THE MITOCHONDRIAC MANIFESTO^2

THE MITOCHONDRIAC MANIFESTO[2]
R. D. Lee

The Mitochondriac Manifesto 2: How Nature Nurtures the Body, and Technology Torments It

The Mitochondriac Manifesto includes many unattributed Dr. Jack Kruse sayings. He taught mitochondriacs most of what we know.

Most unattributed images are sourced from DepositPhotos.com.

ISBN 979-8-9923485-2-1

Enquicken Publishing

For inquires or volume purchases: info@TheMitoMan.com. I plan on doing my own print run maybe mid-2025.

The 2nd Edition is a huge update over the 1st. Includes full coverage of

- Dr. Jack Kruse's story of mammalian evolution, POMC, melanin, UV biophotons, endogenous semiconductors, non-linear optics, dissipative structures, entropy, atomic-molecular optics/organization, topologic insulators and time crystals.
- Methylene blue.
- Photobiomodulation.
- Circadian rhythms, melatonin and red light.
- Near-infrared sauna.
- Sports injuries.

with limited coverage of

- mTOR
- ubiquitination
- I had a chance to try more biohacks and reported my results.

DISCLAIMER

IF YOU'RE NEW TO THE SCIENCES OF BIOPHYSICS, MITOCHONDRIAL BIOLOGY, AND CIRCADIAN RHYTHMS, THIS INFORMATION MAY SEEM A BIT ODD TO YOU… LIKE YOU'VE ENTERED A WEIRD AND WONDROUS NEW REALITY… LIKE EVERYTHING YOU THOUGHT YOU KNEW ABOUT HOW THE HUMAN BODY WORKS NOW APPEARS TO BE FILLED WITH OUTDATED INFORMATION, CONTAINS DANGEROUS HALF-TRUTHS, IS BIASED LIKE YOU WOULDN'T BELIEVE,
OR IS JUST PLAIN WRONG.

ITS TEACHINGS MAY EVEN CAUSE YOU TO QUESTION ALL THE PRINCIPLES, PRODUCTS, AND PRACTICES MAINSTREAM MEDICINE PUTS OUT AS IF THEY ARE THE FINAL WORD ON WHAT'S TRUE OR NOT, AND THAT EVERYONE WHO DARES QUESTION THEIR AUTHORITY IS DEEMED A QUACK OR A CONSPIRACY THEORIST.
…FAIR WARNING.

IT IS NOT INTENDED TO BE USED AS A SUBSTITUTE FOR PROFESSIONAL MEDICAL ADVICE, DIAGNOSIS OR TREATMENT. THE AUTHOR(S) MAKE NO REPRESENTATIONS, AND ASSUME NO RESPONSIBILITY, CONCERNING THE ACCURACY, SAFETY, EFFICACY, OR APPROPRIATENESS OF THE PRODUCTS, PROCEDURES, TESTS, TREATMENTS, SERVICES, OPINIONS, OR OTHER INFORMATION CONTAINED HEREIN.

CONSULT YOUR HEALTH PROFESSIONAL BEFORE UNDERTAKING A NEW DIET, TREATMENT, OR LIFESTYLE CHANGE. IF YOU SUFFER A LIFE-THREATENING EMERGENCY WHILE YOU'RE READING THIS BOOK, RESIST THE URGE TO CONTINUE READING. PUT THE BOOK DOWN IMMEDIATELY AND
DIAL 911.

THIS INFORMATION HAS NOT BEEN REVIEWED OR ENDORSED BY DR. JACK KRUSE. IT IS MY OWN UNDERSTANDINGS OF HIS MATERIAL.

Contents

Additional Material

PREFACE

What kind of document is this?

Readers and critics would like this content to fit neatly into one box or another. Because of unconscious biases, people want to put material in a box with which they're familiar, such as *non-fiction science book with citations, peer-reviewed paper published in a medical journal, blog post,* or *fictional novel.* They don't want to take the time and energy to evaluate the data and decide for themselves what's right or wrong. They don't want to be responsible for separating fact from fiction. Frankly, people are used to (centralized) authority figures telling them what to think and do.

But that's not how mitochondriacs operate. Today, if you want to be informed and independent with your health, you need to be your own investigator because centralized medicine won't tell you this stuff for maybe 25–50 years. So, no, this book does not conform to conventional standards. Instead, it's a hybrid document. It's a series of short assays; written by a layperson with no formal medical or science training; loosely organized into a storyline; with extreme specificity for clarity, utility and engagement; combined with interpretation and opinions. It's as self-validating as possible.

Mitochondriac: (Informal) A fan/follower of mitochondria, biophysics, and seasonal cycles of the body.

If the presentation isn't what you expect, consider the fact that fields of science we think of as being difficult – including quantum physics and relativity – are over a hundred years old and well-examined by now. They're primitive by comparison. Even Einstein had contemporaries to conspire with. Whereas, quantum biology and the biophysics explored within are so far out on the bleeding edge, there's literally only one person in the world who's putting it all together so simple folks can understand it. I'm merely reporting on it (the new stuff is from 2023–2024, by the way).

I'm interpreting the information so both health professionals and health enthusiasts, each with their own interests and expectations, can better understand what Jack Kruse is saying. More than raw information, it offers perspective. Sure, we'd all like quantum biology to be easy to understand. But I can assure you this is as close to that as it will be for a long time. And, trust me, you don't want 500 pages of references (that I didn't read either).

Differences between the 1st and 2nd Edition

The First Edition of *The Manifesto* was aimed at both health professionals and consumers (not an easy task to speak to both groups at once). For the

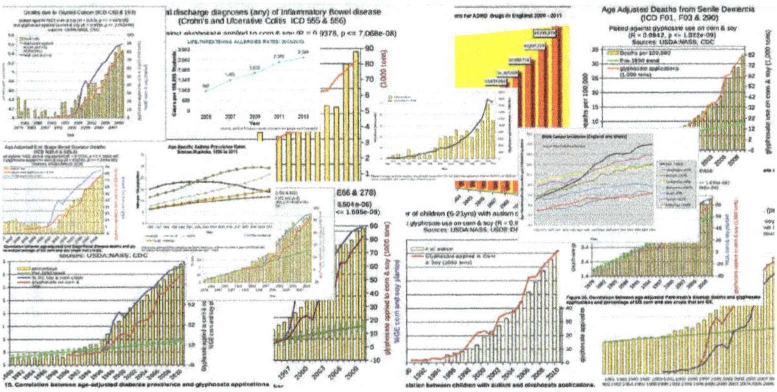

Second Edition, I did not change the sequence in which ideas are introduced that much. Instead, I added about 75 pages of new content toward the back and cut 30 throughout. So, while still aimed at both practitioners and consumers, the new material demanded the target audience be directed more toward professionals. It has to be more technical. The First Edition had a short chapter on the many health crises we now face. However, it's not needed today. By now, everyone knows the threat that disease and dysfunction pose to all life on earth. In its place is the graphic of disease statistics above and the mantra below.

Rallying call

The mitochondriac's mantra says the forces of biophysics that make disease and dysfunctions show up in your life also work in reverse: It undoes them. So however debilitating and irreparable these conditions appear to be, the physics of the mitochondria, and the rhythms of the body, have answers where mainstream medicine offers only lame excuses that keep you ill-informed and dependent. Buckle up. We're going on a whirlwind journey of discovery to the far reaches of human understanding into the physics of life. It will change the way you look at wellness and illness forever. Take its teachings to heart, and the wisdom of Nature will serve you and your family well for the rest of your life.

My sources

The Mitochondriac Manifesto is based on the work of the following thought leaders. For your curiosity, I've named their area(s) of expertise.

1. Dr. Jack Kruse, all major subjects.
2. Doug Wallace, PhD, mitochondria.
3. Dr. Robert O. Becker, polarity of injury and repair.
4. Professor Gerald Pollack, fourth phase of water.
5. Dr. Dean Bonlie, DDS, magnetism.
6. Dr. Alexander Wunsch.
7. Michael Hamblin, photobiomodulation.
8. John N. Ott, light's biological effects.
9. Viktor Schauberger (interpreted by Callum Coats), energetic properties of water, vortexing, and the hydrological cycle.
10. Dr. Jeffrey Friedman, leptin.
11. Brian Richards, near-infrared sauna.

PART 1

Biophysics

(The physics that controls biology)

I

NO ONE KNOWS WELLNESS AND ILLNESS LIKE A MITOCHONDRIAC

(Mitochondriac: A fan/follower of mitochondria, biophysics and seasonal cycles of the body)

Biophysics is more important to human health than diet, exercise, genes, toxins, the microbiome or leaky gut

How so? Because of one simple, but sweeping, principle that holds people back from reversing their chronic disease: **You can't get well when you stay in the environment that made you sick.** This is the mitochondriac's Golden Rule of Wellness. You'd think this concept would be so embedded in our consciousness by now we'd all be applying it consistently as if our well-being depended on it. But no. Quite the opposite. Almost everyone ignores it like the law of Cause and Effect doesn't apply to them.

The more we try to escape its influence, the more we realize wellness or illness are built upon the forces of physics that control mitochondria and our biorhythms – these being light, water and magnetism – not on secondary influences like food, exercise and gut health. To put it more potently, **biophysics is the boss – in charge of all biology – while all other factors are its underlings.** They are administrative assistants that take their orders from biophysics.

That's not to say there's anything wrong with worshipping your favorite diet like a religion, exercising as if you're training for a triathlon, or downing mouthfuls of primal supplements each morning, if that's what turns you on. But if you try to get fully well by adding foods, supplements, or practices to your daily routine, while the real source of your problems is actively contaminating your health, you're deluding yourself with hype and half-measures.

Sure, most of the diet and lifestyle choices in vogue today will give you a little to a lot of improvement. But does your daily regimen take you all the way to 100% in body and mind? Is it sustainable for your budget and lifestyle? And how long will your results last when our electromagnetic environments are getting worse by the week… and the body builds a tolerance to interventions based on emergency response? These are important questions to ask yourself – especially when you measure success by more than just six-pack abs and Instagram followers.

Where the healing arts lost their way

Many moons ago, Western medicine thought it had just about figured out human health because it was successful at improving many people's symptoms, some of the time. Those were the good ole days of the 1970s, '80s and even the '90s. But after people realized pharmaceuticals and personalized medicine weren't half the solutions they were cracked up to be, alternative healers turned their attention toward natural healing techniques like rebalancing gut bacteria, elimination diets, detoxing heavy metals and exercise.

These are the roots of human health. Getting a good handle on these aspects of our biology will keep the doctor away, right? Well, not exactly. Natural approaches like these definitely give us gains in fixing our broken bodies and minds. However, as human technology marches forward – particularly wireless technology – we're beginning to realize there's another dimension to our physical well-being we intuitively sense, but Average Joes deny because centralized science doesn't support it.

We've been taught biochemistry controls our biology

Healers of all kinds have long thought chemicals in the body drive most aspects of health and longevity, while an *imbalance* of said macronutrients, micronutrients, hormones and neurotransmitters bring about dysfunction and disease. Turns out, bio*chemistry* is secondary to the photoelectricity that gives life to plants and animals. Everything that happens on a biochemical level is preceded by an electromagnetic input and our biological programming.

That means signaling molecules, digestion, cognitive function, sleep, and anti-aging efforts – they're all subordinate to light, water, magnetism, and electromagnetic frequencies – both good and bad. Photoelectricity is the real driver – meaning electrons, protons and photons. The ultimate example is ATP production. Crucial to everything the body does, *energy production* in mitochondria is an *electron* transport chain, not a *protein* transport chain, *carbohydrate* transport chain or *fat* transport.

ATP: Adenosine triphosphate is an energy storage molecule made in mitochondria that drives dozens of cellular processes crucial to running the body.

Look past the dogma and it's easy to see a corrupted environment is what really drives dysfunction, disease, and an early demise. The machinations of declining health are merely observable effects of physics acting on our physiology. When they're relieved, disease processes are symptoms that can restore wellness some of the time. However, they are not the actual source. That's why symptom suppression hardly ever fixes underlying problems. For decades now, it's been "our bad." We've mistaken symptoms for source and wondered what went wrong in the process… what we seem to have missed. Stick around, and you'll see why biophysics is more rudimentary to life than any diet, practice, program, or other life science.

We've been told genes control our health, wealth and success in life

For thousands of years, the "haves" of the world (the rich, royal, and privileged elite) have believed they're genetically superior to the "have-nots" (the poor, simple-minded, and sickly masses). "They" fed the plebs that narrative in order to rationalize the differences in health, wealth, and access to the better things in life. And before we understood how environment influences genetic expression, they had everyone convinced that genes predetermine who is going to enjoy good health throughout their life *vs.* which families are going to be mentally and physically handicapped as a consequence of their "bad blood." Mental and physical fortitude would then enable success in other areas of life.

Good genes = good health, and a good life, they posited. And, up until recently, *We The People* bought their story hook, line and sinker. Lacking alternative explanations, we didn't question their storyline that genes and evolutionary advantage plays a bigger role in a person's success than environment, and how well your biology handles it. It's only within the last 10–15 years every socio-economic class has been hit hard enough by chronic illness and cognitive impairment to make us question the *genes cause disease* hypothesis. It's now become obvious to educated observers that's not the case.

Epigenetics: Environmental factors control the way our genes turn into physical traits and behaviors.

Instead, our environment controls our biology much more than genes do. Through epigenetics, our environment controls the switches and programming that run our biochemistry, our gene expression, and the foundations upon which our success or failure in life are built. You'll see in the pages to come just how important the physical forces affecting our mitochondria and circadian rhythms are at every level of our existence from the physical, to the intellectual, to our success in life.

The deficiency-additive paradigm

Drug companies, functional medicine docs, and even biohackers have taught us to think from a "deficiency-additive" mentality when it comes to our health… that we lack something the body needs. So now you have to *take* something, and/or *do* something to get well. When really the best place to start resolving health problems is to *remove* obstacles that prevent the body from doing what it does innately: heal and rebalance itself.

It's becoming ever-more apparent: you have to remove barriers if you ever want to get to immaculate health and longevity. Kind of like fixing your brakes that are sticking before you add more horsepower to your car or even get a tune-up. You have to get out of your own way. Then you want to reconnect with the original, and still best, health support system we call Nature.

Get ready for an ideological reset

Jump down the rabbit hole with us and you'll discover answers and solutions to the greatest health problems that have stumped the world's foremost healers for generation after frustrated generation:

- Why it's so hard to lose weight and keep it off.
- Why so many people feel tired all the time.
- Why kids can't pay attention and focus anymore.
- Why ¼ of all women take antidepressants or anti-anxiety drugs.
- Why so many people have insulin resistance and diabetes.
- Why so many couples struggle with infertility today.

Learn how improving your internal energy production can help you escape the common, complex diseases we all know by name. This is the very cutting edge of human health and healing in the modern world, because it reveals secrets that doctors, public health agencies and media don't know a thing about, and wouldn't tell you if they did. Introducing a brand-new life science that's more foundational to human biology than those symptoms and situations many have wondered about but no one could explain. Introducing mitophysics: the physics of the mitochondria and our connection to sun and earth.

The bioscience that completes human biology: Mitophysics

Mitophysics is a mitochondriac term for the forces of physics that control our mitochondria and circadian/infradian biorhythms (daily and seasonal cycles, respectively). It is an umbrella term to describe the amalgamation of biophysics, quantum biology, mitochondrial biology, chronobiology, and other sciences important to mitochondriacs. Mitophysics is a giant, but previously invisible, piece to the wellness puzzle whose omission has unknowingly undermined our best efforts to get well over the past three plus decades.

More specific to creating wellness or illness than the broader terms of biophysics and quantum biology, mitophysics explains how friendly electromagnetic frequencies, pure water, and the earth's magnetic field give us life, energy, and resistance to disease. Or, conversely, mitophysics explains how foreign frequencies, adulterated water, and non-uniform magnetic fields deplete us of energy and healing capacity on the way to causing disease and dysfunction.

Here are some ways mitophysics influences the life around and inside us:

- **Photosynthesis.** Plants harvest the sun's photonic energy and turn it into sugar, oxygen and excited electrons to feed the entire food chain.
- **Electron transport chain.** Mitochondria harvest electrons, protons, and photons from food to power our biology.

Circadian rhythm: Biorhythm lasting about 24 hours, such as daily cycles of sleep and waking.

Infradian rhythm: Biorhythm lasting longer than 24 hours. Herein, infradian *generally refers to an annual cycle. For example, humans tend to gain weight in late fall through winter, and sleep less in summer, while animals breed, hibernate, molt, and lose/grow fur seasonally.*

Ultradian rhythm: Biorhythm lasting less than 24 hours. For instance, sleep cycles last about 90–120 minutes.

Quantum biology: How light photons and electrons influence biology.

Chronobiology: Time-based biological cycles (e.g., circadian, infradian, and ultradian rhythms).

- **UV and IR light heal you.** UV and IR light control healing and regeneration programs via receptors in the eye, skin, gut and fat.
- **Blue light wakes you up.** Blue light turns on hormone production and the stress response, thereby regulating alertness and stress level.
- **The fourth phase of water.** Water becomes a battery by rearranging itself into areas of net-positive H's and net-negative, gel-state water, thereby creating electrical differential that cells use to do work.
- **Paramagnetism moves materials.** The body uses magnetism to transport blood, hormones, oxygen and DHA – and to re-condense proteins so they can regroup after a day's work.
- **Biology synchs to the seasons.** Our infradian rhythms synch to the seasons through exposure to light, temperature and food.

Paramagnetism: Substance that is weakly attracted to magnetic fields because of its unpaired electron(s).

Docosahexaenoic acid (DHA) is a very special fat that can convert sunlight into DC electricity, and back again.

On the other hand, mitophysics becomes a hero's journey when Nature's processes get corrupted by man's short-sighted technologies, his reckless profiteering, and his need to dominate and control Nature.

Mitophysical challenges we face today:

- **Mitochondria insufficiency.** Weak, inefficient mitochondria make you consume more food (and potentially retain more calories) in order to make enough energy to run the body.
- **Leptin resistance.** The recently discovered hormone called leptin regulates the body's energy production and expenditure programs – including adrenal function, immune function, insulin signaling, fertility, and even our emotions. Can't lose weight and keep it off? Lack of leptin sensitivity (i.e., receptors/reception not working) has probably been sabotaging your efforts this whole time.
- **Deuterium.** Fruits and vegetables eaten out-of-season for your locale can gum up your mitochondria, make you gain weight, and contribute to disease and premature aging. Specifically, *produce* accumulates deuterium and gives it to its consumers.
- **Artificial blue light.** Inappropriate light exposure dysregulates biorhythms and depletes your neurotransmitters. Blue light, in particular, causes hormone and neurotransmitter problems, sleep disturbances, and adrenal issues.
- **Grounding/earthing.** Chronic disconnection from the earth deprives you of energy and healing capacity.
- **Non-native EMF pollution.** Many are surprised to learn how much that non-native electromagnetic frequencies upset human health.

Deuterium: A hydrogen atom with an extra neutron in its nucleus. Nature uses deuterium's different structure and properties to control biological programs such as energy production, food seasonality, and aging.

You can think of mitophysics as the soil that feeds and protects the roots of your health. Or, when you're exposed to the forces of physics in corrupted forms, they become a never-ending source of contamination people try desperately to offset with better food and more supplements. As a result, our electromagnetic environments are so messed up we're no longer shocked to see young children with insulin resistance (formerly called "adult-onset diabetes"), teenagers with cancer, 95% of young women showing autoimmunity to their own thyroid; or 1 in 35 children having autism. We've heard the stories so often nothing fazes us anymore.

Your body isn't defective. The real problem is your environment

Whether learned or assumed, we automatically look inward when our health is interrupted, because **we believe the defect is inside us**. For reinforcement, we run hormone panels or allergy tests, etc. for more evidence our body is betraying us. Any malfunction must mean our body is broken in some way, right? Of course, this reasoning makes perfect sense if you think like a captive consumer (with Stockholm Syndrome). However, in most cases, it's just not true. Your body is not malfunctioning. What's more, it's harmful to think that way because mistakes are inevitable when you experience symptoms and assume they're the source – rather than seeing them for what they really are: distress signals the body sends to your senses, asking for intervention.

For instance, all we can think about when we feel off is the headache, poor digestion, chest pain, weight, foggy thinking, or uncontrolled behavior. But where is it all coming from? That's what should concern us. What's your body really trying to tell you? Important to understand, the body rarely does anything by accident. In most chronic disease conditions, the body is behaving exactly the way it's designed to. Or, at least it's doing the best it can in a difficult situation, using the resources it has available. We tend to forget the body has its own built-in compensation mechanisms and emergency procedures it's running all the time, while we can't think of anything besides *healthy or not* in the moment.

Key point: Look *around* you for the source of your health problems first, before you look *inside*, because **it's your environment that's defective, not you**. In most cases, what you perceive to be your body malfunctioning is just its way of coping with a flawed environment. And most of those influences, good or bad, are actually mitochondrial stories in disguise, because 80–95% of all diseases revolve around the mitochondria, while only 5–20% are primarily genetic. Why? Because mitochondria are profoundly more responsive to the environment than human DNA.

In fact, Dr. Doug Wallace, the world's leading mitochondria researcher, has shown that the vast majority of disease occurs without any changes to the human genome at all. That means for decades now, and tens of billions spent, modern medicine has been studying the wrong genome in an effort to explain disease and treat it. So, instead of looking for the cure to cancer, Alzheimer's or diabetes in human genetic defects, the answer to 95 out of a 100 health questions can be found in the mitochondrial genome – i.e., their state of vitality or degradation.

Every single thing in the universe is made when light slows down

You can visualize how physics creates our material world by observing one simple fact: **When light slows down, things with mass show up.** Our bodies use this miracle of Nature to make all sorts of biochemicals we can't live without. For instance, when light hits an aromatic amino acid in the eye, it slows down and makes dopamine. Dopamine controls higher brain functions, muscles, and our behavioral reward system.

Aromatic amino acid: A building block of neurotrans mitters.

When light hits another aromatic amino acid, that amino acid absorbs UV photons and makes melatonin. Melatonin regulates sleep and mitochondrial recycling. Same thing with a neurotransmitter that makes you feel good, called serotonin. Serotonin is made when light hits the eye and the gut lining (digestion releases light from food). Melanin as well. Melanin is the pigment that turns skin brown under UV light. It's made when one aromatic amino acid turns into another, which turns into another.

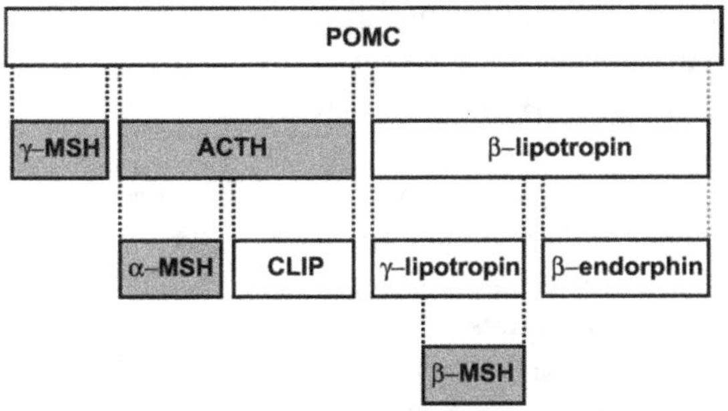

Sunlight also makes a composite chemical in the brain's hypothalamus called POMC (proopiomelanocortin). POMC gets separated into several chemicals, including an opioid called beta-endorphin, which makes you feel good in the sun, as well as ACTH, which is a precursor to cortisol (a stress chemical). Interestingly enough, beta-endorphin is the only opioid that does not increase cravings and dependency on other opioids. So it's the only opioid that's safe and non-addictive to get loaded on.

So, as it turns out, the optical tract is ladened with amino acids like tryptophan, tyrosine, phenylalanine and histidine. They capture sunlight to make neurotransmitters. But the most interesting thing about the energy-to-matter conversion is light controls the pathway – directly, as just described, or indirectly by food electrons that were programmed with light in photosynthesis. Converting light frequencies into material matter – that transformation – is happening in biology all the time. But it all begins as light. Things with mass are made when high-energy light slows down and condenses into lower-energy matter. Then you can touch it, or measure it, as light in its solid state.

What's more influential to our biology: biochemistry or biophysics?

"If I held a dead rat in one hand, and a live rat in the other, I would tell you each of the rats contain exactly the same biochemicals. What's the difference between both of the rats? The amount of energy contained. So if biochemistry really mattered… why is one alive and one dead? Turns out, the issue is all tied to energy. And what is energy on this planet? The entire food web… the entire energy creation on this planet [comes from the sun]."

—*Dr. Jack Kruse.*

Dr. Jack Kruse: Neurosurgeon by training, now the world's leading mitochondriac educator.

'n' is shorthand for the number of people in a study. So instead of a test group representing the general population n=1, in this context, means you are your own test subject. Mitochondriacs use the term "n=1" informally to mean your response to a given diet, supplement, drug, practice, or treatment is unique to you.

Heteroplasmy rate: Percentage of damaged (stretched out) mitochondria vs. healthy and productive. So high heteroplasmy is bad; low is good.

The days of one-size-fits-all treatments are over. It's quickly becoming an *"n=1"* world

Centralized medicine has long used randomized, controlled studies as the gold standard for showing their drugs and devices are safe and effective. The problem is, they presume statistical relevance in a select group (typically designed to exclude undesirable subjects, similar to jury selection) translates well into safe and effective treatment for all. They treat all health consumers as if they're the same. But that's flawed thinking, and people are beginning to wake up to that fact. With its validity diminishing, the presumption of homogeneity contradicts everything we now know about epigenetics and quantum biology. For this reason, mitochondriacs have a right, and a duty, to question traditional safety and efficacy studies. The evidence is piling up: an individual's personal environment, and their biologic response, is as varied as human personality.

Nowhere is this individuality better exemplified then in biophysics. Bio*chemistry*, which is a downstream effect of biophysics, is diverse enough across populations. But when you add exposures of light and electromagnetism to the equation, you can see why the reliability and predictability of Western medicine has been growing more erratic over the last 20–30 years – particularly since toxins such as glyphosate and junk light infiltrate every aspect of our lives.

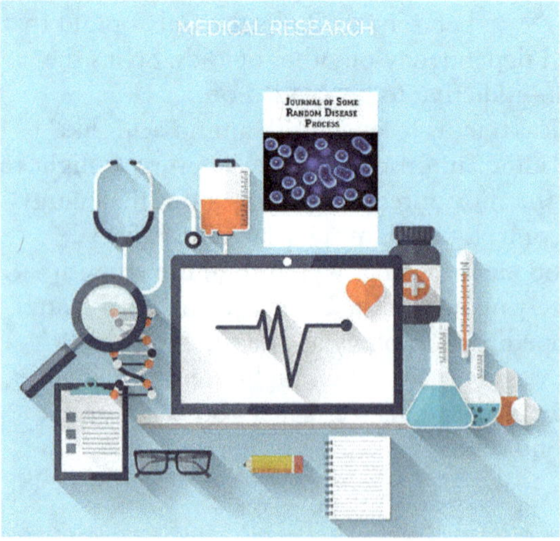

Take light exposure as an example: If you conduct a study near the equator, or even in the summer, human biology acts differently than a study done farther North, or during the winter, under more artificial light. You'll get different results based on the sun exposure each individual gets, because sunlight on the eyes and skin control healing and regeneration programs, hormones levels, and mitochondrial output.

The same variability applies to the amount of fatty acid DHA in the diet, the non-native EMFs a person receives, the amount of deuterium in their water, as well as the individual's magnetic environment. It all makes a difference in your test results. It all matters. However, no clinical trial done in the Western world ever controls for light, water, magnetism, DHA, or heteroplasmy in the design of their studies – not to mention many more factors that could challenge the applicability of standardized tests with real people, in the real world.

Researchers don't take these influences into account. Thus, their results may not correlate well with an uncomfortably high percentage of individuals receiving the treatments coming out of these trials. So, the ugly truth is, drug companies are looking to achieve just one thing in their clinical trials: any statistical improvement, however small, with tolerable side effects and loss of life. It's getting harder and harder to rely on any type of clinical evidence that a treatment will work from one person to the next because our environments have changed so dramatically from that of just a few decades ago. There are just too many factors at play that are never taken into account at any stage of a drug's development – the most influential of which are mitophysical in origin.

Real world example: Studies on the effects of diet have always been done under artificial blue light. That removes a powerful positive influence on our physiology: sunlight. And it introduces potent negative variables such as artificial light and non-native EMFs that can adversely affect each person to a different degree, depending on strength of mitochondria and the mitophysical environment they come from. Conversely, all the dietary research done by Weston A. Price, DDS, on indigenous populations in the 1930s was all conducted outdoors, under lots of real sun exposure and locally-grown food. That gave his research as much scientific validity as modern clinical trials due to their exclusion of incredibly potent variables such as native and non-native EMFs.

Dr. Weston A. Price: Dentist and researcher that travelled the world in the 1930s to study how diet affected the health of indigenous populations.

Bottom line: The healthcare industry has to get used to the idea that the reliability of group studies is becoming increasingly scattered, while the opposite paradigm is taking over. Each person must be treated as if their biology is unique to them, as well as constantly changing, because both are true. A new paradigm we'll call *"n=1"* is taking over, which means there is no such thing as a one-size-fits-all drug, supplement, diet or test anymore. Those days are over. Instead, the response of each individual to any wellness effort is unique to them and their mitochondrial efficiency.

We need to stop assuming the medical industry's results on groups of people, in a clinical setting, are automatically relevant to you on an individual basis. You need to stop thinking the drug companies and your doctors know more about how your body is going to respond to a treatment than it does. Instead, you need to learn a little, be your own test group of one, and *you* need to look out for *you*. That's the new paradigm of $n=1$.

You can't treat your health like a hobby anymore

Before 4G and Wi-Fi came along, you could spend your free time learning how to fight the disease of the day. You'd read "top 10 articles," listen to health podcasts, and scour online forums to find *the* solutions to you and your family's nagging problems. Then you'd try the diet, supplements or practices the guru recommended, based on how much you liked them and their story. You'd try the tips for a while – until you got tired of doing them, they stopped working, or somebody new came along, with a seductive story and something new to sell you. And, like a revolving door, you'd shift your time, expendable income, and hope onto the next greatest thing.

Let's face it: in the old days, the majority of people had time and healing capacity to spare, so they could afford to take a casual approach to their health. And that was fine for most people's lifestyle, biology and bank account. Unfortunately, those days are over. Now, everyone must take their health seriously **or you're going to get sick**. It *is* that simple, because taking your health for granted now has consequences. It may not be common knowledge just yet how symptoms tie back to their source. But trust mitochondriacs when we say to you, '*unfriendly biophysical forces explain the source of disorder better than any health science, period.*' At the same time, putting the *friendly* forces of physics back in your life is *the* best medicine to cure disease and dysfunction today… which is exactly what becoming a mitochondriac teaches you.

Mitochondriacs won't win any popularity contests from mainstream medicine by pointing out new realities like these. But the new path to exemplary health in a 5G world requires thought and reason over "shiny object" syndrome and "seat-of-your-pants" decision-making. That means you have to be more methodical about which recommendations you try out and adopt long-term, and which ones you drop. That will take some proactivity. It will take more weighing of costs *vs.* benefits. It will take more effort than you're probably used to.

As the world's leading mitochondriac Dr. Jack Kruse likes to say, you have to have skin in the game to be a mitochondriac, as we're going to lose a lot of people over the next few decades.*** Getting healthy and staying healthy isn't mere amusement, like it used to be. It's a direction and a journey, not a single product, practice, or event in time. So don't

**** This prediction was given before the recent pandemic. How well do you think it's held up?*

count on it being quick, easy, and convenient to succeed while the health of those around you is sinking to new lows. From this generation of technology forward, it's survival of the best-educated, most proactive, and fully committed. You need to be all-in to stay healthy for good.

Biophysics and mitochondrial biology are multi-layered

The good news is the bad news: Mitophysics has so many layers to it, you won't run out of options to improve your wellness any time soon. There's always another level you can take your efforts to... so many more principles, practices, and products you can add to your routine to get your health where you want it to be. However, if you're coming to *The Manifesto* as a conventional thinker, it can be a sobering to realize almost everything you ever learned about health and disease is more fiction than fact (if not downright fraud), and that the people you trusted to teach you your foundational beliefs have been lying to you out of ignorance, or even on purpose.

It can be hard to wrap your head around the fact that the Establishment has been able to almost completely exclude this information from medical canon for over a hundred years. But I assure you, biophysics is indeed a big deal to medical science, as well as the future of the human race. Mitochondrial biology is crucial to our continued existence. Prepare to be shocked and amazed by what you've always needed to know, but were never taught. Get ready to start your bad information detox.

Be your own health boss. Become a mitochondriac

The mitochondriac learns which forces of physics her mitochondria like or don't like, because mitochondria are the microscopic engines of our wellness or illness. The mitochondriac understands how circadian and infradian rhythms control the body's exertion, recovery, and seasonality programs, because they run our biochemistry cycles. She then incorporates those tips and strategies into her daily routine to fortify her immunity and longevity against the diseases of civilization – let's call them "mitohacks."

Supporting that call for independence and self-reliance, the mitochondriac does not outsource her wellness to anyone. She doesn't let any drug company, three letter agency or clinician dictate what's right for her. Rather, she decides what's best for her and her family. Through experience and intuition, the mitochondriac knows she can't blindly trust what anyone says no matter how educated, experienced or credentialed that source appears to be because each individual's mitophysical environment is unique to them. That makes each person's biology respond differently to a given program, practice or treatment. With that understanding, it's easy to see why there's more information and advice available than ever before, yet no one seems to agree on anything.

This hasn't always been the case. Before electrification, our mitophysical environments were simpler, and more uniform, across population groups, which reduced variation in outcomes. But now, everyone is responding differently to a given healing effort because our mitophysical foundations are more corrupted across the board, and more varied. Hence all the debate about which foods and supplements are good for you, and which ones aren't.

The new reality: it's become an $n=1$ world. And here's your first prescription for prosperity: Don't take the medical industry's literature, marketing materials, and advice at face value any more than you would a used car salesman. Use discernment and don't give your trust away for free. Make your educators and advisors earn it. They work for you, after all. Stop assuming others know what's best for your physiology, because no one does. At the same time, start paying close attention to the clues your body is giving you, because it's always trying to tell you what it likes, what it doesn't, and what it needs. Learn the meaning of its clues and you can take full control of your health and well-being. That's the functional knowledge of the mitochondriac… forged from a modern survivor mentality every one of us is going to need in the wireless era ahead. That's what being a mitochondriac is all about.

2

OUR CONNECTIONS TO NATURE AND DISCONNECTIONS

Our biology is designed to be connected to Nature... not technology

We're meant to live in communion with our environment. Meaning (1) eyes and skin exposed to sunlight at the right time of day; (2) in direct bodily contact with the ground; (3) drinking water from natural sources; and (4) eating food that grows locally. These are the physical forces of Nature that give us energy to power our body's processes, and synch our biology to daily and seasonal cycles.

On the flip side, we suffer in countless ways when we snub Mother Nature and spend our time under the influence of foreign frequencies and modern conveniences. As we're just now beginning to appreciate, the human race (and planet earth) pays dearly when we choose to detach from Nature, and instead plug our bodies and brains into our manufactured technologies. Here are a few seminal moments in that progression of disconnection from Nature, the shift to man-made technology, and why these changes affect our biology so profoundly:

Long ago, plants borrowed a bacterium to power their biology

Archaea: Microbes that are similar to bacteria, but different enough to be considered another life form.

For 3.8 billion years, two types of organism had the planet all to themselves: bacteria and archaea. Life was simple back then, literally, because everything alive at the time was single-celled. But then, life on earth exploded out of nowhere due to a biological collaboration of historic consequence. About 650 million years ago, early plants partnered with a bacterium to get the power they needed to run their biology. Prehistoric plant life made a deal – we'll give you (bacterium) a place to stay inside us. In return, you'll have to modify your genome (DNA) in order to transform the sun's photonic energy into a chemical energy (sugar) that we plants can live on. The deal worked out and the bacteria, now evolved into "chloroplasts," power plant biology using the multi-step process of photosynthesis. Photosynthesis feeds the entire food web on a foundational level – all powered by light, along with CO_2 and water. Or to say it more simply, life begins as light via the photosynthetic process.

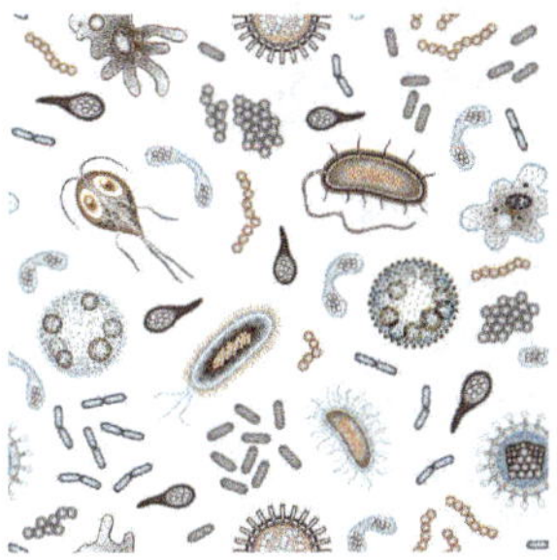

As a by-product of the collaboration, plants also released oxygen and put DHA in the seas. You see around that time

Chloroplast: Primitive symbiotic life form inside plants that uses chlorophyll to convert sunlight into energy. Thought to have evolved from early bacteria, chloroplasts perform photosynthesis.

our sun reached its middle age, which meant increased photonic power. Stronger UV rays hitting ozone (O_3) in the upper atmosphere broke up more of that O_3 into breathable O_2 (oxygen).

So those two things paved the way for animal species to proliferate: the increase of *oxygen* in the atmosphere from less than 15% to then 21%, and the availability of *DHA* to convert sunlight into electricity. Oxygen gave the animal kingdom abundant energy potential by enabling more affluent metabolic pathways like oxidative phosphorylation and the TCA cycle – in contrast to pathways like glycolysis, which are anemic by comparison. Plus, DHA helped multi-celled organisms harvest photonic energy directly from the sun, so they can be in touch with sunlight and earth *intermittently*. In other words, they were mobile.

Oxidative phosphorylation – aka the electron transport chain (ETC): The main process by which ATP (energy) is made in mitochondria.

The TCA cycle: A preparatory process that makes precursors for the ETC, as well as producing a small amount of ATP on its own.

But before that could occur, this had to happen:

A similar partnership was made millions of years later

About 50 million years after the first partnership began, single-celled organisms called archaea made a similar deal with a species of bacterium. Archaea said to these bacteria, 'let's team up so each of us can specialize and benefit from each other's area of expertise.' We (archaea) will give you (bacteria) a place to stay inside our cells and all the food you can eat. And in return, you must modify your genome to get really good at converting food we can't metabolize into an energy source we can.

Over millions of years, those bacteria simplified their DNA to 37 genes in order to specialize at converting fuel stores from the food chain – namely sugar, carbs and proteins – into energy storage molecules that complex creatures can thrive on, which is ATP. Through this partnership, those archaea evolved into complex, multi-celled organisms that science now calls "eukaryotes." And those bacteria became mitochondria.

Eukaryotes: Multi-celled organisms, in contrast to bacteria, which are single-celled.

Having been liberated from the chore of managing thousands of genes, mitochondria were then free to focus on making energy for the cells of the entire animal kingdom in the form of ATP. Meanwhile, the partnership liberated those descendants of archaea, now eukaryotes, from the job of making energy so they could focus on increasing their complexity. They added genes to their DNA to manufacture a wider variety of proteins. That allowed them to build more elaborate anatomy, along with the organs and biochemicals to run it. So archaea became eukaryotic cells, and their DNA became the "nuclear genome." The ultimate eukaryote turned out to be man. The human species has more than 20,000 genes.

Equally important in the deal, the many mitochondria that take up residence inside a eukaryotic cell can use its DNA. You see bacteria, by their very nature, like to exchange genetic information with other cells (called "lateral gene transfer"). And there's tremendous genetic and evolutionary advantage to transferring genes from *many* mitochondria into *one* cell's nucleus.

So that's what they did. Mitochondria transferred all the genes they could into the nuclei of those early archaea. That way, one set of genes in each host eukaryotic cell can serve its hundreds, or thousands, of resident mitochondria, instead of each mitochondrion expending its own energy to support redundant genetic information. To say it more simply, there's a thousand-fold energy savings having each of *your* cells carry genes that each mitochondrion can use to build *its* structure. That's how multi-celled organisms became more elaborate over time: mitochondria specialize in making energy, while eukaryotic cells focus on fancier things like building

- a brain that can figure things out;
- muscles, nerves and a skeleton/exoskeleton that can move us about;
- and systems that adapt to dynamic environments.

But mitochondria did something revolutionary in this whole partnership we shouldn't overlook: They brought the sea into the cell. The animal kingdom may have left the watery confines of its ancient ancestors. But water is still vital to life in so many ways. So to cross that chasm, mitochondria basically do the photosynthetic process in reverse by turning food and oxygen into CO_2 and water. That way, animals don't have to live in the sea if they don't want to. Instead, they can bring the sea with them wherever they go, as water inside and around cells. Pretty neat, huh?

Through these two revolutionary partnerships of energy and anatomy, called "endosymbiosis," countless species of plants and animals sprung up anywhere and everywhere life could possibly exist on earth. The Cambrian Explosion was born thanks to endosymbiosis.

Endosymbiosis was like an archaeon kidnapped a bacterium, or a bacterium hijacked an archaeon, but they both got Stockholm Syndrome because they both received benefits from the other. They then decided to continue living together as a eukaryotic organism.

The first circadian mismatch in human history occurred when man started to wear clothing

Conquering the cold with clothing made far more of the earth's surface habitable by allowing people to live outside the warmth of the tropics. In fact, it was such an important development in human history one could argue that clothing advanced civilization faster than fire. However, the unintended consequence we've always overlooked is clothing blocks light from hitting our skin.

Sensed on a visceral level but never taught to us in school, our skin is literally a solar panel that collects light in order to make DC electricity. It's DC electricity from this source, and others, that power most cellular activities that run and repair the human body. So covering our solar panels virtually our entire lives cuts back on the power we're able to harvest from

the sun. Fortunately, Nature gave us several backup methods to get energy from our environment when we lack sun exposure. Stay tuned for more on that.

The second circadian mismatch in our history is that we learned how to harness fire. Use of fire helped early man cook his food and keep warm. Later, we used it to make tools and scare off wild animals. But the unintended consequence of heating with fire is that it circumvents the need for the body to generate its own heat from within. This weakens mitochondria over time by preventing them from exercising their thermal "plasticity," as well as shrinking their proteins to keep them in-tune.

In the late 1800s, man invented alternating current electricity, which led to machinery and mass production

From then on, every sort of disconnection from Nature industrialists could make money off of was brought to market – from artificial light, to artificial heat, to rubber-soled shoes, to UV-blocking glass, glasses and sunscreen, to fruit shipped in from all over the world in winter. So now, for just about every limitation or inconvenience we might face in our lives, industry has invented a solution that makes money for someone, or serves someone's agenda. With fewer basic living needs left unsolved (i.e., new things to sell), we've reached an inflection point in our history. Today, wireless technology, including smart devices and 5G phones, have become High Tech's one big meta-opportunity for long-term growth. But while industry profits, human health is collapsing.

It's getting harder to disagree: Human health is being sacrificed for the sake of convenience, consumerism, progress and profits. Our technology is literally killing us. And we have unbridled capitalism, hidden agendas, and our own frailties to blame. Indeed, our disconnection from Nature is nearly complete. And our transition to living in the Matrix full-time has begun in earnest. So what will it mean for our health, and the future of mankind? Well, we'll just have to wait and see how far these, and other commentator's, warnings carry in waking us up from our blue light-, EMF-induced state of dependency and zombification. We shall see.

How light creates life

We learned in biology class that the food chain begins with photosynthesis in plants. Photosynthesis takes the sun's photonic energy, combines it with carbon dioxide and water, and converts them into chemical energy in the

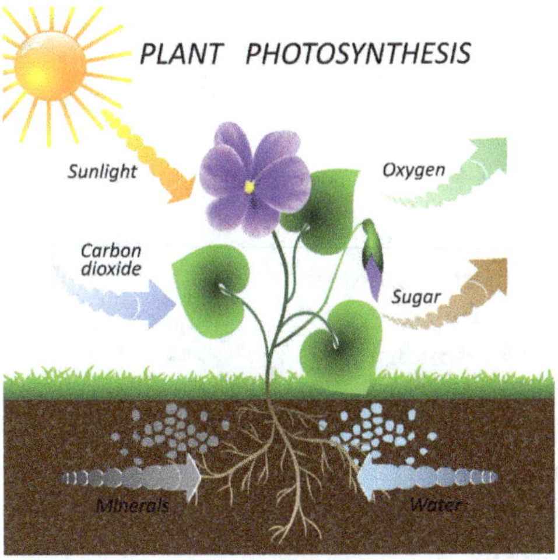

form of sugar. These sugars feed organisms up and down the food chain – from soil microbes below the plant, to the plants themselves, to the animals that feed on those plants, and so on. Most important to appreciate, photosynthesis turns an intangible energy – light – into chemical energy – sugar. That's right; photosynthesis turns light into tangible matter – the happy by-product for animals being oxygen. And the big "a-ha" for inquisitive minds is, **when you eat food, you're eating condensed sunlight and you didn't even know it!**

But the lesson we didn't learn in school is that our mitochondria do the same process in reverse: They take food (sugar and fats) and oxygen, and turn them back into CO_2 and water. The happy by-products are ATP and light (much in the form of infrared heat). Also neat to know, food electrons carry energy and information from the sun, which are released as those electrons hop along the electron transport chain. So you see, Nature wastes nothing. Plants and animals form opposite ends of an interdependent, coupled system, where one organism's trash (waste) is the other one's treasure (sustenance).

Image of Albert Einstein by Oren Jack Turner. Public domain work.

The Photoelectric effect

In 1922, Albert Einstein won the Nobel Prize for his work in describing the Photoelectric Effect. The Photoelectric Effect explains how light makes certain materials emit electrons. By inference, this tells us that light (photons) can only interact with one subatomic particle – electrons – and not with protons, neutrons, or any other particles. It's a foundational phenomenon of physics that comes into play whenever and wherever light needs to be captured, transported and used. And it's another reason electrons are so important to health and healing: they capture and deliver light.

On a subatomic level, the Photoelectric Effect tells us that when light hits an electron, it makes it spin in certain ways, based on the frequencies of light absorbed (e.g., visible, IR and UV), and their intensity. Electrons retain that spin until mitochondria reverse the process by turning food into ATP. Through cellular respiration, mitochondria convert those spin states back into forms of light our cells can use. What's more, those spin states give our mitochondria information about the environment that food was grown in, which the body uses to control metabolism and infradian rhythms.

In other words, light *programs* the electrons in a plant's cells with spin. In living systems, mitochondria later decipher and convert those spin states back into a variety of frequencies of light, but mostly IR and UV. For what purpose? The "reconstituted" light energizes respiratory proteins. And it builds exclusion zone water – or "e-zone" as I call it – which cells use to power their processes.

Respiratory protein: Workstations in mitochondria that make ATP using electrons and protons.

Exclusion zone water, or e-zone: The 4th phase of water, between a liquid and a solid. Full explanation in chapter 8.

By extension, the Photoelectric Effect also explains why many plants struggle to live outside their native light environment, while most animals can live anywhere that temperature and food permit. The difference lies in the number of electrons carried in the core of their most basic compounds. Chlorophyll, which fuels photosynthesis in plants, looks almost the same as the hemoglobin that transports oxygen in animals. The biggest

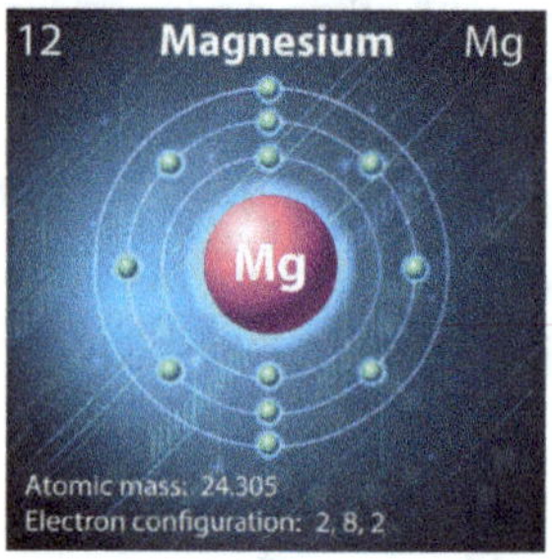

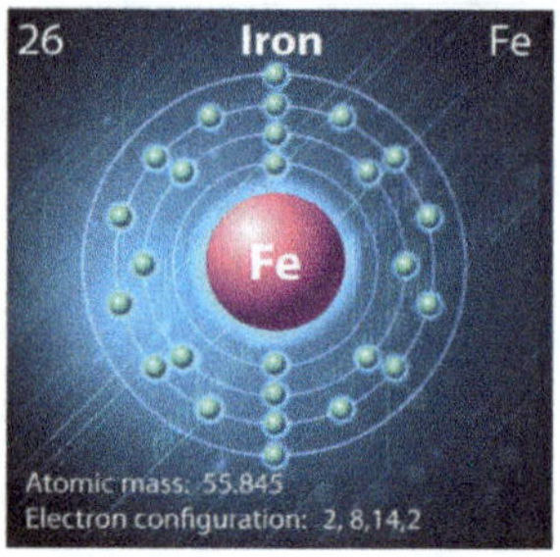

difference is that the chlorophyll molecule has a magnesium atom at its core, while hemoglobin is built around an iron atom. Magnesium, being lighter, has 12 electrons that can absorb light, while iron, being heavier, has 26. (See the molecular structures of chlorophyll and hemoglobin on page 184.) Fewer electrons mean chlorophyll in plants can absorb and use a narrower band of frequencies – mostly blue and red – to power photosynthesis. That makes plants less tolerant, and less adaptable, to altered light conditions than higher life forms utilizing hemoglobin. Meaning, too much light and the plant fries; too little and it can't make enough food to feed itself.

Each ring in a Bohr model of an atomic element represents an electron orbital. When hit by photons of certain frequencies (only), electrons jump to a higher orbital.

In the animal kingdom, the iron in hemoglobin has more than twice as many electrons as magnesium. That gives it the ability to absorb and use a much broader spectrum of light, including IR, UV, and everything in between. Hence, animals are much more tolerant of less-than-ideal light exposure, and still get by okay. It's because they have more electron density in their hemoglobin to harvest a wider variety of light, and more of it. To put it plainly, animals have bigger and better storage tanks for light. Which is to say, the more electrons you have, the better equipped you are to assimilate light. But that also means you're more dependent on it.

Ideal health depends on our ability to use light

Light is the alpha and the omega of terrestrial biology.

1. Light slows down and forms all matter in the universe via Einstein's equation $E=MC^2$.
2. Light combines with CO_2 and water to lay the foundation for the entire food chain via photosynthesis.

But as influential as those foundational phenomena are, you can't fully appreciate light's importance to multi-celled organisms until you understand what light does after it enters your body. You see, a healthy body is good at capturing light, storing it, transforming it, using it, and reabsorbing it – including these processes:

- Photoreceptors in the eye, skin, gut and lung capture light information. These "opsins," as light-sensitive biochemicals are called, are sensor/switches that activate our internal pharmacy to make the hormones and neurotransmitters that control our metabolism, sleep/wake cycles and regeneration.
- Cells in our body store light energy when water in and around them is exposed to (mostly) IR and visible light. That is, light turns regular water into charge-separated e-zone.
- DHA converts light into DC electricity (and back).
- Respiratory proteins on the inner mitochondrial membrane take light of one frequency and convert it into another. For instance, mitochondria take light borne on electrons, and release it as infrared to heat and shrink water around respiratory proteins, increasing their efficiency.

In contrast, bacteria (and archaea) have a completely different relationship to light than we do

They're not designed to harvest light like us. Having developed along a different evolutionary tract, single-celled organisms (prokaryotes) don't have any DHA in their cell membranes. That means they can't turn the sun's energy into DC electricity. They can't store light as fat. And they don't have respiratory complexes to transform one wavelength of photonic energy into another so it can be used for other purposes. All of which means bacteria can't build complexity. And they have to stay single (mono-celled) their entire lives. In fact, bacteria and light don't always get along. Large doses of UV light can kill them. To their disadvantage, they release almost all the light they absorb not long after collecting it (i.e., they can't store it).

A leakage of light causes disease

In the animal kingdom, organs and whole organisms become dysfunctional when they're not good at retaining light. This happens when they lose too much light to their environment when they should be holding on to it and using it to do cellular work.

Key concept: When a cell is stressed, it emits minute amounts of ultraweak

UV light, often called biophotons. The more stress a cell is under, the more ultraweak UV it releases – whether that stress is from physical injury, toxin damage, nutritional deficiency, infection, environmental extremes, emotional duress/crying, orgasm or exertion. However, this type of emission isn't inherently unusual or harmful. Instead, abnormally rapid loss, and/or slow replenishment, is what creates problems.

In other words, every life form on the planet (onions, grasses, monkeys, bacteria – you name it) releases a certain amount of biophotons. The more light an organism can retain and use, the more complexity it can build, and the more vitality it has. So a hallmark of good health is the efficiency with which your body is able to harvest light from the environment, convert it into different frequencies of light, and store it for future use.

In simplest terms, light is the currency of life. To illustrate this in action, when the sun makes a photon, that photon is retained for 100,000 to a million years before it's emitted in a ray of light. The physical forces that contain that photon are strong electric and magnetic fields. Interestingly, the same thing happens in us. Strong electric and magnetic fields around our mitochondria help us retain our photons so we can put them to good use. Conversely, illness and obesity are unhappy consequences of a loss of light due to weak electric and magnetic fields made in mitochondria.

One way we lose light: The double helix of DNA needs to unwind to let cellular machinery read the instructional code to make proteins. The strength with which photons are held in place when your DNA unravels determines how many of them escape in this process. Whatever the reason may be for these weak electric and magnetic fields – whether that's worn-out mitochondria, chronic hypoxia or lack of DHA – light can't be held in place. So it escapes as ultraweak UV, IR (heat) or other forms of light. The informational aspect of light also gets mismanaged when light is squandered, which means the body isn't able to read the information contained in electrons about seasonality. This messes up the way food is metabolized, which can make vegetables and fruits raised in strong sunlight fattening and damaging to mitochondria when eaten under weak sunlight.

So when, where, and how will an individual be affected by a leakage of light? It all depends on which mitochondria are weakest, as that will be the area of the body you'll find the lowest electric and magnetic fields. For instance, people with anorexia leak light from the brain, while obese people lose light predominately from the body – especially the liver and pancreas. In time, Dr. Jack Kruse believes research will prove each disease has its own distinct "light loss signature." But whatever science ultimately finds, the rule is simple: The more light you release, the sicker you are; the more light you retain, the healthier you are. That's how loss of light connects to disease.

Electrons are the *currentcy* of life

All things life- and health-related revolve around harvesting, utilizing, and retaining electrons. Our biology runs on electrons because electrons "turn the gears" of ATP production, they hold electrical charge that drives our biochemistry, and they power regeneration efforts as redox potential. We depend on electrons to live and thrive on, because electrons supply energy, healing and fluency of cellular communication. On the other hand, electron deficiency produces positive charge, which presents itself as acidity, inflammation, low voltage, dehydration, and weak regeneration of cells and mitochondria. Of course, this accelerates aging. To say it more simply, electrons support life, while a lack thereof creates weakness, disorder, aging and premature death. We get electrons from four sources: (1) grounding; (2) sun exposure; (3) food; and (4) burning stored fat. Water quality/quantity, and how tightly your mitochondrial proteins are coupled, then increase or decrease energy yield from those electrons.

Redox is short for "REDuction-OXidation." Redox reactions involve oxidation and reduction.

Redox potential: Pools of electrons and their net-negative charge (and, in some cases, pools of positively-charged protons) that the body uses to move materials and perform chemical reactions (think battery power*).*

Herein, "redox" usually means the latter, because redox potential describes the ability of charged particles to accomplish redox reactions.

Mammalian battery: Informal, general term describing stores of electric charge and photonic energy that cells can use to do work. (1) E-zone is the biggest cache. (2) ATP holds electrical energy in its chemical bonds. (3) Cell membranes hold electrical charge. (4) Muscle movement releases electrons – as piezoelectricity mostly from bones, ligaments, cartilage and tendons – into the acupuncture meridians. (5) DNA is its own battery, powered by a spiraling coalescence of cosmic energy, often called "scalar energy."

Plants are like batteries that are always plugged in

Plants are designed to be connected to the earth 24/7 their entire lives. That allows their roots to gather electrons from the soil, and their foliage to harvest photons from the sun. Those two things – grounding and sunlight – combine to make plants like batteries that are always plugged in. That's why plants don't need to eat food, or store fat, the way we do. Instead, they harvest whatever energy they can from the sun, soil and air through the Photoelectric Effect. Their chloroplasts, which are like mitochondria for plants, then convert the light, water and CO_2 they collect into sugar via photosynthesis. They store relatively little of that food (sugar) for leaner times because, being connected to a source of sustenance full-time, they expect their next meal to be served each time the sun comes up in the morning. On the other hand...

Humans are designed to be unplugged and mobile

Our biology is designed to collect energy from more sources, to get it sporadically, and store it for later use as fat and muscle mass.

1. We can harvest energy second-hand by eating plants.
2. We can get it third-hand by eating the animals that ate those plants.
3. We also convert our own stored fat back into usable energy.
4. We can gather electrons directly from grounding and sun exposure.
5. And our mammalian battery gets charged up by exposure to native light frequencies.

Macronutrients such as fat, carbs and protein are basically storage containers for the sun's energy until we eat them and our mitochondria release those energy resources as electrons, protons and light. This is why plants need to be plugged into Nature their entire lives to get their energy needs met (i.e., chlorophyll- and chloroplast-based metabolism), whereas animals can charge up their batteries whenever and wherever they can from the food they eat. We've basically got energy storage tanks in the form of fat and muscle mass, conversion machinery in the form of mitochondria, and processes like ketosis (fat-burning) that let us move around and survive the feast and famine of daily and seasonal cycles.

The more electrons you get from Nature, the less you need from food

When you get your daily dose of electrons from the environment, you don't need to eat as much, you don't get hungry as much, and it's easier to lose weight. And the less food you eat, the lower the calorie, sugar and insulin burden on your metabolic pathways. So doing things to improve your use of electrons – such as hydrating better, grounding, getting more sun, and biohacks that increase mitochondrial efficiency such as Cold Thermogenesis – all these things reduce your need for electrons from food. And that plays a big role in weight management, and metabolic disorders such as diabetes, because you aren't eating as much. We'll talk more about losing weight without dieting and exercise in the energy mismanagement section, Part 2.

Modern living steals electrons

Living in the world of today steals electrons from us, putting us in a state of electron deficiency that decreases our health and healing capacity. Not only are we getting fewer electrons coming in from traditional sources, but our lifestyles and technologies are using up our electron supplies faster than ever. Indeed, every facet of our lives in which we replace Nature with technology, we dig our powers of renewal a deeper hole to climb out of – challenging our ability to recover ever more. Just as bad, most people don't even realize the stress that electron deprivation subjects their body to. So while the entire wellness industry talks about diets, supplements and fitness regimes, the most fundamental building block of optimal health – electrons and voltage – seldom receive any attention. Here's a quick overview of how electron deficiency drains the health out of you and me.

Disconnection from Nature

We're doing pretty much everything possible to divorce ourselves from Nature: not getting direct sun exposure, not directly touching earth, not drinking pure water, and not eating seasonal, whole foods. These are the traditional sources of electrons that used to fill our tanks with energy to support cells, brain function and resistance to disease.

Inflammation and oxidation

Modern living is brimming with foods, toxins and lifestyle choices that increase inflammation and oxidative stress. At the center of it all, electron deficiency keeps inflammation and oxidative stress going after their usefulness has run out. Persistent inflammation and oxidative damage consume more electrons – electrons that, under more healthful circumstances, could be used to keep you looking younger and feeling better. Instead, those electrons are being used to fight imbalances and emergencies.

Technology and our environment

Just about every modern technology and societal exposure we encounter steals electrons from you because of the way they eject electrons from their sphere of influence, and/or reduce mitochondrial efficiency. From artificial blue light, 4G and 5G, to Wi-Fi, fluoridated water, and processed foods... most inventions for communication or convenience interfere with the way our bodies collect and use electrons.

EMFs. The non-native electromagnetic radiation emitted by our communication devices jostle electrons loose from tissues chronically. Meaning, small amounts per device, times copious devices, equals significant sub-ionizing electron loss. Non-native EMFs also oscillate mitochondria at frequencies that directly impair fat-burning and ATP production – basically wasting a portion of our electron supply that should be going into, and out of, energy production.

Water. Fluoridated and chlorinated water is bad because it reduces water's ability to separate its positive 'H's' from its negative 'OH' groups. Both limit the amount of light energy that water can store for future use (as e-zone water). This spreads the body's supply of electrons thin.

Processed food. Most packaged, high-carb food becomes more acidic in manufacturing, because electrons are lost in processing. That makes it relatively more proton-rich, positively-charged and acidic. ...Which is a major reason why processed foods are less healthy for you: they have fewer electrons. And it's why *whole foods* from Nature are more wholesome: they contain more electrons.

Moving air steals electrons from you. Ever notice how fatigued you get after sleeping under a fan or air conditioner for many hours; or riding long distances on a motorcycle, bicycle or convertible with skin exposed? One reason these activities can be so tiring when done for several hours is they steal electrons from you. Other factors offset this effect (like sunlight), but fatigue of this kind is caused by more than just wind buffeting.

To sum up the electron-deficiency dilemma, technology and modern conveniences restrict our supply of electrons coming in, and they use up our supply of electrons faster than in previous generations. Our health and healing capacity declines as a result.

Seasonal cycles

How come wild animals don't freeze to death when it's cold out?

It's not just that their fur or feathers insulate them from the cold like a jacket. Simply *retaining* more heat would not be enough to keep them warm when it gets chilly out – resting especially. The forgotten factor in surviving winter in the wild is that animals have built in mechanisms to *generate* more body heat, combined with retaining more of that heat with fur, feathers or fat. You see, mammals have built in seasonal programming that cranks up heat production from their mitochondria to keep their body temperature livable in the freezing cold. We'll call this relatively new science of seasonal biology "infradian rhythms," referring to longer than a 24-hour period. The infradian system in the brain basically tells mitochondria to consume brown fat, food and other light/energy resources to turn more of that energy into heat.

Brown fat: A specialized type of fat whose dense mitochondria populations burn it to make heat when you get cold.

Even less widely known, it's not just cold temperature that activates winter programming in mammals. Dwindling ultraviolet frequencies in late fall also tell photoreceptors in the eye, skin and gut to turn on this winter programming. These sensor/controllers tell the infradian system to grow a thicker coat of fur in anticipation of cold weather, as one element of this winter programming.

Still another force of physics that helps animals stay warm in winter is grounding. The simple act of being in direct contact with earth gives animals an endless supply of electrons. And, as we discuss throughout, being alive, and well, and warm has a lot to do with how many electrons you have. To illustrate, when chickens and other livestock are raised in ungrounded pens, they're more sensitive to cold than their wild counterparts. Supplemental heat may then be necessary to keep them from freezing to death – often with heat lamps that radiate IR light.

Conversely in nature, those same animals get electrons from constantly touching earth. More electrons means better circulation, because grounding thins the blood by keeping cells from sticking together. This helps animals stay warmer without the need for added heat. More electrons also result in better light assimilation, because electrons bear light energy for living systems. These are some of the unrecognized reasons wild animals can tolerate freezing temperatures that we can't... Or can we?

But modern man has never faced a winter

To this day, we still have adaptive programming built into our biology that helps us survive cold weather. But we as modern humans – living a climate-controlled existence – have invented ways to avoid being cold at all costs. We can't stand the cold, so we live in our comfort zone between 65° and 80°F. But comfort and convenience come at a price, because our mitochondrial fortitude atrophies when we never allow our bodies to feel real coldness. In effect, many people have lost the seasonal temperature variability that beneficially stresses mitochondria into making heat the way we're supposed to. That turns our mitochondria into weaklings that have scarcely had to lift a finger to help us survive.

To try out this thermal plasticity on myself, I went out multiple times per day over a winter, purposely underdressed for the high 30s to low 50s. And you know what? I was uncomfortably cold at first. It was definitely a shock to the system initially. But something purposeful happens when you exercise your thermal regulatory muscles: Your mitochondria kick in, and each time you go out in the cold, they respond quicker and more vigorously by releasing more IR light. So you just don't get cold as easily.

But the real benefit beyond not *feeling* cold is that your mitochondria stay younger and fitter over time because heat shrinks water around the respiratory proteins, making them more efficient. On the flip side, if you never use your mitochondria's ability to make heat, respiratory proteins in your mitochondria get stretched out and stay stretched out, you lose energy production efficiency, you age faster, and disease sets in sooner and more easily. For these reasons, athletes like Michael Phelps have learned to swim in 50°F water to strengthen the productivity of their mitochondria. Similarly, football trainers have long soaked their players in ice water to reduce inflammation. However, what Phelps and the trainers may not realize is that the cold tunes up your mitochondria.

In summary, many adults in the modern world can't stand the cold because they never use their body's built-in thermal regulatory flexibility, whereas many children can play quite comfortably in cold weather because they still have the natural temperature variability humans are born with. In fact, humans can even tolerate soaking in ice water for a really long time, given special training, as Wim Hof has demonstrated. He set a world record by staying submerged in an ice bath neck-deep for nearly two hours.

Wim Hof (Nicknamed "The Iceman"): Developer of The Wim Hof Method of controlling the autonomic nervous system to heal.

Simply put, energy-making efficiency drops across the board (chronically) when mitochondria expand. So you can either use the heat production capability your mitochondria were born with, or you lose mitochondrial fitness over time. That's the unfortunate sacrifice you make if you never leave the comfort of your climate-controlled bubble. Think of it this way: Many of us have heard the body gets sluggish making a hormone or biochemical when you take it as a supplement. Well, the cold is no different. Our bodies are meant to generate heat from within. It's built into our biology. But our mitochondria have gotten lazy in modern society, because we're letting clothing and technology do all the work our mitochondria used to do.

You pay a biologic toll eating foods out of season

Humans are the only organism that can radically change the world around them to suit their every fancy. And probably the second most important change we make today (after shelter) is with our food. Westerners import fruits, vegetables and high-carb foods from sunny climates so we can eat them in our fall and winter. But just because we *can* eat fruits and vegetables in winter doesn't mean we *should*.

Nature had a plan when it designed our metabolism around light cycles of the seasons. In a word, it's energy *balance*. When the body has too much energy coming in or going out, both create problems. So our biologic programming has elaborate mechanisms to match energy coming in with energy leaving, as a coupled system. In practice, that means you can safely eat fruits, vegetables and carbs in the summer when the sun emits a lot of UV. However, when you eat those same foods in winter, they detune your mitochondria and promote weight gain. The difference is, when you eat sweet fruits and starchy vegetables in the *summer*, the biologic consequences of eating those carbs are offset by strong UV that minimizes fat storage, and charges up the water in your body.

Summer programming is designed to maximize instant energy, while minimizing fat – at a time when you're likely to be more active and there's more food around. On the other hand, winter biorhythms do the opposite: They make less ATP, more heat, and more fat when you're designed to be in survival mode and less active. So you're actually losing mitochondrial efficiency, and retaining more calories, when you eat sweet and starchy foods in winter. Most people then erode mitochondrial function even further by chronically avoiding cold exposure, which adds insult to injury.

To observe this wisdom of Nature, have you ever noticed your craving for fruit goes down when it gets cold out, while fat cravings increase? That's Nature's way of telling you you're not supposed to be eating fruit at that time of year. The energy density of fat is better for generating body heat and maintaining mitochondria. On the flip side, have you ever noticed your craving for fruit increases for no apparent reason when it gets hot out? That's because your body knows the photonic energy contained in fruit is now matched to your local light environment – which means mostly your latitude, partly altitude, as well as cloud cover and temperature.

Unfortunately, people assume incorrectly that because a fruit or carb is available in a grocery store that its nutrients will benefit them the same year-round. Not true. Eating a banana, pineapple or even sweet potato in winter represents a photoelectric mismatch that metabolizes those carbohydrates using the wrong seasonal programming. Adding fuel to that fire, carbohydrate-rich processed foods amplify this effect, because the whole food ingredient is refined and concentrated in manufacturing. At the same time, complimentary elements that help you assimilate nutrients are removed, such as fiber and moisture content.

Type of chlorophyll is a seasonal signal

At its source, food seasonality is based on how a plant's particular type of chlorophyll captures the light spectrum of its growing environment, and encodes that energy and information onto the electrons of the plant's sugars and fiber. For example, *chlorophyll-A* likes strong UV environments like you would find at the equator, and doesn't mind long hours of exposure. Chlorophyll-A is good at harvesting intense light. So that's what tropical fruits are predominately made of.

On the other hand, plants grown at higher latitudes, under weaker sunlight, need help to gather their daily dose of sun. So they have an additional accessory pigment, called *chlorophyll-B*, that helps them absorb more high-frequency blue light to make up for the weaker spectrum they live in. In this way, different mixtures of chlorophyll change which biological programs are activated when food is broken down and its light hits photoreceptors in the gut.

To sum up the seasonality of food, the information encoded onto a plant's electrons needs to match the photoelectric environment in which that plant is eaten or else sensors in the gut get decoupled from sensor/controllers in the eye and skin, causing chaos. That's one of the main reasons so many people today struggle with seasonal weight fluctuations. As a result, it is possible to violate these rules and still lose weight in winter using conventions like *eating less and exercising more*. But you're paddling upstream when you do that, because you're fighting built-in ancestral programming that's designed to do the opposite of what you want. Look back at everything you've witnessed and perhaps experienced, and you'll notice this phenomenon acting against you.

The take-home message: Carbs in particular are not uniformly bad for all people, all the time, as paleo, Atkins and keto fans would have you believe. Rather, it's a mismatched light environment that makes them *appear* bad, because infradian sensors in your eye, skin and gut are confused about the difference in photonic energy between your eye and food. But, interestingly enough, if you were to change your light environment – like going to Hawaii, for example – in very short order, you could eat that pineapple or papaya and it wouldn't cause problems, because it *is* matched to the light exposure you're getting when you eat it.

In this way, people are like plants: Wellness increases when you get your light intake right; it suffers or fails when there's a mismatch. Taken to the extreme, you wouldn't expect a coconut tree to grow in Siberia, or a cold weather pine tree to grow in the tropics. Likewise, you're stressing your body's adaptive mechanisms by eating foods grown in a different light environment.

These are all aspects of infradian biology that many nutrition educators call "seasonal eating." Fortunately for us, it all makes intuitive sense, it's simple to follow, and it gives you big benefits. Just learn how Nature synchs your biology to the seasons. Do what you'd do naturally if you were living in the wild. And you can stop fighting invisible forces that sabotage your best efforts to lose weight. That, right there, delivers a body-blow to the once-indisputable paradigm of *eat less and exercise more.*

However, there's another layer of metabolic programming operating underneath our seasonal biology we can't do anything about. Our "haplotype" is racially-inherited ancestral programming that influences how we metabolize our food like a baseline, or a source code.

Ancestral groups are programmed to eat according to the light, temperature and altitude of their environment

People are programmed to eat according to where their ancestors came from. Food compatibility is determined by how tightly or loosely-coupled your mitochondria are from birth – called your "haplotype" individually, or your "haplogroup" collectively.

Population groups that grew up near the equator are programmed from a mitochondrial haplotype perspective to make more ATP from the food they eat, and produce less free heat. Living under strong sunlight, they didn't need extra heat. Instead, they're designed to make lots of energy to run away from predators that want to eat them, and vice-versa. On the other hand, for Inuit Eskimos and Scandinavians, their biggest threat was the cold. So they have mitochondrial programming that innately makes more heat, and less ATP, than equatorial haplotypes. This "loose coupling" of electron transport efficiency is the basic predisposition upon which the preceding story of seasonal eating acts.

As a result, people from Northern climates do not tolerate high photonic energy foods as well as people from the tropics because Northerner's mitochondria are built for the high flow, higher ATP yield of *fat*, more than the low flow, low ATP yield of *carbs*. In other words, Northerners tend to yield less energy out of foods grown in strong sunlight, no matter where the individual currently lives, or their heteroplasmy rate (more about heteroplasmy in Chapter 3: Mitochondria). Thus, they can eat more and not gain as much weight, because nutrients are extracted or made in the gut first (e.g., vitamins, minerals, fiber, moisture and antioxidants), before calories from food constituents make ATP or heat at the cell/mitochondria level.

To put it plainly, Northerners have a larger percentage of their calorie intake going into heat production, so they have a built-in advantage when it comes to weight management. But it also means they're more susceptible to metabolic problems like diabetes when they eat summertime foods… as well as more free radicals, more inflammation, and lower redox than a high fat diet. That's because carbs are dirtier and less energy-dense as a fuel source, compared to fats and proteins. Which means Northerners overdose on carbs more easily.

See page 21 sidebar for definition of "redox."

At the other end of the spectrum, equatorial haplotypes don't handle high-fat diets as well because they can't uncouple their mitochondria as easily. Too much fat overloads their mitochondria's circuitry with a stronger electron- and proton-flow than their electron transport chains were designed to handle. Energy excesses like this can flood their systems with free radicals, which can make them sick because their mitochondria are better equipped to handle lower octane carbs.

This explains why elite runners from countries close to the equator, like Kenya, arrive a week early to marathons held in the cold. Their mitochondria are used to being tightly-coupled. So when their respiratory proteins try to decouple themselves in cold weather and lower UV, their free radical production skyrockets, which cause even the fittest athletes in the world to get sick for a few days, until they acclimate to the cold. In other words, their systems take a few days to adjust to the photoenergetic mismatch between their innate biology and that of their new environment.

This explains **why African-Americans have a harder time maintaining a healthy weight than other races, as well as being more susceptible to cardiovascular disease and other conditions.**

- The Standard American Diet contains more fat than their haplotype is built for.
- They don't get the sun exposure they're meant to.
- Many under-exercise in relation to their infradian programming. It's designed for lots of activity.
- And inflammation is habitually higher because their electron transport chains spend far more time in "adaptive mode" than they do operating at their most efficient settings.

They're basically living a mismatched lifestyle, in mitophysical terms, compared to their ancestral programming. This predisposes equatorial haplotypes to easier weight gain and more chronic inflammation, given less sunlight, less physical activity, and an altered diet. So the darker your skin, the more attention you need to pay to seasonal eating, mitochondrial fitness, leptin sensitivity and redox potential.

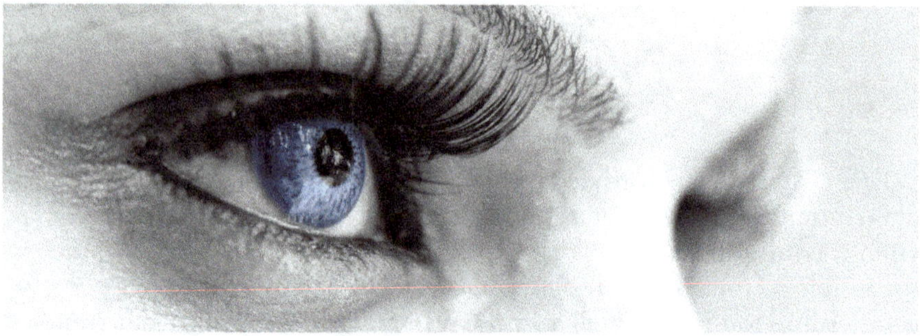

What runs our biologic programming?

The eye is much more than a camera

We all know the eye is a camera through which we see the world around us. But it has another job that's at least as important, if not more. The eye's function we were never told about is it senses which individual frequencies of light are in a mixture of full-spectrum sun. It uses that information to regulate almost every biological program and process inside us – including energy production, stress level, sleep cycles, digestion and repair processes. That's right, the eye is designed to:

1. decipher exactly how much of each wavelength of the electromagnetic spectrum is present in sunlight – including invisible frequencies ultraviolet and infrared;
2. tell the brain what time of day it is, based on the presence or absence of colors in sunlight;
3. regulate daily and seasonal cycles, growth, metabolism and regeneration accordingly.

And, fundamentally important to mitochondriacs, the eye controls these biological programs *photo-electrically*. That means photoreceptors (light-sensitive biochemicals) in the eye collect light and turn that mixture of frequencies into electrical signals that the brain understands to be a certain time of day, at a certain time of year. For instance, one blend of light frequencies, after it's deciphered, tells the "suprachiasmatic nucleus" (SCN) in the brain it's high noon in the middle of summer. Biological programming built into the brain then makes the necessary adjustments to sleep/wake cycles, weight loss/weight gain, detoxification and more.

Suprachiasmatic nucleus (SCN): One of the brain's primary control centers that run daily and seasonal cycles (circadian and infradian, respectively), based on the information it receives about your environment.

How the brain loses control of organs and systems

Circadian and infradian rhythms are controlled by a tiny region of the brain where the optic nerves cross and lead into the brain. About the size of a pea, this suprachiasmatic nucleus (pictured on page 148), as it's called, is one of the brain's primary control centers of daily and seasonal cycles. Most impressive of its duties, the SCN helps control every metabolic and growth program in the body – including (1) the speed of your metabolism via the leptin receptor (Chapter 10: Leptin), (2) sleep and recycling of mitochondria through melatonin, and (3) fertility through testosterone and estrogen. Unfortunately, the SCN is not working so well in people today, and all sorts of unexplained disorders show up as a result.

What makes the SCN malfunction? There's a time-keeping mechanism in the SCN that coordinates signaling throughout the body – quantum signaling. You can think of it as a master clock in your brain for your biorhythms, or a timer/regulator for tissue-to-tissue communications. Below that – in location and pecking order – every gene in every mammal has a circadian clock gene in front of it through which the SCN controls gene expression and biologic activity of cells and organs. Most important for circadian signaling, the master clock above in the SCN has to run several milliseconds ahead of the clocks that run genes down below, or else the message to increase or decrease activity of peripheral organs and tissues arrives late. This imperfect timing confuses organs and tissues, and makes them operate at reduced functionality at best, or perhaps not at all (think "safe mode" in computers).

Dr. Jack draws this analogy to help mitochondriacs grasp the concept

GPS satellites 22,000 miles above the earth need their atomic clocks to run 38 milliseconds faster than geolocation devices at ground level, or else the GPS app in your smartphone will be off by 10–100 kilometers by the time the signal reaches you. That's because gravity bends light through a physics principle called "gravitational lensing" – satellite signals included. In other words, gravity creates an extra 38 milliseconds of lag, so engineers advance the broadcast signal's timing to stay perfectly synchronized, not its trajectory (as you would try to hit a moving target with an arrow).

If you didn't advance the timing of the master clock above, devices on the ground would get confused by the lag time – much like a concert musician would in receiving directions from their conductor after a delay. Therefore, your master clock has to operate with a little lead time in order for the signal to arrive in time to be executed on schedule. That's important to the physics of life because cells and mitochondria are releasing light all the time, converting it, recapturing it and storing it. Any asynchronous delivery in that process – any leakage of light – is energy wasted and communications lost. Point being, biophysical communications in the body happen at the speed of light (new paradigm), not chemicals (old paradigm). And electricity is no exception. Electricity is simply light oscillating at a slower frequency, but still travelling near the speed of light.

In summary, when the clock genes in your peripheral organs fall behind the master clock in your SCN, the communication and operation of cells in your liver, pancreas, gut and feet become uncoupled from each other and lose their coherence. When they fall out of synch, regulatory glands then can't talk to cells underneath like they need to, and tissues misbehave. Organs disobey important instructions, creating inflammation, which turns into dysfunction and then disease.

How is this disconnection happening?

The #1 reason timing errors happen is because of weaker electric and magnetic fields in your brain's SCN than in your body. Lower electrical charge in the SCN slows down your brain's master clock so it falls behind that of distal organs. It's that simple. Nature's solution is DHA. As we'll talk about later, DHA is a very special fat that converts sunlight into DC electricity (and back). More electricity makes your SCN run faster than peripheral cell clocks down below. And as long as the suprachiasmatic nucleus has enough DHA and "juice" to run ahead of peripheral clock genes, orders get received on time and obeyed. Unfortunately, blue light destroys DHA in the eye. That muddies the connection between the SCN and peripheral clocks that rely on the SCN for instructions. The liver then has trouble controlling blood-sugar levels, as one example, which makes the individual susceptible to diabetes and metabolic dysfunction.

A second type of timing error is caused by exposing your internal organs to nnEMFs. For instance, using a tablet computer on your lap vibrates cells in your lower abdomen at billions of oscillations per second. That speeds up clock genes in your reproductive organs faster than your SCN, which make it appear as if you've got hormone imbalances, when the real problem is signaling errors that make organs malfunction. This is one of the leading causes of infertility and sexual dysfunction in women and men today. To get your organs back in synch, try this mitohack: Soak your bottom half in cold water, and let the sun shine on your top half. This slows the clock genes in your reproductive organs, while speeding up your SCN, leptin system and pituitary.

A third timing error happens due to elevation. Every foot you ascend above ground level slows your SCN a little bit. The smallest effect would be on a mountain top, because you still have positive influences like grounding and the Schumann resonance to offset the slow-down. However, those who live in tall buildings are indeed mildly affected. Air travel is even worse. It's more disturbing to our brain clock than being in a skyscraper. Fortunately, exposure time for passengers is usually limited to just a few hours, whereas pilots and flight attendants aren't so lucky. They're exposed to timing errors both acutely and chronically. And, let's not forget, being irradiated by microwaves in a giant metal tube for hours on-end can really hit those with heightened sensitivity hard. After being exposed to these influences in-flight, the majority of people feel a little fatigued, but they get over it quickly. Others get wiped out, even within one time zone. While an unfortunate few possessing a low biophysical safety margin can suffer pulmonary embolism, stroke or heart attack from the dreadful electromagnetics aboard.

Schumann resonance: Earth's natural resonant frequency, which is slow and gentle at 7.83 cycles per sec.

Finally, the most disruptive timing hazard of all is space flight. (Whew, most of us don't have to worry about this one.) Long stays in high earth orbit decimate circadian timing because, up there, you're disconnected from pretty much every mechanism that keeps your biology tightly-regulated – including light cycles, magnetism, grounding, gravity, good water and the Schumann resonance.

How tissues and organs talk to each other: entangled particles

Biochemists would say regulatory glands in the brain control organs below primarily through biochemicals such as hormones and neurotransmitters. For example, they say: (1) adrenals control stress level through cortisol and adrenaline, (2) the pituitary controls cell growth with human growth hormone, and (3) the thyroid controls metabolism through thyroid hormones T3 and T4. All true. But the link between endocrine glands in the head and organs in the body doesn't start there. Nor is it the main way they stay connected.

Hormones are merely the chemical messengers we observe between glands and organs. They're biochemicals released into the blood as a result of coherent communications around the body. Most important, hormones are simple one-way activators, not two-way connections offering fast and accurate feedback. That job goes to "entangled particles" – a well-established physics phenomenon going way back. Regulator glands talk to organs around the body through free radicals and their once-connected electrons. Together, science calls them *entangled particles.*

After an electron is removed from an O2 molecule to form a free radical, the electron and resulting free radical maintain a bond between them through a physics phenomenon Einstein called "spooky action at a distance." Like twins that can each sense what's happening to the other, *spooky action at a distance* says a particle (in this case an electron) that was once part of a whole atom or molecule will continue to react as if they're still connected. So when you stimulate one, the other will react as if it experienced that stimulation itself, no matter how far apart they are in space and time. They're forever yoked on a quantum level.

A superoxide free radical may go to the liver, for example, while its now-liberated electron may travel to the brain. All the while, they know what their former particle-mate is experiencing through quantum entanglement of particles. This is how regulatory glands in the brain collect information about the state of organs throughout the body. It's an important way Nature keeps bodily systems on the same page, singing the same tune. Therefore, the more entangled particles you have in your body, the better that organ systems communicate, and the more coherently they operate.

On the other hand, when the body can't make free radicals due to weak electric and magnetic fields, lack of electron flow, or shortage of transition metals such as iron, copper and manganese, then *spooky action at a distance* also breaks down. That breaks the bond between endocrine glands and the organs they're supposed to regulate. We call that chaos, which presents as inflammation, hormone imbalances, malfunctions of unknown origin, and disease.

1. Take **obesity**, for example: Obese individuals have lost coherence in their body, but their brain may still work fine. The circadian clock in their suprachiasmatic nucleus is decoupled from the clock in their liver and pancreas. That impairs metabolism and a whole lot more, putting you at-risk for obesity and diabetes.
2. The opposite is true for **Alzheimer's.** The dysfunction is happening from the neck up. The SCN gets disconnected from the rest of the brain, but their body organs may remain connected. So brain function deteriorates, while weight remains stable.
3. **Fertility**. When clocks in the leptin receptor and pituitary system get unyoked from the ovaries or testes, you can become infertile, lose libido, or become impotent. In case you hadn't noticed, disconnection problems like this are sweeping the population like a plague to end all man, with nothing but silence coming from the Establishment to explain why it's happening.

In basic terms, when your biology is uncoupled due to a lack of free radicals in the body, organ systems can't talk to each other. That's what quantum entanglement does for you in biologic systems. It's the secret signaling mechanism that keeps organ systems perfectly synchronized which, unfortunately, is failing or broken in most people today. And that contributes mightily to diseases of modern civilization for which modern medicine has more questions than answers.

Mitochondria and cell membranes are environmental sensors

Mitochondria pay close attention to pretty much every biophysical force around you in order to regulate your energy level, cell function, and seasonal changes that run your circadian rhythms. Based on energy and information exposures, mitochondria regulate the production of chemical energy (as ATP), free radicals (for restoration), body heat (as infrared), metabolic water (to store energy), magnetism (to move materials), and redox potential (for healing).

Crucial to every last thing the body does, mitochondria oversee the collection of energy, storage of energy, when and how energy stores are spent, and the efficiency of energy production (conversion, really). Mitochondria are in the middle of *all things energy* in the body. As we'll discuss in more detail next chapter, the electron transport chain is Grand Central Station for unpacking and analyzing all the energy and information they get from electrons and your environment.

Mitochondria are responsive to many stimuli in energy production

- **Electron density.** Fats produce a stronger electron flow than proteins and carbs do. Whole foods tend to have more electrons than protons. They're more alkaline, while processed foods have fewer electrons and are more acidic. Mitochondria adjust efficiency of their electron transport chains accordingly.
- **Oxygen level.** Mitochondria are acutely aware of oxygen levels – whether at rest, while exercising, from poor circulation or at altitude.
- **Water quality.** Deuterium gums up the ATP synthase. Conversely, low deuterium water helps you store more energy.
- **Temperature.** Cold temperature turns your furnace on by burning brown fat and releasing IR light.
- **Red light.** Infrared light absorbed through the skin makes the ATP synthase spin faster.
- **Magnetism.** Magneto-responsive chemicals that we call free radicals regulate beta oxidation, urea cycle and other metabolic pathways.

The main way mitochondria measure efficiency of energy production is by examining free radical emissions from their electron transport chains (ETC). It's this free radical signal that tells mitochondria how efficiently they're burning (oxidizing) input materials to make ATP. The main mechanisms by which mitochondria then optimize ATP production are: adjusting input rates of materials such as oxygen, respiratory protein "stretch" (casually referred to as heteroplasmy rate), and how much each metabolic pathway is employed. The end result being, input materials, reactants and oxygen combine to burn best.

In response to the free radical signal, our epigenetic *software* then dynamically alters the gene expression *hardware* (mitochondrial DNA and nuclear DNA) in order to regulate heteroplasmy rate of cytochrome proteins. In other words, heteroplasmy rate and energy efficiency change in real time by free radicals turning genes up or down. Unfortunately, exposure to blue light and nnEMFs scramble the free radical signal beyond normal operating limits. Hence, mis-tuned mitochondrial engines suffer decreased energy production and increased free radical formation – both of which lead to cell and mitochondrial damage, which develop into dysfunction and disease.

Cytochromes: The electron transport chain (that makes ATP) is comprised of 5 cytochrome complexes, aka respiratory proteins, or just 'cytochromes' for short. Diagram on page 51.

Mitochondria help harvest the sun's energy from excited electrons

Orbital: Path in which an electron might be found orbiting around its nucleus – not necessarily a linear path, but a path of probability.

In plant photosynthesis, photons from the sun hit electrons in chlorophyll. If the photon has enough energy – meaning a specific frequency of light – the electron will absorb that photon and jump to a higher energy level around the nucleus, called an "orbital." The flight path of orbitals, and their energy level, are incrementally stepped, with nothing in-between.

The photon then becomes pure energy in the form of the electron's higher orbital energy state, corresponding to the frequency of light absorbed. The more energy a photon has, the higher and more powerful the resulting electron orbital.

However, if the photon has too much energy, the electron overshoots the atom's highest orbital, escaping its orbit (and the atom) entirely. Science calls the escape of electrons from atoms "ionization." Radioactivity is the most potent ionizer. On the other hand, if the photon does not have enough energy for the electron to reach a higher orbital – meaning the wrong light spectra – nothing changes. The electron ignores the photon and stays in its present orbital. These phenomena explain the absorption spectrum of elements, as well as their emission spectrum.

Plant electrons maintain this excited state until an ETC in a eukaryote recaptures the light in increments. Picture this: As an electron jumps from spot to spot along the respiratory chain, the energy previously captured in photosynthesis is released. Respiratory complexes collect this light energy on light-sensitive proteins, called chromophores or porphyrins, and use it for whatever the body needs at that time.

Watch YouTube videos entitled "Physical Science 7.3h - Atoms Absorb and Emit Light" and "Electron excitation, emission and absorption spectra" to see how mitochondria harvest light.

This is how mitochondria harvest photonic energy from food. And this is the code that tells mitochondria the conditions in which that plant was grown. They then use this information to tell your infradian system which seasonal programming to run. In this way, food is a ferry for light. Food electrons haul light energy around in their orbital state, until mitochondria transform it back into light proper. At that point, respiratory complexes will often transform higher frequencies, like UV, into a type of light that mitochondria need at that moment, such as IR.

Voltage-gated calcium channels: Extremely sensitive, electrically-powered valves that let calcium into the cell.

Add it all up and we can see why biology's special skill in the universe is that it's learned how to take excited electrons and capture the sun's photonic energy as the electron falls back to its ground state (lower energy orbital). Fundamental to our connection with light, this is how biology harvests both energy and information from food electrons.

Membranes are electromagnetic antenna

Cells and mitochondria also collect data about your environment from the way vibrational frequencies make their surfaces oscillate. Specifically, proteins in cell membranes oscillate when they're exposed to electric and magnetic fields, light, food constituents, water, supplements/drugs, and even person-to-person "vibes" (via biophotons). In other words, **cell membranes are electromagnetic antennae** that continuously sample the vibrational frequencies of everything around you. For example:

- **Voltage-gated calcium channels.** Voltage-gated calcium channels (VGCCs) on the cell membrane detect the tiniest electrical currents. Cells then pass information like this back-and-forth to mitochondria inside them, so together they can make the necessary adjustments to metabolism, seasonal cycles and cell operations.

- **Oscillation of inner mitochondrial membrane.** Respiratory complexes normally oscillate at 100 Hz. Many countries have electric power grids that operate at 50 Hz, which is the second harmonic of 100 Hz. That's a problem for fat and protein burning. Our tech devices then complicate mitochondria and cell function further with their alien frequencies.

A good way to describe this partnership is that membranes are frontline sensors for your internal *biology-management computer.* Cell nuclei, mitochondria, and the brain then respond according to intrinsic programming developed over eons. The take-home message for mitochondriacs: Membrane oscillations, plus the orbital state of electrons and protons, are the code that biology uses to decipher what's going on in your environment, so it can take the appropriate steps to run your circadian systems. That's chronobiology for you.

Chronobiology: Time-based biological cycles (e.g., circadian, infradian, and ultradian rhythms).

Unfortunately for modern humans, when our built-in programming conflicts with our present-day situation that spells trouble for our physiology on every level – big trouble, from metabolism and hormones to stress, sleep, cognition and restoration. Basically, when our environment changes, but our programming stays the same, it's a disaster for our biology on all those levels and many more.

In summary, mitochondria and cells receive inputs and information about your environment in the form of frequencies, food, toxins, and even the emotional state of those around you. The worst forces we have to contend with on a daily basis are blue light, microwave EMFs, foods eaten out-of-season, high-deuterium food and water, and stress. To the best of their ability, mitochondria and cells respond to those stimuli by increasing or decreasing ETC efficiency, among countless other adjustments. But, most important, when your respiratory proteins get stretched out, and stay stretched out, past your fitness level and age, that's how/when you lose energy, vitality and lifespan.

And that's what the next chapter lays out for the biology geeks among us. It's a mini-masterclass on how mitochondria make energy for the body, what causes them to break down, and what happens when they're unable to meet all the body's energy needs. Prepare to get down into the weeds of the *what* and the *why*, because this is the very inception point of health or sickness. Pick up as much as you can from your present level of knowledge, and don't worry about leaving some understandings still on the page and not in your brain.

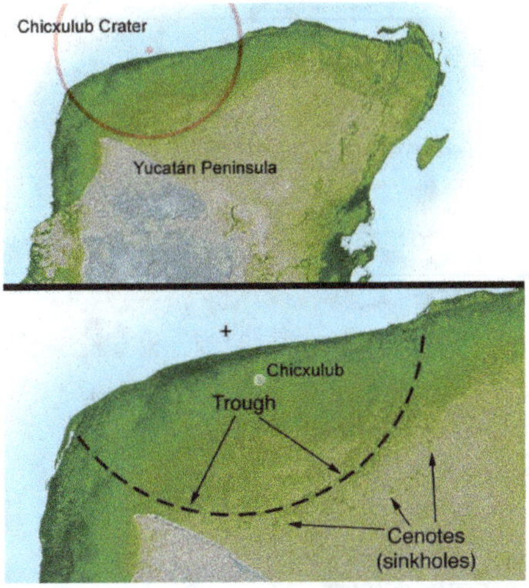

Public domain work. Author: NASA/JPL-Caltech, modified by David Fuchs.

3

MITOCHONDRIA: WHERE HEALTH OR SICKNESS BEGIN

Excess mitochondrial capacity enabled two types of creature to survive The Great Extinction

Sixty-five million years ago, an asteroid over six miles wide hit shallow water off the coast of present-day Mexico's Yucatan Peninsula. The Chicxulub asteroid, as it's known today, threw so much debris into the atmosphere it blotted out the sun and depressed photosynthesis in plants for 1 to 1000 years. The tsunami and superheated ejector cloud were so cataclysmic to the earth's ecosystems they're thought to have wiped out the dinosaurs. Bad for the dinosaurs, but good for us, it was the extinction-level event that's believed to have launched the age of mammals.

So what made it through that seminal event? Two classes of creature had something special that allowed them to make it through The Great Extinction: *Flying dinosaurs* and *mammals that hibernate* both had excess mitochondrial capacity that helped them get by with less. They basically had bigger mitochondrial "gas tanks" designed for flight or hibernation. That means they could "fill up" less often and scrape by while adapting to new conditions. It's thought that flying dinosaurs evolved into birds, while ancient mammals evolved into early hominids.

Whether that story is true or not, the biology of modern man still has links to ancient ancestors in the form of reflexive survival behaviors – switches encoded deep within our reptilian brain that activate survival programming. One such behavior is called the "mammalian dive reflex," in which human babies – like marine mammals such as whales and dolphins – wait to take their first breath after being born into water. This reflex is activated by cold receptors on the face. But when the "leptin-melanocortin pathway" malfunctions in us, you can get sleep apnea.

Another trait we have in common with our ancient ancestors is we've taken hibernation programming still present in some bears and have shrunken months into hours when we sleep. In sleep, we're supposed to go through cycles of ketosis and renewal. But when our sleep is disturbed, our bodies can't get into a parasympathetic resting state deeply enough to recover from a day's exertion. That shorts us on cell repair, rebalancing of biochemicals such as dopamine and serotonin, detoxification and, important for this discussion, restoration of mitochondria.

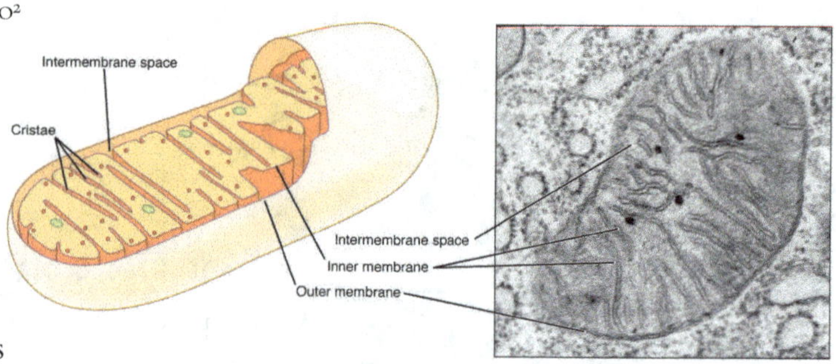

Image used under Creative Commons 4.0 license. Author: OpenStax.

What are mitochondria?

Mitochondria are double-membraned organelles (a cell's organs), often described as being the power plants of the cell. Mitochondria are essential to every member of the animal kingdom because they make a biochemical that the body uses to meet its energy demands: ATP. Through ATP production, mitochondria also produce signaling molecules that allow cells, mitochondria and the immune system to talk to each other. We call these signaling molecules free radicals, reactive oxygen species (ROS), oxygen radicals, or redox molecules... all the same thing. In addition, mitochondria make deuterium-depleted "metabolic" water.

About 200 to 5000 of these mitochondria live inside each of our cells – most cell types anyway – equaling 30% of a person's dry weight. They take care of the cell in so many ways (or not). However, they are their own (dependent) life form and possess their own DNA. So they ride along inside each cell. Yet they live, reproduce and die separately from their host's cells. Well, what a peculiar pairing of independent entities. Why, do you suppose, did they end up joining forces?

Of epoch consequence, mitochondria and multi-celled organisms made a deal to work together millions of years ago in what would become the most successful partnership in earth's history. Mitochondria agreed to take up residence inside the cells of eukaryotes and get fed on a regular basis by the organism. In exchange, they **make ATP** to power many of our cells' activities (1). They also make the best kind of water for our biology, which is **deuterium-depleted water** (2). They generate **magnetic flux** (3). They produce **redox potential** (4). They're **environmental sensors**, as explained in the previous chapter (5). And they **liberate light** from electrons (6). **These are mitochondria's 6 most important functions.**

Deuterium: A hydrogen atom with an extra neutron. Nature uses deuterium's different shape to control biological programs such as energy production, food seasonality, and aging.

Basically, the more mitochondria you have, and the healthier they are, the more energy you can produce, and the better your cells operate in many ways. Conversely, depressed mitochondria equate to chronic low energy, poor cell-cell communication, reduced cell repair and accelerated aging. Unfortunately, mitochondria populations weaken (1) as you age; (2) when toxins and disease mutate their DNA; (3) when you lack sleep and exercise; (4) when their communication network breaks down; and (5) when the both of you are malnourished, among other reasons.

Reactive oxygen species (ROS, aka free radicals, oxygen radicals or redox molecules) metabolism makes dozens of different ROS molecules characterized by oxygen atoms with one or more unpaired electrons.

But ATP isn't what we think

Biologists say mitochondria make the only form of energy our cells can actually use – a storage molecule for electro-chemical energy called adenosine triphosphate, or ATP. But, as it turns out, that's only partly true. ATP is indeed essential to the operation and maintenance of the human body, just not for the reasons we think.

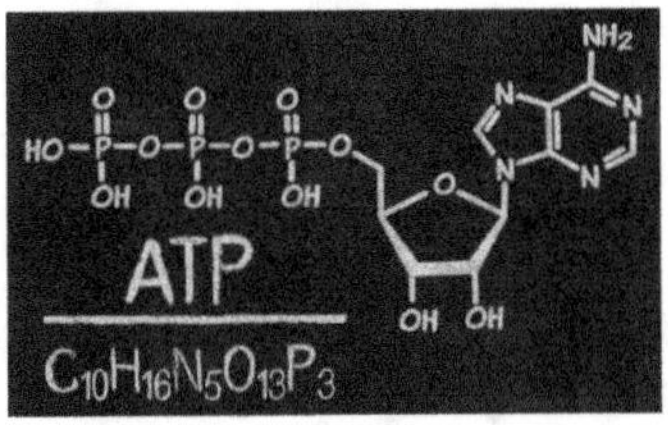

Conventional wisdom says ATP is used to activate ion channels like VGCCs, drive chemical reactions, help transmit nerve signals, control DNA synthesis, and even organize cell expiration. All true. However, leading minds now know ATP's primary role in biology is it unfolds proteins to expose their binding sites to water. Proteins then function correctly when they're properly shaped, which is hydrated. At the same time, water in a special form provides power to the cell with net-negative electrical charge.

The more water that surrounds proteins in cells, the bigger the battery that cells have to power their processes. That battery is a particular form of water – now called an "exclusion zone," or what I call "e-zone" when referring to the substance, or "e-zone layer" when describing its formation. E-zone is basically water that's chemically restructured into a honeycomb shape so some of its positively-charged hydrogens are separated from its negatively-charged oxygens. This charge differential is what gives cells the DC electricity they need to run their operations.

As part of this process, ATP helps e-zone form around proteins by relinquishing one of its phosphate groups (left arm of the ATP molecule above) to release electron energy contained in its bonds. The adenosine triphosphate molecule (charged-up) becomes adenosine diphosphate (partly discharged). And those electrons are voltage that make enzymes and proteins change their structural posture, thereby enabling them to do what they're designed to do. Now you know why mitochondriacs think of ATP as one of the body's main mammalian batteries: **ATP is portable electricity that switches proteins into a different "conformational" shape. The shifting of shape is the mechanical action by which many proteins accomplish their work – pumping, for example.**

Conformational: An object's shape or structure.

In the use case described above, ATP *helps* proteins harvest and store energy from the environment by opening their structure to charge-separated water. Therefore, ATP is important, but for different reasons than that claimed by orthodox medicine. This is a hard pill to swallow for almost every academician and researcher who learned ATP is a fuel – no more and no less. But how can we be so sure, despite 50+ years of believing that ATP is the biofuel of the body?

Dr. Peter Mitchell: Scientist famous for discovering how mitochondria make energy (ATP): electrons jump along the ETC, pumping protons as they go, finally the fifth cytochrome adds a phosphate group to adenosine diphosphate (ADP) to form adenosine triphosphate (ATP).

Well, in 1950, Dr. Gilbert Ling showed that if ATP really powered all 100,000 biological processes taking place in a cell, it would break the second law of thermodynamics by 500-fold. That means there's no way on earth mitochondria can make enough ATP to accomplish all the things that popular science thinks it does. As a result, some of Dr. Peter Mitchell's conclusions about ATP must be wrong – brilliant as they were and still are – despite the fact he won a Nobel Prize in 1978 for describing this process of "chemiosmosis" (aka oxidative phosphorylation) that makes ATP. The numbers just don't add up. Problem is, no one was listening to Ling, and the misconception about ATP persists to this day.

So is there any validity to the idea that ATP powers almost all of the body? Could so many scientists and researchers have been so wrong, for so long, about such a fundamental principle? A: I believe time will show ATP is an essential *enabler* in the processes by which proteins operate. You see, researchers observed ATP performing mechanical and chemical operations around the body, but could not see that which they were not looking for: ATP's electrical energy making proteins bend so they functioned correctly. Honest mistake, because ATP does carry electrical energy, but important distinctions to make, nonetheless.

A 155-pound person needs to make 170 pounds of ATP per day

That's at rest. When you exercise, you need to make closer to 185 pounds of ATP per day. Think about that for a moment. You actually need to make more ATP each day than you weigh! But here's the crazy part: Food is only supposed to provide ⅓ of the electrons to make that ATP, while ⅔ of those electrons are meant to come from sunlight, e-zone and grounding. Unfortunately, most of us don't get anywhere near that volume of electrons from traditional sources living the way we do. And that, believe it or not, is a primary cause of energy shortages that turn into modern diseases – mostly due to lack of sun, pure water, and direct earth contact.

Overview of cellular energy production

Cellular respiration. The body employs three major, multi-step processes to make energy, in addition to several intermediate sideline processes. The whole energy production process, called cellular respiration, takes chemical fuels such as glucose and fat, combines them with oxygen and burns them to make ATP, metabolic water and carbon dioxide. Meshing like the gears of a watch, the major metabolic pathways involved in making energy include glycolysis, the Krebs cycle, and the electron transport chain. The pyruvate pathway is one of the

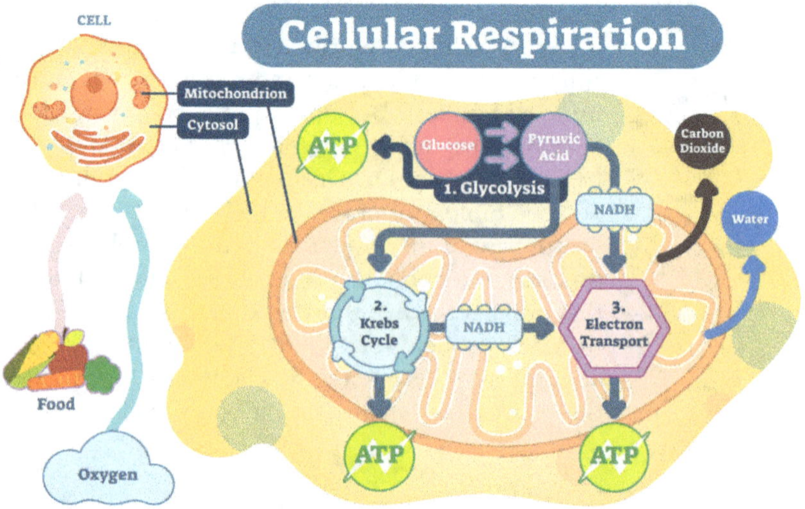

The pyruvate pathway is explained on page 50.
Beta-oxidation is the process by which long-chain fats are broken down to move ATP production forward.
The Krebs cycle (aka citric acid cycle, TCA cycle) is covered on page 50.

main gears in the middle that connects glycolysis to the Krebs cycle (aka TCA cycle, citric acid cycle). And beta-oxidation (aka fatty-acid oxidation) is one of the sideline processes that breaks carbon groups off long chain fats to make acetyl-CoA to feed the Krebs cycle.

Warning: This section gets technical. It's intended to help biologists and health professionals understand how energy is made in the body, and where breakdowns occur. **Health consumers reading for their own education can safely skip the next five pages and not miss anything you need to know.** Indeed, cellular respiration is so ridiculously complex, yet powerful and highly efficient, you'd think Germans had (over)engineered the entire process like a fancy luxury car. Instead of envisaging the electron transport chain from words, I recommend everyone just watch the YouTube video noted (left).

To see how mitochondria make ATP through their ETCs, watch the YouTube video entitled "Electron Transport Chain (Oxidative Phosphorylation)": https://youtu.be/LsRQ5_EmxJA.

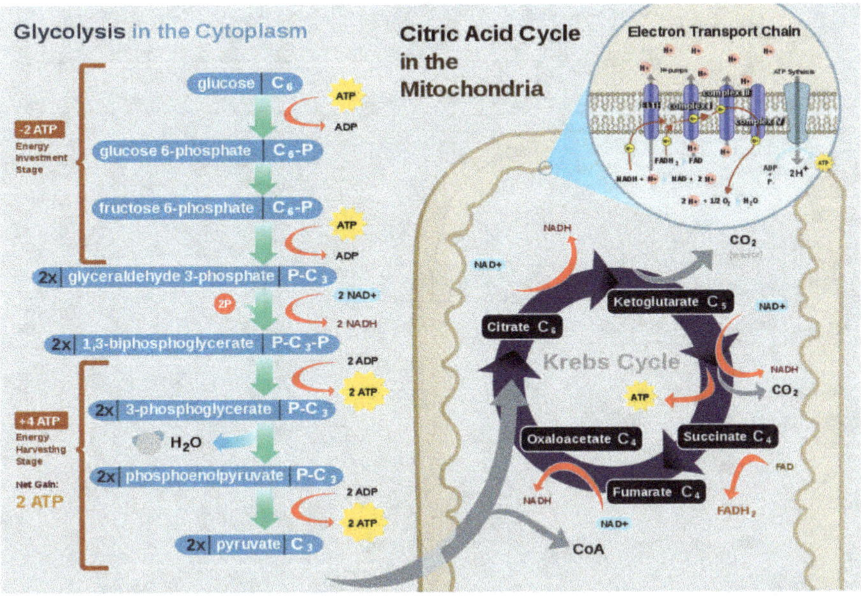

Image used under Creative Commons 3.0 license. Author: RegisFrey.

Glycolysis. Glycolysis happens in the interior of the cell, called the cytosol, not in the mitochondria. The main purpose of glycolysis is to make pyruvate to feed the Krebs cycle, and a small amount of ATP. Glucose goes into the cycle. ATP are invested as fuel to power the first three stages of glycolysis. And, through some seven-odd stages, sugar-based biomolecules are converted from one into another, to arrive at pyruvate, which powers the Krebs cycle. And the investment of two units of ATP through this glycolytic process yields four units of ATP at the end, for a net gain of two. In short, glycolysis is a weak way to make energy. But the process almost virtually unbreakable.

Pyruvate pathway. Pyruvate's most important function is that it gets converted into acetyl-CoA – the chief reactant that drives the ensuing TCA cycle. That is, acetyl-CoA is the chemical compound that, like pedaling a bicycle, propels the 8-plus enzymatic steps of the Krebs cycle forward… over and over again. The interesting thing about this step is that lactic acid, which is part of the pyruvate pathway, gets made as a by-product of ATP production in conditions of over-exertion, hypoxia/ischemia (lack of blood flow), infection and heart failure. In relatable terms, that means lactic acid is the chemical that makes muscles sore when you exercise too hard.

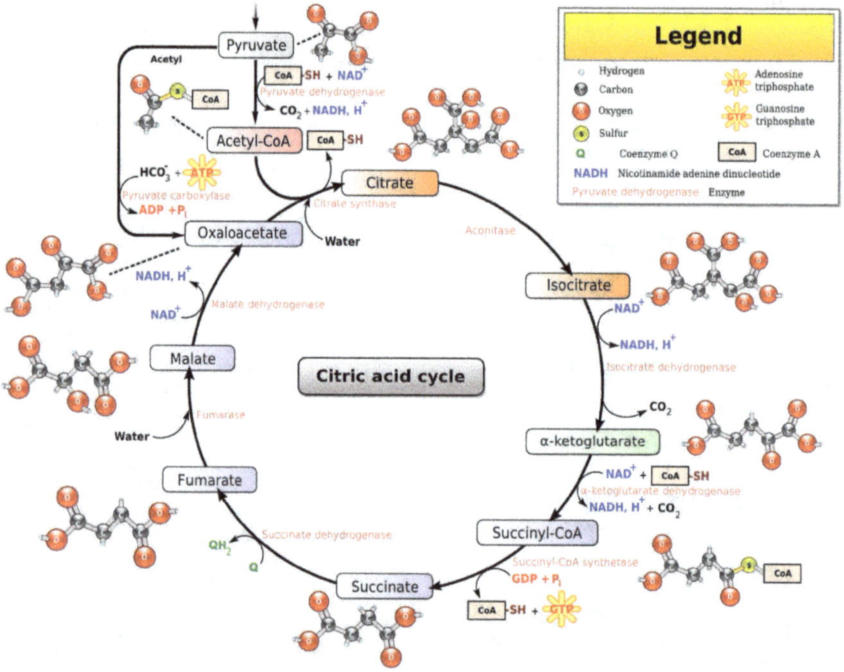

Image used under Creative Commons 3.0 license. Author: Narayanese.

Tricarboxylic acid cycle (aka the TCA cycle, citric acid cycle or Krebs cycle). The main purpose of the TCA cycle is to make input materials for the electron transport chain, in the form of NADH and FADH2 molecules. Specifically, respiratory complexes need hydrogen protons to produce a voltage differential (i.e., proton pools of different concentrations). And they need electrons to pump those protons from a mitochondrion's interior into its intermembrane space. The TCA cycle makes NADH and FADH2 molecules through nine enzymatic steps you don't need to know about unless you're a pre-med student studying for a biology exam.

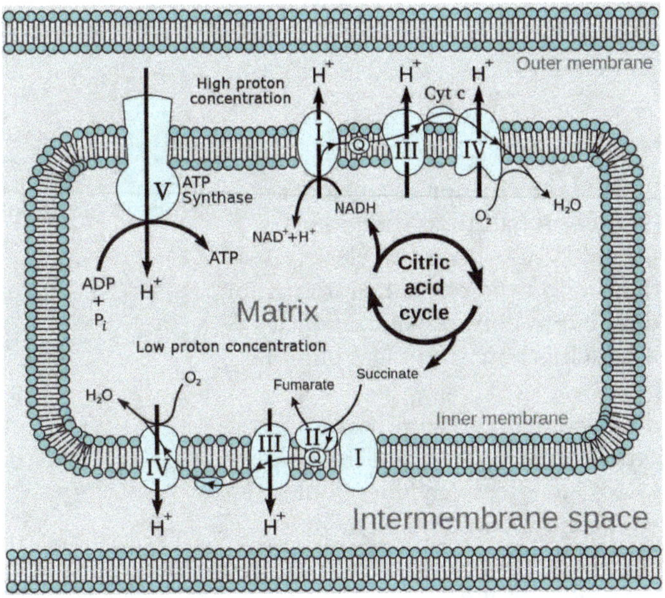

Public domain work. Author: Tim Vickers.

Note: Roman numerals I-V, which denote the respiratory protein complexes, or "cytochromes."

FYI: This graphic shows us that there are two different pathways for electrons to enter the ETC to pump protons: NADH at Complex I or FADH2 at Complex II, though the ETC is often perceived to be one sequential chain.

Electron transport chain (aka oxidative phosphorylation). This is the "main event" process of energy production in a multi-celled organism from the animal kingdom. Mitochondrion have two cell membranes to contain a pool of protons between them (called the intermembrane space). The electron transport chain (ETC) is a set of protein structures that sit on the inner membrane.

The ETC uses a flow of electrons from food and other sources (via chemical storage molecules NADH and FADH2) to pump hydrogen protons from the interior of the mitochondrial "matrix" into the intermembrane space (between the mitochondria's inner and outer walls, aka its "phospholipid bilayer"). The high concentration of protons in the intermembrane space (relative to the matrix) offers an electrical gradient, or voltage differential, that drives the rotating head of the ATP synthase, like water turning a hydroelectric turbine.

Technically, the electron transport chain only refers to Complexes I–IV, because Complex V (ATP synthase) only deals with protons; it doesn't transport electrons. But it's often included in the ETC for the sake of simplicity. Here are a few noteworth details about the ETC, included for their special significance in energy efficiency and mitophysics in general.

- Electrons from carbs tend to make the molecule NADH. NADH deposits its electrons into Complex I. While fat electrons tend to make FADH2, which deposit its electrons into Complex II. For those interested in seasonal eating, this means if you eat carbs when your respiratory complexes are uncoupled (e.g., when you generate heat in the winter), then more free radicals will pour out of Complex I and damage the mitochondrion's DNA, which is nearby. Conversely, eating carbs in summer, when your respiratory proteins are more condensed, makes just the right amount of free radicals to induce mitophagy (mitochondria recycling) in a controlled manner.
- Once inside, electrons navigate the respiratory complexes through a series of stations, called "redox centers." As electrons jump from one redox center to the next, they release photons of different color frequencies. These photons represent both energy and information, which was encoded onto them by sunlight in photosynthesis, resulting in a certain orbital state of electrons. Respiratory complexes

then capture this photonic energy and information as an electron falls from an excited spin state (higher orbital) to its ground state (lower orbital). Complexes I, III, and IV use electrons' electrical energy to pump protons. And they collect the photonic information about the plant's growing environment to control your seasonal biorhythms – most important of which is how many UV photons the ETC harvests from food electrons.

- When Complexes I and II are done extracting energy and information from electrons, the molecule CoQ10 loads them up and transports them to Complex III. CoQ10 effectively does double duty moving electrons, which is why taking a CoQ10 supplement is known to enhance energy production (particularly in the heart), and why it's a good antioxidant (i.e., electron donor).
- Electrons from Complex III move to Complex IV via cytochrome C.
- Complex IV makes water. This is where most metabolic water comes from. It's the best kind of water to build e-zone, and harvest light energy, because it's naturally low in deuterium.
- Oxygen molecules receive almost all the electrons coming out of the ETC, which ends at Complex IV. Without oxygen to carry electrons away from the ETC, the ETC gets backed up, ATP production drops, and free radicals have a feeding frenzy on your mitochondria's DNA.
- Lastly, Complex V, the ATP synthase, uses the mass of protons between the two membranes to spin its rotating head. Its machinery adds a phosphate group to adenosine diphosphate (ADP) to become adenosine triphosphate, or ATP. **The electro-chemical energy held in this third bond is one of the body's main power sources to move materials and drive chemical reactions. So ADP, collectively, is like a discharged battery, while ATP is a battery that's fully charged, though not confined to one area.**

Note: Dr. Jack Kruse and others commonly refer to Cytochrome V as the ATPase due to incomplete understandings recorded in the literature. It's only fairly recently that the respiratory complex which makes ATP (by attaching a phosphate group to ADP) have come to be known unequivocally as the ATP synthase. Conversely, ATPase describes an enzyme that breaks down ATP into ADP + P.

Let's recap cellular respiration. Glycolysis and pyruvate metabolism make input materials for subsequent processes, and small amounts of ATP. The TCA cycle further contributes to energy production processes by making NADH and FADH2. But the big payoff of ATP happens in the ETC. Oxidative phosphorylation (ox/phos) uses protons and electrons from NADH and FADH2 to make two to three times as much ATP as all the earlier steps combined. Combine that with fatty acid oxidation (it's complicated), which feeds acetyl-CoA into the TCA cycle, and that's how mitochondria make 140+ molecules of ATP from ox/phos of fat, compared to 30-something from carbs.

Amusing but useless factoid: We've examined a single electron transport chain and its respiratory complexes in isolation. But in reality, each mitochondrion's entire inner membrane surface is densely-packed with about 3000 complete ETCs. So if we add up all the cells in a human body (~37 trillion), times the number of mitochondrion in each cell (200–5,000), times the number of ETCs each mitochondrion has, that means you have about **111 quintillion electron transport chains in your entire body** (37,000,000,000,000 x 1,000 x 3,000 = 111 quintillion)!

That's what makes your inner mitochondrial membranes unimaginably powerful at making ATP, electric fields and magnetic fields. In fact, despite their tiny individual size, you can even measure mitochondria's combined electric and magnetic fields several feet from the organs that have the highest densities: the heart and the brain. We know these tests as EKGs and EEGs.

Oxygen accepts electrons exiting the electron transport chain

Everyone knows you need more oxygen when you exert yourself. And everyone knows that oxygen, in general, is good for your health. But how many people know why oxygen is good for you, exactly? And how many people know what happens when you're deprived of O2? A: Your respiratory rate increases when you exert yourself because mitochondria need more oxygen to receive the electrons coming out of mitochondrial metabolism. That is, electrons from food and fat-burning need to be carried away from the "tailpipe" of the electron transport chain or else the ETC gets clogged, masses of reactive oxygen species (ROS) get made, and mitochondria can't make as much energy (ATP).

In other words, oxygen supply *at the end of the electron transport chain* needs to match electron flow *going into it and through it,* or else you get a bottleneck in the manufacture of ATP, and the system starts making excessive amounts of mischievous molecules called free radicals. That's why we breathe harder when we exercise or exert ourselves: O2 is the taxicab that accepts those electrons into its molecule, and carries them away safely, without creating harmful by-products or imbalances.

Free radicals are made by shortages or surpluses at any point in the electron transport chain

Reactive oxygen species (aka free radicals or oxygen radicals) are like smoke from a fire. They are the result of too much of one input, and not enough of another, anywhere in the ETC's oxidation combustion process. These imbalances produce "hiccups," or small imperfections, in the oxidation and reduction of molecules, which become molecules with one or more unpaired electrons. However, it's normal and healthy for the ETC to make a small amount of free radicals as they go about their business, because these signaling molecules control how the nucleus and mitochondria fine-tune metabolism and regeneration.

To give you some idea how prevalent free radicals are, 2–5% of electrons tunneled through the ETC become free radicals. And the oxidative phosphorylation process will normally turn about 0.4–4% of oxygen consumed into superoxide free radicals. If you're on the low end, the efficiency of your mitochondria rock. On the high side, that means your mitochondria are struggling to make ATP, and free radical damage has likely turned from acute and helpful to chronic and toxic.

Current research indicates the majority of free radicals are made at Complexes I and III. Complex I seems to make more ROS when the ETC receives more input materials than it can handle. While further down the chain, it looks like Complex III makes more free radicals when ATP is not being used up fast enough, so not enough of ATP's breakdown product ADP is available to feed back into the ATP synthase to remake ATP.

In real life, that means eating high-energy fats in a hypoxic state (lacking oxygen) produces lots of electron flow through the ETC, but not enough oxygen to carry those electrons away. Like a fire with too much fuel and not enough air, this creates incomplete combustion products at various stages of the ETC, which we call reactive oxygen species. On the other hand, eating carbs in a hypoxic state is a better match because carbs produce a smaller stream of electrons. Thus, the "fuel-to-air" mixture of combustion is balanced so you make a healthier amount of both ATP and oxygen radicals.

Another too-common situation of special significance is eating foods out of season. You see, metabolism pathways such as oxidative phosphorylation, the TCA cycle, urea cycle, and glycolysis are precise and particular in their tuning. So when you eat foods that contain more photonic energy than your circadian programming is expecting – because light on the eye and skin is different than the light released by your food – that mismatch ruins the ETC's pre-set programming. The program that's running is mis-calibrated for the fuel/food you're taking in, basically. Rough running combustion then creates excess free radicals, more positive charge/acidity, more inflammation, and a lower redox state.

In case the point slipped by you, this is the single biggest reason eating fruits, vegetables and carbs out of season is worse for you than you might have thought. It's because you suffer extra free radical damage, more mitochondrial degradation, inflammation goes up, and you lose healing capacity because your redox potential drops. A perfect example you may not have considered is refined vegetable oils. One reason omega-6 vegetable oils from soy, corn and canola are inflammatory is because their electrons contain the energy signature of the strong sunlight they're grown in. Meanwhile, our metabolism wants to run a different program when our eyes, skin and gut are getting the message that it's fall or winter. This means man-made vegetables oils present a concentrated circadian mismatch to our metabolism, thus raising ROS, positive charge and inflammation. Your sources ever tell you that?

What happens when mitochondria lose their mojo?

When mitochondria are inefficient at producing energy, you tend to gain weight or lose weight. You can't regenerate cells as well. And, most important for this discussion, the diseases of aging accelerate in whatever organ or system has the weakest mitochondria from bad genes and toxic exposures. In fact, the vast majority of disease and aging is a function of mitochondria's potency and population density. So when your mitochondria fail to produce energy efficiently, first organs and systems go into power-conservation mode (which you usually don't notice). Then they malfunction. Then they shut down. And then you're in real trouble. We call that disease and death.

This decline in mitochondrial productivity through genetic inheritance, toxic environment, and/or age shortens the time your body is able to maintain energy, health and youthfulness. Historically, the process of going from a low heteroplasmy rate (strong mitochondria) to a high heteroplasmy rate (weak mitochondria) took 5–8 decades to deteriorate to the point of incompetence. But, now, we're seeing more and more cases of mitochondrial insufficiency developing over the first few decades of life, as it does in childhood cancers. In extreme cases, we even see mitochondrial collapse happening in the womb before birth, as with some birth defects. But whatever age it happens at, this process of mitochondrial degradation is equivalent to having the life sucked out of you and decay setting in.

One coping mechanism employed by the mitochondria is this: Cells that can't make enough ATP revert to a much less efficient fallback mechanism of fermentation (aka glycolysis), which is sugar-based energy production. Energy-starved cells basically resort to consuming sugar to try to sustain themselves. And that manifests as sugar and carb cravings, which contributes to gluten sensitivity and candida overgrowth.

"Mitochondrial coupling" controls efficiency of ATP production

Mitochondrial coupling efficiency explains why some people are born with a "fast metabolism" and have no trouble maintaining their weight, while others have a "slow metabolism" and tend to gain weight easily. A fair amount of it (its basis) is rooted in where your ancestors came from – specifically the women in your family tree, because mitochondrial DNA is passed exclusively through mom's side.

Before modern times, people born closer to the equator didn't need to generate as much heat internally to maintain their body temperature. They also led more active, outdoor lifestyles in lots of sun. So their mitochondria made more ATP and less heat/waste. This type of mitochondria is described as being "tightly coupled," referring to the spacing of respiratory proteins, which enables efficiency. Tightly-coupled mitochondria expend a lower percentage of calorie consumption out of their ETC as heat.

On the other hand, people who lived in colder Northern climates gained a survival advantage when their mitochondrial DNA mutated to make more heat and less ATP. Their mitochondria leaked more protons out of the ETC to make heat. Those happy accidents survived to be passed on. We say their mitochondria are "loosely coupled," which equates less ATP and more heat. These people have an advantage in losing weight and keeping it off, because a higher percentage of their calorie consumption exits the ETC as heat, rather than ATP.

Now, at face value, you'd expect this respiratory complex heteroplasmy to weaken mitochondria. But, to offset this effect, the release of IR heat condenses the water around the respiratory proteins, which helps maintain their efficiency so mitochondria don't go lame. Also important to note, nutrient extraction (e.g., minerals, antioxidants, probiotics, moisture content, fiber) happens at the gut level, prior to ATP production. That's why loosely-coupled people don't necessarily just eat more to make up for the drop in ATP. In reality, most of them probably feel compelled to change *what they eat,* more than they change *how much*, because their mitochondria do operate better on fats and protein than carbs.

Again, the difference in coupling efficiency is a function of how expanded or condensed the respiratory complexes are. The closer the respiratory proteins of the mitochondria are to each other, the easier it is for electrons to jump from one spot to the next (redox centers) along the electron transport chain. Conversely, when respiratory proteins are more spread out, electrons can't move as easily, more heat escapes from the system, and ATP production drops. However, as we all know, body heat from metabolism goes up when you exercise, regardless of where your coupling efficiency started out, or its present state.

One mechanism that mitochondria use to fine-tune their coupling efficiency is through use of a protein called "uncoupling protein 2." In the same way that a carburetor adjusts an engine's fuel-to-air mixture, *uncoupling protein 2* controls how much deuterium is allowed into Krebs Bicycle (not to be confused with the Krebs *cycle*, which is part of Krebs *Bicycle*). You see, deuterium stymies the enzymatic steps of metabolism. So, by controlling efficiency of steps in the citric acid cycle and urea cycles, efficiency of metabolism is increased or decreased. That's how metabolic programming is designed to work anyway. As we'll discuss

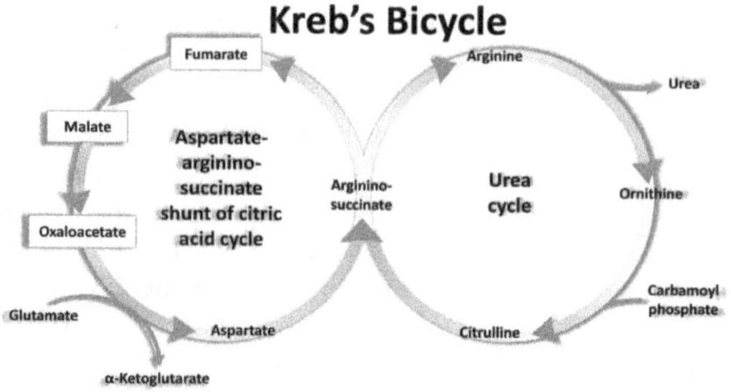

further, excessive deuterium fouls up energy production in mitochondria.

Interestingly enough, coupling efficiency can be traced back along your maternal lineage many tens of thousands of years to the migration patterns of your ancient ancestors. In fact, Dr. Doug Wallace has traced mitochondrial lineages back a quarter million years to today's central Africa – now called the "L-Zero" haplotype.

Dr. Doug Wallace: The world's leading mitochondria researcher. He taught the field most of what it knows about mitochondria, including how they are generationally inherited, and how local food and climate controls metabolism, epigenetic expression, and aging (through free radicals).

Mitochondria haplotype

We call the mitochondrial genes of your ethnic lineage your "haplotype" or "haplogroup" (think *ancestral*). Your haplotype controls how tightly- or loosely-coupled your mitochondria are made to be straight "from the factory" – meaning, how spread apart your five respiratory complexes are at birth. Coupling efficiency then morphs slowly over time on a daily and seasonal basis in response to temperature, food, solar yield, disease/infection, altitude (O_2 level) and toxins.

Your haplotype comes from what area of the planet the women in your family tree came from. This tells you something about mitochondrial DNA mutations that occurred in an area. When environmental adaptations proved to be useful, the changes were passed on to future generations. When the changes made your life worse, the mutations eventually were corrected by their genetic machinery, or the person didn't make it, thereby removing those genes from the gene pool.

Ideally, the distance from the first cytochrome to the fifth should be about 36–40 angstroms (super-small distance). That's where mitochondria operate most efficiently. But environmental exposures cause the respiratory proteins to expand or contract, altering both efficiency and potential side effects. *Unfriendly forces* working against you stretch the respiratory proteins out so they might be 60 angstroms apart in an extremely obese person with run-down mitochondria. On the other hand, when *favorable exposures* work to your advantage, respiratory proteins contract closer to the 35–40 angstrom ideal, and ATP production improves.

The takeaway lesson: It's inaccurate to think of either loosely-coupled or tightly-coupled mitochondria as being bad or unhealthy. The right way to think about it is how closely, or poorly, matched your coupling efficiency is to *your* environment, *your* biology and your age. That is, when you go from where you started, and you lose gobs of coupling efficiency inordinately fast from smoking crack, or even a highly mismatched diet, that's how resulting ATP shortages become dysfunction and disease. In other words, it's how much looser your coupling efficiency is from where it should be. That spells trouble, because that's what causes organs to malfunction and your health to decline. On the flip side, when you optimize your coupling efficiency to its theoretical maximum, perhaps even exceed it, you'll live forever and never get sick ;-)

Visualize energy inefficiency in stars

When a star passes its prime, it becomes energy-inefficient. No longer able to burn as hot due to weaker combustion, it expands into a red giant and loses thermodynamic efficiency. At that point in the star's life cycle, its energy production become wasteful, and so the star literally gets bloated. Later, as it runs out of fuel toward the end of its life, it shrivels up and dies. Well, thermodynamic laws are the same at the microscopic level as it is at a galactic level. So the same thing happens in humans when their mitochondria become inefficient at making and using energy: they tend to get fat. Excess weight in people is a strong indicator that their mitochondria consume too much fuel per unit of energy made – in this case, calories per ATP. Basically, their mitochondria are decoupled and damaged, so the body expands. The same thing happens with the heart. When the heart weakens, it expands to try and compensate. We call that congestive heart failure. In summary, stars, proteins and hearts are more thermodynamically efficient when they're compact *vs.* bloated. So to get that sleeker physique, focus on fine-tuning your engines.

Bad genes don't cause disease... human genes, that is

In his over 40 years of ground-breaking research, Dr. Doug Wallace discovered that modern disease conditions such as obesity, heart disease and neuro-degeneration occur with no genomic changes. That means maternally-inherited and accumulated defects to mitochondrial DNA contribute more to disease than your own human genes do. He's even been able to correlate mitochondrial mutations of various haplogroups with the prevalence of major diseases such as diabetes, heart disease and autism. Simply put, disease is caused mostly by feeble mitochondria and the epigenetic changes they exert upon our gene expression.

Cancer research is focused on the wrong genome

"We need to be looking in the mitochondria, not the human genome, because when we study cancer, we're looking at genes that have already misbehaved. So to really, truly understand cancer we need to know what happened to those mitochondria before the cancer came. That's where we get into the heteroplasmy rate. That's where we get into the distance between respiratory proteins."

—*Dr. Jack Kruse.*

Aging and illness are functions of "percent heteroplasmy rate"

When mitochondriacs talk about heteroplasmy rate, generally speaking, they mean the global amount of degradation your mitochondria have accumulated to their DNA, and thus their output. The cause: respiratory

Complex I sits next to a mitochondrion's DNA. So high heteroplasmy there besieges a mitochondrion's DNA with free radicals, which is how mutations occur. Technically, percent heteroplasmy rate is defined by the percentage of healthy mitochondrial DNA (mtDNA) in the genome *vs.* mutated. In other words, heteroplasmy is the mixture of normal *vs.* altered DNA. But, interestingly enough, mitochondrial DNA only makes thirteen of the protein 'subunits' that form respiratory complexes, while nuclear DNA makes the other sixty-one.

Acting something like the wiring diagram for a power plant, the respiratory complexes determine ETC efficiency. **The greater the uniformity of the construction, the better your mitochondria function. The more disparate their DNA, the more dysfunctional their productivity will be.** That's because if any respiratory complex were to tunnel electrons at a different speed than the rest, due to mismatched mtDNA, you'd get an electrical short in the system. Electrons will leak out of the electron transport chain where the mismatch occurs and produce loads of reactive oxygen species.

However, not all mutations are bad. Some alterations end up helping the recipient adapt to their environment, while others disadvantage the individual in their ability to thrive locally. Thus, those alterations are less likely to be passed on. Dr. Wallace sums up the *advantages* acquired over time as "ancient adaptive polymorphisms," and the *disadvantages* as "recent deleterious mutations." A practical example: Moving from the tropics to the Arctic Circle can make a once well-adapted haplotype (what you might call "mito-type") person suddenly become poorly-adapted to their climate, and vice-versa.

The starting point for your heteroplasmy rate comes from how much defective mitochondria you inherited from your mother. Mitochondrial biologists put these defects in the category of **"maternally-inherited mutations."** From that bioenergetic baseline, your percent heteroplasmy rate typically rises as a result of mutations that alter protein synthesis. You can put these changes in the category of mutations due to environmental exposures and disease. Then, your heteroplasmy rate naturally increases as you age from mutations in both the mitochondrial genome and the nuclear genome.

These increases in mitochondrial diversity fit in the category of **"somatic mutations"** (meaning after birth). Problem is, mitochondrial DNA has 10–100 times greater propensity to mutate than our own DNA. And these naturally-occurring mutations often get unequally distributed when the host cell divides and mitochondrial populations have to split up. So Nature installed powerful mechanisms to prevent differences in mitochondrial DNA from happening at birth, and increasing over time.

On a generational scale, you inherit as low a heteroplasmy rate as possible by having mitochondrial DNA inherited only from the maternal side, not from dad. Nature could not mix mitochondrial DNA from both parents, because doing so would potentially create respiratory proteins with different electron tunneling speeds. That would start you off in life loaded with free radicals and a high heteroplasmy rate. After that, heteroplasmy rate is kept in check on a day-to-day basis, and over a lifetime, by autophagy, apoptosis, heat shrinking the respiratory proteins, and heat shock proteins (explained on next page).

Autophagy: How the body reconditions mitochondria

Autophagy is the controlled breakdown of cells so their contents can be recycled and reused – called "mitophagy" in the case of mitochondria. It's one of the primary processes that keep your mitochondria young and your internal energy production high. Autophagy requires good sleep, intermittent fasting/ketosis, and regular exercise in order to maximize oxygen level, electron flow, magnetic field, and the controlled production of free radicals. In order to make new mitochondria, you first need free radicals to break down the old, worn out mitochondria to make room for replacements. This is the job of the oxygen-based free radical called superoxide.

Free radicals are made when a stream of electrons from the electron transport chain energizes electrons in metals such as iron, copper and manganese in the mouth of cytochrome complexes. It's these transition metals, and their excited electrons, that hold oxygen to the inner mitochondrial membrane by electrostatic cling and strip it of an electron, thereby making a free radical. Unfortunately, if you have a reduced electric and magnetic field due to low electron flow through the electron transport chain (e.g., stretched out respiratory proteins), you can't hold oxygen in place and remove an electron. When that happens, autophagy suffers because you can't make free radicals that pave the way for new mitochondria to replace old.

To sum it up, when you're low on oxygen, DHA, voltage, water or solar yield, you can't make free radicals when you're supposed to. Specifically, if those things are substantially lacking, you can't make superoxide. And when you don't have superoxide to break down old mitochondria and signal for new ones to be made, you can't make new mitochondria. That forces your body to operate on years-old mitochondria that should have been retired long ago. When that happens, your worn-out mitochondria can't produce enough energy to regenerate tissues and organs. Hormonal signaling can suffer. And lots of bad things happen health-wise. Unfortunately for current and future generations, a lot of people with type 2 diabetes, Alzheimer's or

Parkinson's suffer from this poor mitochondrial recycling. Measure their superoxide level, and you'll find it's absent.

Some famous food gurus are stuck in this dilemma too. Because fats and proteins make more ATP than carbs, they assume the longer they're in ketosis, the better. But little do they know, being ketotic all the time is like rowing a boat with one oar: You are indeed moving, so at first it seems like you're making progress. Unfortunately, avoiding carbs entirely means you don't make any superoxide radicals. Hence, you don't trigger autophagy to freshen up your mitochondria. So mid-to-long term, you and your mitochondria are rowing around in circles. Another half-truth becomes a full lie.

Apoptosis

In a related process, when it's time to cull the herd, ATP actually helps mitochondria self-destruct in a controlled manner. Programmed cell suicide is called "apoptosis." It's the process of self-sacrifice when a cell is damaged beyond repair. Unfortunately, when you don't have enough ATP and UV biophotons, apoptosis falters. When apoptosis malfunctions, mitochondria stick around past their expiration date and limit the energy you have available. Point being, even the proper expiration of mitochondria requires energy.

Heat Shock Protein 70

Mainly affecting people of Northern descent, cold stress (or heat stress) releases something called "Heat Shock Protein 70" (HSP 70) in the cell. HSP 70 stabilizes the size and shape of proteins in mitochondria, so the respiratory complexes stay the same and don't get too stretched out over time. It's one of Nature's ways of keeping a Northerner's mitochondria efficient by limiting the degree to which their mitochondria uncouple themselves, which is their natural inclination.

This then minimizes heteroplasmy, ATP loss, and excessive free radical production. In other words, if you never let yourself get cold (or particularly hot) – as a Northerner would in the wild – you lose the benefit of this protective programming. Your mitochondria are then free to swell, divorced from Nature's normal system of checks and balances. Sound like any modern humans might be stressing out their mitochondria with this very situation today?

Let's review: You can slow down the rise in heteroplasmy rate, and even reverse it to some extent. But time ultimately makes weaklings of all our mitochondria. Barring an effort to reverse accumulated damage, your percent heteroplasmy rate rises as you're exposed to toxins, diseases, environmental factors (e.g., excessive carbs), and never getting cold, as well as passage of time. In fact, heteroplasmy and aging are virtually one and the same because, oddly enough, heteroplasmy causes *biologic* aging, and *chronological* aging causes heteroplasmy. (Whoa, that's deep.)

But why does heteroplasmy affect one organ and not another?

Each organ has different energy needs. So a loss of mitochondrial efficiency in a high-energy consumer disproportionately affects it more than other organs. The brain, for example, has the highest energy demand, with low reserves. It's 2% of the body weight, but uses 14% of the blood flow and 20% of the oxygen. So a 10% decrease in systemic energy has huge effects on the brain, without affecting your toes. The heart, muscles, kidneys, hormonal system and liver also have high energy needs. But they also have greater reserves to cope with deficits.

To sum up the importance of heteroplasmy rate, **when mitochondrial energy production drops below minimum threshold levels, organs fail and disease shows up.** So, like a flashlight with a low battery, first you get a brownout, then your light doesn't shine at all. Indeed, cutting-edge thinkers believe heteroplasmy rate will soon become one of the most accurate measurements of just how much time you have left before disease shows up. To illustrate, let's just make up some numbers and say diabetes shows up when your heteroplasmy rate crosses 45%; cancer and Alzheimer's show up when your heteroplasmy rate hits 65%. You can think of this as your "aging clock."

Until those numbers are worked out, percent heteroplasmy rate represents how productive your mitochondria are at producing ATP, electrical charge, magnetic field and water. These fundamental mitophysical forces power our biology when plentiful and pure. So their absence invites sickness when the body is forced to make compromises. This is why most super-centenarians have a brain that works well, but you never see huge bodybuilders at 100+ years old. It's because maintaining enormous muscle mass is a big mitochondrial expense the body can't afford in old age. At some point, there isn't enough energy to pay for luxuries like that. So the people you see that make it to the century mark have strong mitochondria in all the right places.

Is there a test to measure heteroplasmy rate?

The field of mitochondrial biology does not yet have a test to directly measure heteroplasmy rate. But you can look at a test done by SpectraCell Laboratories that measures ROS, and correlate it with another measuring TCA metabolites, to extrapolate heteroplasmy rate indirectly. Ideally, when your heteroplasmy rate is low, you tend to burn more fat than glucose. And you make a manageable amount of free radicals from the ETC. But when your heteroplasmy rate rises, your metabolism shifts away from fat-burning, toward glucose. That makes ROS and TCA metabolites shoot through the roof because they aren't being processed. In other words, when you see more free radicals and TCA metabolites, you know metabolism has shifted to glycolysis and Warburg (weak), which tells you indirectly how heteroplastic your mitochondria are.

Mitochondria convert one wavelength of light into another

Far more than just power plants, mitochondria have the fascinating quantum ability to turn one type of light/energy into another. They use this special skill to do for us what the sun does for plants: produce light to do cellular work. In doing so, **each mitochondrion acts like a mini-sun inside the cell.** For instance, the Q Cycle in the ETC (which transfers electrons from Complexes I and II to Complex III) takes the vibrational frequencies of light contained on electrons, it cuts up the shorter frequencies of purple and blue – slicing up their sine waves like a stick of pepperoni – spaces the slices out, and puts them back together to make longer wavelengths of red for use by Complexes III, IV and V. Recall that red light energizes the water around proteins, and the ATP synthase directly.

Unfortunately, when the respiratory proteins get too stretched out due to blue light exposure, nnEMFs and other mitochondrial toxins, then electrons traversing the ETC are spaced too far apart for the slices of light to fit back together properly. Unable to fabricate red from the purple and blue slices, light then leaks out of ETC, and ATP production drops. This is how a lot of energy is lost in the system, and a contributing factor in many/most diseases of civilization.

The most mischievous mitochondrial toxins

- **Non-native EMFs.** Notably, blue light and microwave EMFs.
- **Statins.** Cholesterol-lowering drugs inhibit CoQ10 production. This blocks respiratory complexes I and II from passing electrons to Complex III on the electron transport chain.
- **Antibiotics.** Mitochondria are believed to have evolved from bacteria. So anything that kills bacteria also poisons mitochondria.
- **Smoking.** Breathing toxic gasses like carbon monoxide, can't be good for your ATP and free radical load.
- **Anti-inflammatories and pain killers.** Including aspirin and Tylenol.
- **Anti-anxiety.** Including Valium and Xanax.
- **Antidepressants.** Including Prozac.
- **Antipsychotics.**
- **Artificial colors.** For example, one artificial blue color used in some candy and shaving gel inhibits oxidative phosphorylation.

We're doing everything possible to screw up our mitochondria

Unfortunately for our health and that of future generations, we're doing everything humanly possible to screw up our mitochondria:

- non-stop stress response of Wi-Fi and cell phone signals;
- circadian mismatches of artificial light and avoiding temperature changes of the seasons;

- gumming up energy production with processed foods, high carb diets, deuterium-bomb fruits, and fluoride in our water supply;
- and lack of real sun exposure and direct earth contact.

It's clear: Modern living assaults our bodies like a mitochondriac's nightmare. That's the real reason disease rates are going from terrible to tragic. Crippled mitochondria simply can't make enough energy to keep us well anymore. Without an ample supply of (1) ATP, (2) intracellular water, and (3) DC electricity to power our processes, cells can't repair and replace themselves, we detox slowly, disease and dysfunction hits you harder and more frequently, and we age faster. That's why the nexus of wellness or illness is the health of your mitochondria.

Rejuvenating your mitochondria

Most of this manifesto is dedicated to reviving your mitochondria, so we'll just point out the main pillars of mitochondrial restoration here.

- **Get more full-spectrum sun.** IR and UV light make more ATP, and they activate regeneration programs.
- **Make more melatonin**. UV light – particularly in the eye – makes tryptophan, which makes melatonin, which controls mitochondrial DNA. mtDNA controls mitophagy and apoptosis, which keep mitochondria in shape by limiting heteroplasmy rate.
- **Reduce your stress level.**
- **Avoid nnEMFs.** Cut your exposure to non-native EMFs – especially blue light, wireless microwaves, bipolar magnetic fields and dirty electricity. Blue light in particular destroys melatonin via dissociated melanopsin-vitamin A.
- **Drink good water.** Make sure the water you drink is not fluoridated or chlorinated, and is low in deuterium.
- **Avoid deuterium in food.** Eat fruits and vegetables in season – preferably grown inland. Avoid processed foods, refined carbs and refined vegetable oils.
- **Get some cold exposure.** Let yourself get cold, whether by under-dressing for the weather, or by doing Cold Thermogenesis therapy.
- **Avoid mitochondrial toxins** such as cholesterol-lowering statins, antibiotics and cigarette smoke.
- **Take supplements.** D-ribose and CoQ10 supplements can help mitochondria make more ATP.

Redox reaction with electron transfer

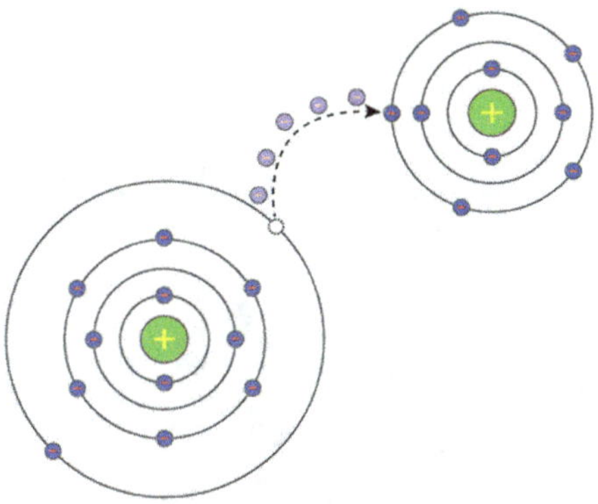

4

REDOX

How oxidation, reduction and inflammation keep the body tuned-up, or cause it to break down

Oxidation-reduction overview

For the last 20–30 years, healers and marketers have taught people free radicals and antioxidants potently affect our wellness and longevity. They are the yin and yang of the "oxidation-reduction" cycle in the field of oxidative medicine. Unfortunately, partial-truths have crystalized into many flawed beliefs contaminating millions of mindsets today. Let's deconstruct the core tenets of oxidation and reduction so we can get to the bottom of redox reactions.

Oxidation-reduction: Simply put, oxidation and reduction describe the exchange of electrons.

Electrical charge: Electrons are negatively charged. So *removing* electrons through oxidation increases positive charge and acidity, while *adding* electrons through reduction does the opposite: it increases negative charge and alkalinity. This is important because most pathogens, toxins and free radicals are in their comfort zone when positively charged and acidic, while antioxidant activity fights those threats by donating electrons, thereby increasing a tissue's negative charge and alkalinity. Electrical charge is synonymous with redox potential.

Oxidation is the "stealing" of electrons from a molecule – or an *increase* in the state of oxidation. So molecules with a propensity to oxidize substances are called "oxidants." To illustrate, when oxygen takes electrons slowly from iron, that's a form of oxidation we call rust. When a flammable material burns or explodes – again, with oxygen – that's oxidation happening rapidly right before our eyes.

Oxidants: In biologic systems, oxidation destabilizes the matter inside cells by stealing their electrons, which makes them blow apart in a hurry unless they find an electron to pair up with to zero out their charge. Employed for their ability to obliterate, oxidants are used by our immune systems as the ultimate antimicrobial, detoxifying agent. Most oxidants, as the name suggests, are predominantly oxygen-based molecules, so they're called reactive oxygen species or oxygen radicals. Some of the best-known oxidants in the functional medicine field are oxygen, chlorine dioxide, ozone, and hydrogen peroxide. Nitrogen and sulfur also form their own less-diverse reactive species, such as nitric oxide.

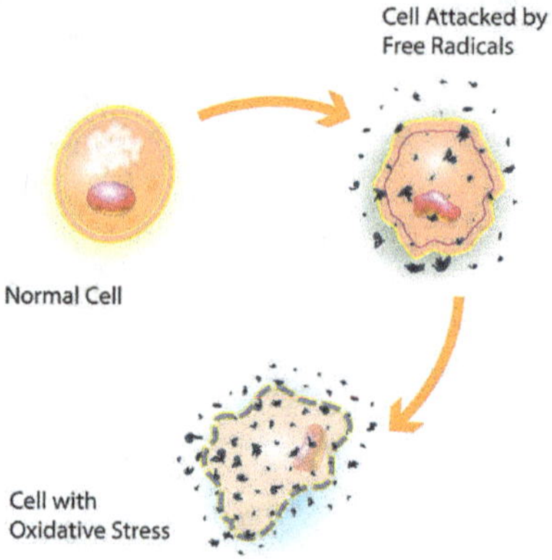

Free radicals: When a molecule with a balanced pair of electrons gains or loses one of those electrons due to oxidation or reduction, the resulting molecule can become a free radical because of its extra, unpaired electron, which is highly reactive and potentially damaging to cells. Its destructive effect on cells is why the public was taught to fear free radicals starting in the 1980s, and why we were instructed to get plenty of antioxidants to combat free radical damage. Most, but not all, reactive oxygen species are free radicals.

Reduction is the opposite of oxidation. It's the giving of electrons, or a *decrease* in the state of oxidation – hence "reduction."

Reductants: Molecules that give up their extra electron in chemical reactions are called "reductants" (even though that sounds backwards), or reduced species (RS).

Antioxidants (stricter definition) are tiny molecular catalysts that make oxidants give their extra electron(s) to reductants, thereby neutralizing them both of electrical charge and biological reactivity. The body's naturally-produced antioxidants such as glutathione can perform tens of millions of these reactions per minute. Reactive oxygen species and reductants then turn back into salt water (from which they came). More loosely defined, antioxidants are any substance that reduces free radical damage.

Redox: Not too long ago, redox scientists realized that saying the words *oxidation, reduction,* and *reactive oxygen species* out loud made them sound like nerdy chemists who use big words to confuse or impress people. So they came up with a cooler sounding umbrella term to describe the players and processes. The field swapped and shortened oxidation-reduction to "redox," which is short for REDuction-OXidation. And it seems to have a ring that's right for the mainstream. However, when mitochondriacs say redox, they're usually referring to just electrons or protons, and their electrical charge, not redox signaling molecules.

Redox signaling molecules: Reactive oxygen species/free radicals (ROS), and reductants/reduced species (RS), are collectively called "redox signaling molecules" or just redox molecules. In particular, mitochondria's *oxygen redox molecules* are by-products of metabolism that form a major communication network between mitochondria and human cells. Their signature trait is they have one or more unpaired electrons. Similarly, bacteria make "carbon redox molecules" as by-products of their metabolism, which they use to communicate between themselves and cells of the gut lining. Carbon redox molecules are the communication medium of the microbiome (see *Gut-Brain Secrets* to learn more).

Redox potential, redox power, or simply "redox" (less formal). Not to be confused with redox signaling molecules, *redox potential* is a general term referring to how many electrons, and their negative charge, are available to do work in the body. In most cases, redox potential describes the condition of having more electrons available than protons, resulting in a net negative charge, which can drive chemical reactions, reduce acidity and inflammation, and move materials. Basically, redox potential describes oxidation-reduction capacity.

Oxidative stress: Oxidative stress is the amount and duration in which oxidants outnumber reductants. Oxidative stress can be damaging when you don't have enough antioxidants and reductants available to neutralize oxidation – particularly, in chronic, uncontrolled circumstances. On the other hand, oxidative stress can be beneficial when used therapeutically.

Why oxidation and reduction matter

Oxidation and reduction are fundamental to life as we know it, because biochemical reactions revolve around the exchange of electrons. Redox reactions are integrally involved in energy production, healing, immunity, detoxification, hormonal response and anti-aging… when they're working properly – a lack of cell repair, many chronic, degenerative problems and accelerated aging when not. So redox reactions are more than important to all life on the planet; they're essential. Life would cease to exist very quickly without oxidation and reduction. Oxidation and reduction, or redox signaling (they're the same system), is one of the biggest, most important stories for medical science to study today, and you to learn about, in our ongoing effort to get healthy and stay healthy.

We used to think free radicals were harmful by-products of exercise

Until just a few years ago, wellness experts believed *free radical oxidation* harmed cells and accelerated the aging process… and that was all there was to it. Researchers and educators basically thought inefficient energy production and aerobic exercise make reactive oxygen species by accident. Because free radicals are good at destroying things, they believed the resulting oxidation is responsible for randomly injuring cells and making us age before our time. Therefore, we needed to protect ourselves against this sort of cell damage by eating foods high in antioxidants or take them as a supplement.

But science has since come to its senses as new research has revealed we shouldn't think of reactive oxygen species as harmful by-products of metabolism that you need to eliminate, else they harm cells and cause premature aging. Instead, they're made for a very good reason – many beneficial reasons, actually. Researchers have come to realize that the body uses oxidants in a controlled manner to selectively repair or replace injured cells, depending on the extent of damage. Even more pertinent to the discussion, redox molecules are created to communicate what's happening inside the cell.

So redox molecules are both the communication network of the mitochondria, and the disinfectant that wipes out unwanted material to make way for healthy, new cells. But, up until just 15–20 years ago, most scientists didn't realize that a healthy system purposely makes an equal amount of destructive oxidants and restorative reductants to maintain balance. In other words, good health is a balanced redox system. In fact, large amounts of both reactive oxygen species and reduced species are made when you exercise. So, oddly enough, taking a redox molecule supplement does many of the same things that exercise does, whereas ingesting antioxidants right before you exercise blocks some of its benefit because you neutralize some of the cleansing and restoration.

We used to think antioxidants only came from food

We now know our own cells make large quantities of antioxidants that are far superior in reach and capacity to those supplied by foods. In fact, most antioxidants that come from foods can't get inside cells, so they don't do what scientists thought they did (vitamin C is an exception). That said, antioxidants like vitamin A and E do offer some benefit quenching free radicals in the blood and extracellular matrix – colorful vegetables being their richest source. However, that's only half the story.

Extracellular matrix: Supportive structures and biochemicals outside of cells (e.g., collagen, enzymes, glycoproteins, and minerals.

As a compensation mechanism, raw vegetables increase alkalinity in the body through their calcium, magnesium, potassium, sodium and iron. The compounds formed from these minerals become a reserve of alkalinity, or buffer, the body uses to raise pH level when the blood becomes too acidic (low pH). It's a standby method to adjust pH level that overlaps in purpose with a vegetable's antioxidant effect.

That is, **redox reactions and pH regulation aim to achieve electron balance.** Together, they are big reasons why raw, naturally-grown vegetables can be good for you. Unfortunately, food companies tend to exaggerate the benefits of antioxidants taken in isolation – similar to how drugs try to mimic a medicinal herb's active compounds, and the way some vitamin supplements are promoted. Whichever way it's presented, redemption in a pill or a potion *sells* with a good story behind it.

In any case, antioxidants from food can be good for you when used intelligently. They're just not as all-powerful as those the body makes internally. To give you some perspective, dietary sources contribute about 4% of the body's total supply of antioxidants, while internally-produced glutathione, by itself, represents about 85%, and is much stronger.

Molecular structure of glutathione. Public domain work. Author: Ben Mills.

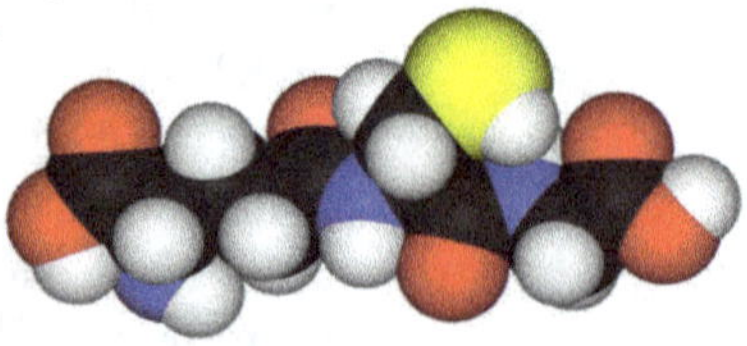

Our new understanding of oxidation and reduction

We now know oxidation and reduction is a carefully orchestrated, essential process the body uses to rectify injury and degradation so repair processes have a pristine environment in which to rebuild. Without oxidation and reduction healing you, you'd have mere hours to live – mere minutes without redox happening in the ETC. Oxidation and reduction are fundamental to human life because the body uses oxidation to destroy everything it deems undesirable – including microbial pathogens, toxins, heavy metals, dead cells and unrepairable cells. However, oxidants are like wild beasts in that they're useful in controlled circumstances, a destructive nuisance when they get out of control.

Normally, when phase one of inflammation is done disintegrating its target with oxidants, phase two brings in antioxidants to clean up the mess and restore the peace. Antioxidants bring excess oxidation products together with reductants, and make them exchange electrons. This neutralizes the oxidant of its destructive capacity, and turns both of them back into harmless salt water. The body's own antioxidants such as glutathione can do this tens-of-millions of times per minute.

Exercise is oxidative stress. Oxidative stress is like exercise

Most important of all its functions, exercise is controlled oxidation. It produces a cleansing and renewal effect at a cell level. That's *why* exercise is good for you, and *how* it benefits the body. It leads to very important mechanisms of cell regeneration, aerobic capacity and speed of recovery. It goes like this: Exercise makes mitochondria burn fatty acids or sugar along with oxygen to power cells. Through this process, energy production in the mitochondria makes large amounts of reactive oxygen species that act as both universal cleansing agents (that need to be neutralized) and immune-system messengers.

Exercise, oxygenation, and the resulting capacity for oxidation and reduction also help improve electric charge and pH balance. Electrical potential improves cell wall permeability, hydration, nutrient exchange and detoxification. In conjuction, breathing from exercise balances pH because O_2 alkalizes, while CO_2 acidifies. These factors promote either a high state of health and healing when present or disease conditions that favor pathogens and breakdown of cells when absent.

Point being, exercise, oxygenation and oxidative activity are precious to your wellness. But, the other side of the story is, it takes time to build a capacity for exercise, which is largely dependent on antioxidant capacity and completion of the redox cycle. Otherwise, you experience extended recovery time and overuse injuries pushing yourself too hard when you're out of shape. When well-controlled, oxidative reactions represent the bulk of the body's mechanisms for healing and anti-aging. Conversely, when those systems are overwhelmed or fail altogether, like ROS not being neutralized properly, then oxidative damage becomes the mechanism of imbalances, exposure to genetic weaknesses, breakdowns and rapid aging.

In summary, exercise is oxidation. Exercise is cleansing and rejuvenation on a cellular level. But exercise needs to be matched up with an equal amount of aerobic fitness, which is determined by your mitochondrial strength, your body's capacity to neutralize free radicals, and balance between oxidation and reduction. You need all that to sustain aerobic exercise. Otherwise, you get too much breakdown and not enough repair. So basically, aerobic capacity is brought about by a redox signaling system that performs well under load.

Oxidative therapies use oxidation, then reduction, to heal and renew

- hyperbaric oxygen
- chlorine dioxide
- ozone therapy
- hydrogen peroxide
- exercise.

High blood sugar causes a burst of oxidation

Basic biology says the hormone/endocrine system regulates blood-sugar level by releasing insulin to let sugar into cells. Insulin also causes the liver to store sugar for future use. But did you know that eating sugar makes mitochondria crank out free radicals as a by-product of ATP production? The problem is, the more broken-down your energy production systems get to be, the more ROS tends to be made relative to reductants. Those extra reactive oxygen species increase chronic oxidative stress, which can threaten cells all over the body with oxidative damage, including insulin-producing beta cells in the pancreas.

To protect itself from this rapid burst of oxidative damage, the pancreas is designed to pump out insulin as fast as it can in order to store that sugar in the liver, instead of burning it immediately and overwhelming the system with free radicals. Unfortunately, it has the opposite effect. What ends up happening is, insulin drives the sugar into cells faster and more furiously, which only increases oxidative stress and inflammation as by-products of metabolism.

So basically, eating lots of refined carbs and sugar is like pouring gasoline on a fire. **It's a sugar-induced explosion of oxidation** that can only be controlled (hopefully) by flooding the system with insulin to stash sugar in the liver. This isn't a problem when antioxidant supplies are capable of handling the increased oxidative load. But, more and more often, that's not the case as our population loses its healing capacity, and sugar consumption rises.

Thus, you've got yet another factor, among many difficult to avoid, that increases inflammation… another coping mechanism stretched to the limit. This is one of several vicious cycles involving redox reactions that increase inflammation. In this case, with antioxidant and redox reserves exhausted throughout the body, insulin conspires to keep the destruction phase of inflammation going longer than it's supposed to. Frequent high levels of insulin then cause the battle between inflammatory *destruction* and *repair* to rage on continuously. Elevated blood sugar from blue light and nnEMFs also help keep the wheels on this cycle spinning. Of course, constant demand for insulin also contributes to diabetes (another pathway whereby poor redox signaling turns into disease).

Redox signaling

Through the communication network of redox molecules, our cells and mitochondria read conditions inside and outside of cells to run their operations. Conversely, when communication breaks down, they can't respond to threats when, where, and how they should. Made by mitochondria in cellular respiration, *oxygen redox molecules* help the body

- detect when cells are under stress
- tell DNA to activate coping mechanisms
- call in the immune system to fight off threats
- repair mildly damaged cells
- hit the self-destruct button on badly damaged cells
- regulate hormonal response
- fine-tune mitochondrial metabolism
- and turn off coping mechanisms after a threat subsides.

When cells become damaged for any reason, they enter a state of oxidative stress. Free radicals are the "S.O.S. distress signal" that tell cells and systems there's a problem that needs to be fixed, where it's coming from, and how dire the situation is. Like smoke from a burning building, the more oxidants the immune system finds leaking out of cells – because they were not neutralized – the more emphatic the distress signal is deemed to be. After the situation has been resolved, the lack of free-floating redox molecules inside and outside cells prompts the nucleus and immune system to turn off their coping mechanisms and go back to business as usual (i.e., homeostatic balance).

So it's through mitochondrial redox signaling that the nucleus reads the state of oxidative stress occurring in the cell and can activate genetically-controlled coping mechanisms to deal with the threat and restore balance to the system. Of course, most of the time, a cell reads its status report and realizes it's fine and doesn't need repair. Still other times, cells look at their neighbors and realize they're different from that of the host organism. They realize they're a cancer cell and need to be sacrificed to protect the health of the whole.

But the thing is, without a broad vocabulary with which to communicate, cells don't even realize what a healthy cell is supposed to look like *vs.* a cancer cell. They don't talk to each other fluently about their status and needs. And they don't relay that information to organ systems that need to know such things, like the immune system. All of that are jobs of redox signaling molecules – known to most of the world as free radicals. Summary: In the same way that analyzing a car's exhaust tells a mechanic how well an engine is running, the ratio and volume of ROS to RS indicates whether the cell is happy or distressed. Mitochondria, cell nuclei, and organ systems can then use this info to activate healing or turn it off.

List is from Gary Samuelson's paper "The Science of Healing Revealed," pp. 45–46.

Here are some of the "buttons" the nucleus can push to call for help

1. **The DNA Repair** "button" mobilizes the DNA damage detection-and-repair crew.
2. **The Antioxidant Boost** button makes more antioxidants to neutralize the potentially harmful surplus of oxidants.
3. **The Intercellular Communication** button strengthens lines of communication between cells.
4. **The Increase Blood Supply** button dilates blood vessels to cells that need more resources.
5. **The Stronger Cell Adhesion** button makes cells hold more tightly to each other.
6. **The Inflame Tissue** button stops damage from spreading.
7. **The Secrete Antibiotics** button deploys antibacterial substances to fight foreign invaders.
8. **The Stop Cell Division** button prevents distressed/damaged cells from replicating.
9. **The Send Distress Call** button sends a distress signal to the immune system.
10. **The More Energy to Repair Crew** button brings in more energy for repair processes to work with.
11. **The Prepare Cell for Shutdown** button places the decision to euthanize a cell with its neighbors.
12. **The Master Shutdown** button kills and demolishes the cell.

That's healing at a cellular level

If you didn't notice, those buttons are basically healing on a cellular level. Genetically-controlled processes activated by the nucleus in response to redox signaling are the mechanisms by which cells are instructed to activate repair processes when repairable, or make way for replacements when gravely injured. These are the mechanisms that heal cells and stave off the aging process. A perfect example is button #12: The body is supposed to shut down attempts to repair cells after two hours if they were unsuccessful, then turn on cell suicide (apoptosis). But poor redox signaling disables this programming, allowing damaged cells to persist and replicate. In a word, that's aging.

Both cell-cell communication networks come from mom

As Dr. Doug Wallace taught us, you inherit all your mitochondria from your mother. Each human egg has some 100,000 mitochondria, which is more than any other cell in the body. On the other hand, the 200-or-fewer mitochondria in sperm are seen as foreign and selectively destroyed after the egg is fertilized. Mitochondria then reproduce as welcome guests inside most new human cells made. Which means all the mitochondria you will ever have descend from that original batch in the egg.

A curious consequence of this process is that you acquire half your redox communication ability from your mother *at conception* in the form of mitochondria and the oxygen redox molecules they make in cells. And you inherit the other half of your cell-cell communication from your mother *at birth* in the form of bacterial microbiota and the carbon redox molecules they make in the microbiome. That is, when a baby is born conventionally, friendly bacteria from the mother's birth canal seeds baby's digestive tract with starter cultures. Over a baby's first few weeks, those initial bacterial colonies populate their GI tract and make the carbon redox molecules that form the communication network of the microbiome. So, as it turns out, fathers are almost useless in the world of redox communications. Their role happens on the sidelines of all the action.

Rate of injury *vs.* rate of repair

Cells in your body are constantly being injured and killed every minute of every day. At the same time, the body's healing mechanisms continuously repair damaged cells, and replace dead cells with healthy new ones. So there's always some rate of injury happening versus some rate of repair. That ratio determines your ability to stay well. It greatly influences your energy level – physically and mentally. It controls your recovery time from exertion, injury and illness. It has a lot to do with your rate of detoxification. And it's a primary factor in the rate at which you age biologically.

Both of those states – damage and repair – can change quickly through interventions, or gradually through natural processes. But you seldom notice any of these changes because the human body has a built-in buffer of coping mechanisms – a reservoir of healing capacity – that's designed to favor life and healing, until your compensation mechanisms are over-extended. So you only notice three conditions:

1. When your rate of **new injury increases rapidly** due to some event or circumstance (e.g., a major infection or massive toxin exposure).
2. When **repair has fallen far behind** and injury is clearly winning (i.e., the basis of disease).
3. When your rate of **healing improves quickly** through a lifestyle change such as a healthy new diet, an exercise program, becoming a mitochondriac, or getting out of a bad relationship.

Complex organisms are designed to start their lives with a full bucket of life force, if you will. They (should) spend the majority of their lives with healing capacity to spare in the form of compensation mechanisms, emergency procedures, and various workarounds the body uses to cope with sub-optimal conditions. It's only when injury exceeds the body's ability to repair by a wide margin that you see acute disease manifest, or you die. In daily life, however, you typically don't notice a gradual expansion or contraction of healing or repair, because your buffer of healing capacity absorbs any slow changes you encounter. Your body just does its normal healing thing and you're none the wiser.

But our multitude of sins against Nature have stretched our coping mechanisms to the limit

Our excesses and irresponsibilities have exhausted our healing capacity at all levels – cell, organ system, individual, society and planet. So now the majority of people are living their lives on a razor-fine edge between health and sickness, with dangerously little safety margin to spare. That's an important concept to internalize, because the slightest new insult in EMF exposure, diet, lifestyle choices, or internal milieu can push a person over the tipping point into disorder. In case you hadn't noticed, that very situation is occurring all around us in every health statistic you'd care to measure. Almost everyone is on the verge of manifesting a chronic, degenerative disease. Those in jeopardy just can't see it. So as soon as the body uses up the last trick it has in its arsenal to keep you running problem-free, serious disease appears out of nowhere, and we blame it on unfortunate genetics, crummy diet or tough luck.

Inflammation

Chronic inflammation is integrally involved in creating most disorders involving the immune system, hormones, heart, brain, digestion and cell repair – basically, everything that happens in the body. That's because **the majority of damage a disease causes is not done by the pathogen or problem itself, but by inflammation the body employs to fight the threat.** In other words, most infections, toxins and disease processes are not the thing that does the most harm. Instead, the body's own threat annihilation system – inflammation – is far stronger at wiping out friendly cells than the menace itself. When it persists unchecked, inflammation causes symptoms we collect into categories and call disease.

In fact, learned practitioners realize that modern diseases are nothing more than a variety of ways and places that chronic inflammation exhibits itself in the body, as repair fails to keep up with injury. Therefore, a failure of cells to repair themselves completely is what causes modern disease. And that's caused by inadequate energy production, combined with poor redox signaling, which results in chronic inflammation and unfinished repair. They're partners in degeneration, with chronic inflammation being the culprit that actually "pulls the trigger." Cancer cells are a quintessential example of what happens when inflammation lingers and cell's distress signals go unanswered. Cancer is believed to take place after a cell has borne the brunt of some 20,000–25,000 unrepaired injuries to its DNA (i.e., failure to repair/replace).

How inflammation heals

The body uses inflammation to repair pretty much any sort of damage that can happen anywhere in the body, including:

- cuts, bruises and overuse injuries;
- damage to bones, muscles, tendons, blood vessels and internal organs;
- microbial infection;
- radiation damage;
- chemical and heavy metal poisoning;
- as well as ordinary wear and tear.

Becoming amoeba-like and squeezing through a vessel wall (as white blood cells do) is called diapedesis.

It does this by first increasing permeability of blood vessels so immune cells such as white blood cells can pass through the vessel wall to fight invading pathogens. This also lets more blood into an area, which we call swelling. Swelling makes movement painful, thus limiting mobility and further damage. Barring outside intervention, inflammation then launches an oxidative attack on everything in the area that the immune system does not recognize as your own healthy cells. Through oxidation, the first stage of inflammation kills infectious microorganisms, neutralizes toxins and heavy metals, and it destroys your own damaged, diseased and dead cells.

When inflammatory processes are working correctly (i.e., acute inflammation), the first phase of inflammation eventually shuts down. In the second phase of inflammation – the repair phase – unhealthy cells and

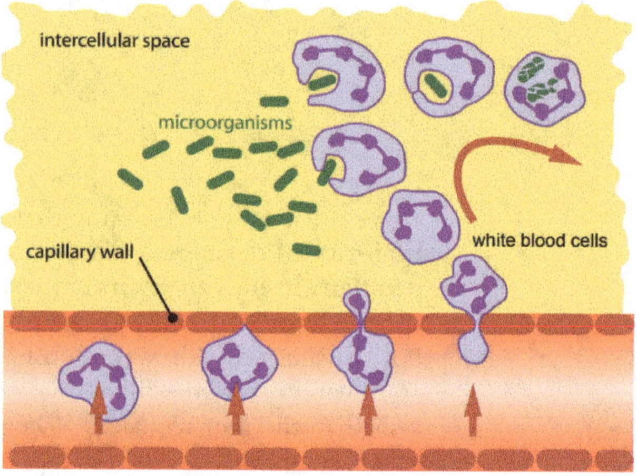

debris are replaced with new cells and collagen fibers (made from protein and cholesterol). On the other hand, when inflammation doesn't let up, both phases stay active at the same time – destruction and creation. That's the leading cause of chronic, degenerative disease today – inflammation that doesn't stop destroying tissue – which is itself caused by poor redox signaling and a lack of energy from the mitochondria.

But the mind-blow you need to know about is **pain and swelling are integral parts of the healing process. If you prevent pain and swelling from running their course – like with anti-inflammatory painkillers, icing, elevating, compresses or fever reducers – then full healing may never take place because you shortcut the healing process.** Believe it or not, pain itself actually stimulates healing by kicking the immune system into action. This is why so many people today suffer from chronic pain issues that linger too long, or never end: they've interrupted the healing process so many times throughout their life – with drugs that block inflammation and pain – that injuries and inflammation become more or less permanent.

Acute inflammation *vs.* chronic inflammation

There are two kinds of inflammation: acute inflammation, which is temporary and beneficial, and chronic inflammation, which is protracted and harmful. Acute inflammation heals and protects you on a daily basis from threats that can damage tissue, such as cuts, bruises, infectious organisms, ordinary wear and tear, oxygen deprivation and toxin exposure. Everyday occurrences like exercise and sun exposure also turn on inflammation without you even knowing it.

Through acute, fast-acting inflammation, the immune system gets called into action. It wipes out the threat in a matter of days to months using a four-step process whereby: (1) oxidants *destroy*; (2) antioxidants team up with reductants to *neutralize* the oxidant; (3) redox signaling partners with the immune system to *repair or replace* damaged cells; and (4) when that process is finished, if everything's working properly (i.e., complete healing and the redox signaling to match), inflammation shuts down and the body resumes normal operation.

On the other hand, chronic inflammation is where the same series of processes get activated, but they're unable to turn themselves off. *Step 4 shut down* kind of starts, but is never completed. Instead, the first three stages get stuck in a vicious circle of destroying and repairing cells repeatedly, which can smolder along unbeknownst to you for years or even decades.

Reasons the immune system stays activated

Your antioxidant system is overwhelmed. First, some significant insult takes place requiring the immune system to use its oxidative stress tool – often between the ages of 10 and 30. This could be a sports injury, a nasty infection, or an acute poisoning event (such as a major round of vaccines). In this situation, cells don't have antioxidant and reductant reservoirs large enough to neutralize the oxidation being done. So the repairing and replacing of damaged cells is done slower, and less completely, than it should.

Cells then eke by and reproduce themselves in a partially-damaged state, which is frustrated by lingering attempts by the immune system to simultaneously *destroy* with low-grade oxidative stress and *heal* using redox signaling and repair processes. This smolders along underneath your conscious awareness until later in life your cells are being injured by oxidative stress faster than they're able to heal. At that point, you get symptoms that bother you. But, all the while, you've had chronic inflammation you only noticed occasionally (like an old knee injury that only hurts when it starts to get cold out).

Pro-inflammatory exposures sustain the situation

- You eat pro-inflammatory foods, such as refined vegetable oils.
- Leaky membranes cause poor digestion, food sensitivities, immune system hyper-activation and autoimmune conditions.
- You continue to stress that old knee injury with the help of anti-inflammatory painkillers or other medications.
- You come down with a long-term illness such as Lyme disease.
- You continue to take in toxins faster than you release them.

Fewer reductants, diminished redox signaling capacity, and a shortage of energy all prolong inflammation. As we age, we lose mitochondria. Scarcity of mitochondria means fewer reductants get made, diminished redox signaling capacity, and not enough ATP for all bodily processes. Plus, the mitochondria that do stick around age with us. Their DNA accumulates damage just like ours does. As our mitochondria age, our once-balanced blend of oxidants-to-reductants tends to tip toward a surplus of oxidants, and a deficiency of reductants. In other words, lower redox potential. That pushes us toward oxidative stress as we grow older. However, too much of either one – oxidation or reduction – results in unregulated oxidation and cell damage. To complicate matters further, inefficient redox signaling sends unclear messages to the nucleus, which then makes fewer antioxidants.

In this situation, any sort of severe or prolonged damage can cause the immune system to dump oxidant into an area that it's not able to clean up properly due to reductant and/or antioxidant deficiency. So as oxidative stress, positive charge and acidity build up in the cell, the clarity of message conveyed by redox molecules declines. From then on, the immune system's oxidative response becomes its own worst enemy. Oxidant is dumped in an area to start the clean-up process, but cells don't have antioxidant reserves large enough to clean up the oxidation. Full healing never takes place.

The immune system senses something's wrong, but it's confused by the mixed signals, and failure of its tools to do their jobs properly. So it continues promoting the destructive side of inflammation, as it rebuilds as best it can. Neither side – destruction or healing – is able to prevail while the immune system is caught in a vicious cycle of antioxidant/reductant deficiency and imbalanced/unclear redox messaging. That's what's happening when acute inflammation turns into chronic inflammation. And that's how chronic inflammation causes disease. It's the failure of cells to repair or replace themselves effectively. Cells then continue to live and reproduce in an unrepaired state. That's most of what drives modern, degenerative diseases today – including heart disease, arthritis, Alzheimer's, cancer, diabetes and GAPS conditions.

GAPS: Coined by gut-brain health pioneer Dr. Natasha Campbell-McBride, Gut and Psychology Syndrome conditions are gut imbalances, such as leaky gut and gut dysbiosis, that cause impaired brain function, such as ADD, autism, anxiety and depression.

Redox potential

Redox potential is a general term mitochondriacs use to describe concentrated zones of electrons and their net negative charge and, in some instances, concentrated pools of protons and their (isolated) positive charge. Not to be confused with redox molecules, *redox potential* represents electrical charge itself, and what the body can do with that voltage. It's largely a function of how well you're able to turn sunlight, grounding, food and magnetism into DC electric charge, while at the same time minimizing the loss of electrons from modern living. Broadly speaking, redox potential is the energy behind health and resiliency, healing capacity, movement of materials, chemical reactions and life itself. When you're loaded with redox potential, your biology is more full of life. On the other hand, when you lose redox potential, your body lacks energy to maintain itself.

Negative charge and alkalinity are synonymous with health and healing. Positive charge and acidity create inflammation and disease

If you had to pick only one metric to assess your relative state of health or sickness, electrical charge and pH belong near the top of that list. You can think of the two as pretty much the same thing in the body because they're like dance partners. When oxidation steals electrons, cells and atoms gain positive charge. That means more corrosion, greater instability, and inflammation that doesn't know when to quit. It means your cells and atoms are literally falling apart and not repairing themselves efficiently, which is imbalance, sickness and aging, in a nutshell.

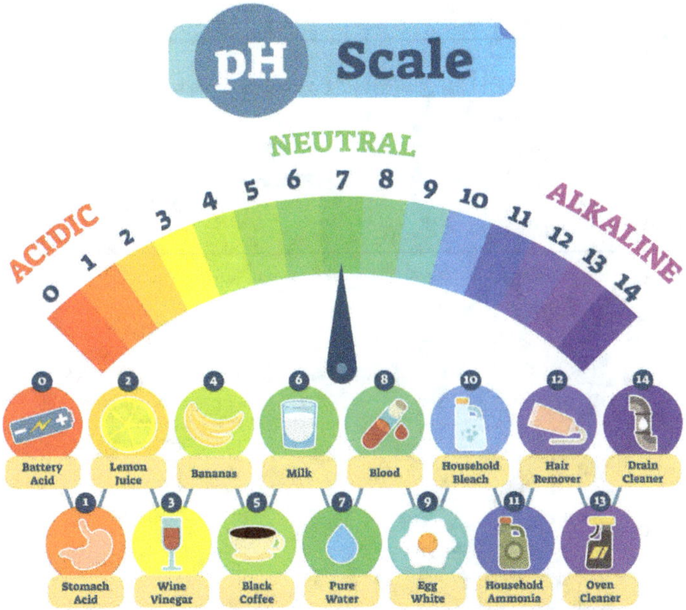

On the other hand, when your immune system is good at making antioxidants, and it has all the electrons it needs for reduction, the atoms in your cells stay happy and whole, instead of falling apart. We call this *neutralizing, quenching,* or *extinguishing* oxidative stress. This translates into better cell-cell communication, inflammation that starts and stops appropriately, and more healing capacity. On the other side of the scale, acids have more protons than electrons, which gives them a positive charge. The more protons a substance has, the more acidic it is. And the opposite is true: the more electrons something has in relation to protons, the greater its alkalinity. Hence, many alkaline foods and substances are famous for their healing/antioxidant effects.

In short, extra electrons are *negative charge* and *healing capacity* through oxidation quenching, and helping the immune system know when to shut down inflammation (alkalinity supports cell-cell communication). While at the other end of the spectrum, electron deficiency is *positive charge* and *poor healing ability* through corrosion, lack of light energy (electrons hold light energy), and failure to realize that inflammation is not working the way it should. Acidity aids and abets breakdown.

Redox potential in action

When the water in cells has a lower net negative charge, it can't hold as much oxygen, because voltage dictates how much oxygen will dissolve into water. With less oxygen to support mitochondrial metabolism, ATP production declines. Metabolism then switches from aerobic respiration in mitochondria to <u>an</u>aerobic glycolysis in the cell. Unfortunately, glycolysis is 20 times weaker than the aerobic respiration of fats.

So not only do you produce less ATP when low redox makes you hypoxic (adding insult to injury), but most pathogens thrive in anaerobic conditions. Invigorated by a low-oxygen environment, pathogens that had been idle wake up and get hungry. They release digestive enzymes to dissolve cell walls to access their nutrients. You've experienced this as an irritated throat. Pathogens can also change form (called pleomorphism) from, say, a bacterium to a fungus, based on conditions such as low oxygen and acidity in their environment. Three consequences of low redox potential: metabolism shift, pathogen surge, and pleomorphism.

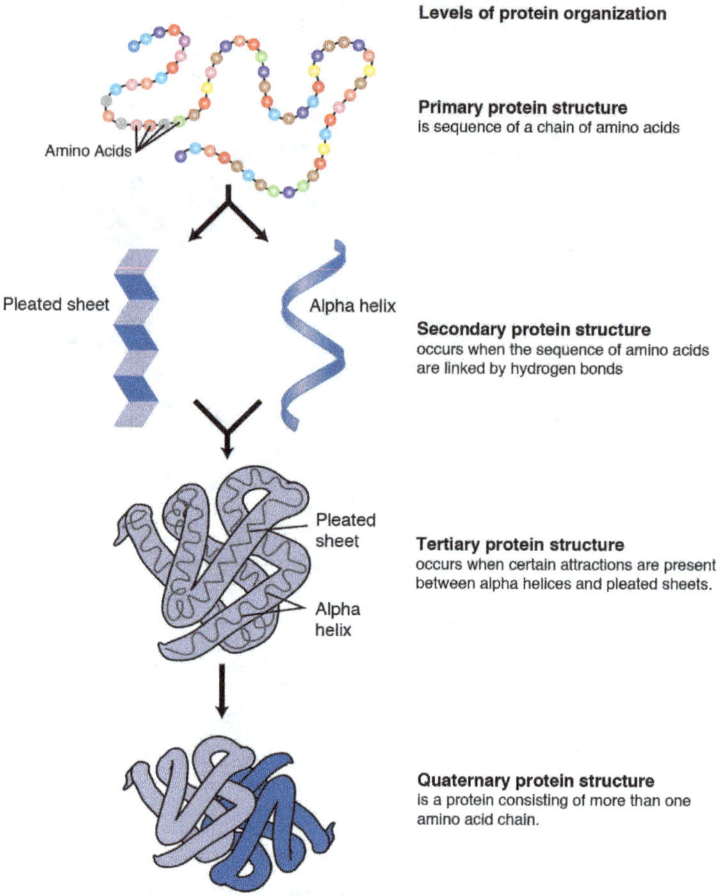

Another scenario that's greatly important to good health is voltage differential through the electron transport chain. A hearty redox potential is present when the transport chain repels electrons from one spot, and attracts them to the next, with a force of 250–400 millivolts from beginning (Cytochrome I) to end (oxygen). On the other hand, weak mitochondria measure under 250–200 millivolts between redox stations in the ETC. The forces of reduction and oxidation attract and repel electrons through the ETC so mitochondria can power the body.

A third example happens in neuro-degeneration. Mis-folded proteins in the brain are a primary cause of Alzheimer's, dementia and Parkinson's. That's because proteins have four levels of structural bending to them. The first two are controlled by DNA. The third and fourth are controlled, or botched, by redox potential. So when your redox is low, errors occur in their shape, which alters the way energies make them resonate and perform. This is an important way that redox power helps build either functional or dysfunctional proteins.

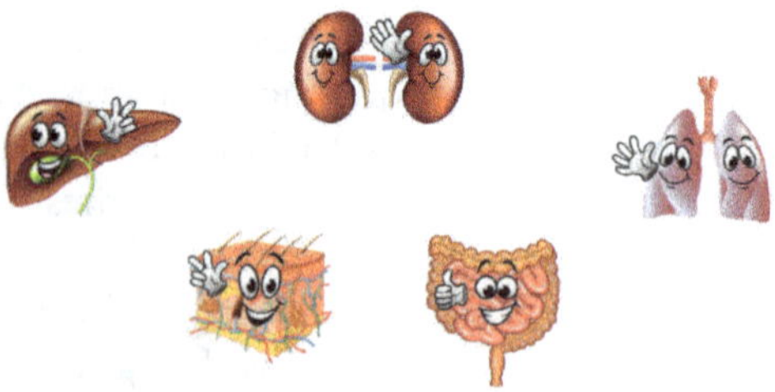

Redox before you detox

Detoxing has become a hot topic in natural healing over the last few decades – and for good reason:

- **Mercury** in amalgam fillings, seafood, coal (for power plants) and vaccines is believed by many to be a major cause of autism, ADD, neuro-degeneration, and now heart disease.
- Industrial **chemicals and pollutants** such as bisphenol-A (BPA) and glyphosate have been shown to mess with the endocrine system through hormonal disruption.

- Severe emotional/psychological distress (e.g., rape, abuse, armed combat) can cause brain cells to hold on to **heavy metals**, which can hinder attempts to heal from that trauma.

These concerns are very real. However, detoxing through a product or a program may not be the best way to avoid toxin injury. Instead, think about detoxification as a function of your redox potential.

Detoxification systems are controlled by redox potential

Cells and detox organs need large caches of electrons, protons and antioxidants at their disposal to denature and remove toxins from the body. In particular, the single most important element in how well or poorly the body detoxifies is the availability of electrons and their negative charge, as well as pools of protons and their positive charge. The body uses these charged particles to power chemical reactions and move materials around. For example:

1. Electrons repair oxidation damage through the reduction process.
2. Electrons enable you to harvest the sun's energy for better everything.
3. Electron flow through the electron transport chain makes ATP. ATP operates the sodium-potassium pumps. They pump positively-charged sodium and potassium into cells, which pushes positively-charged toxins out through electrostatic repulsion.
4. High electrical differential at the cell wall helps remove decommissioned toxins by bringing more water into the cell.
5. More electrons moving through the ETC also improves magnetic fields in the body, which increases blood flow, oxygenation, DHA delivery and hormone transport.

So electrons and protons help break down toxins, mobilize them in water, push them out of cells, and transport them away. For these reasons, raising your redox potential is better at fixing toxicity issues than a detox product or program. Electrical charge helps the body do what it's designed to do naturally, instead of applying an external force to support one particular step in the detoxification process. In this way, interventional detox products and programs are like a drug: As soon as you stop doing it, you lose the effect. They also tend to target fewer issues, whereas improving your redox potential straightens out a wide variety of problems that lead you to better health and longevity.

The lesson for the day: "Redoxing" is more potent and comprehensive in its benefit. It's like a shotgun approach that fixes a wide range of toxin-related issues, while accomplishing many things that expand overall health. On the other hand, *detoxing* is closer to a rifle approach in that it often targets specific toxins, or specific organs, for specific reasons, over a finite period of time. It's for these reasons Dr. Jack tells his followers that heavy metals are only a problem when your redox potential is really low. That's why he says you should *redox before you detox*.

The field of redox science has only just begun

The last 15–20 years of research into redox reactions have propelled medical science into new paradigms of healing, energetics and optimal aging. As a result, genetics is now known to play a much smaller role in health and sickness than previously thought. On the other hand, the end products of mitochondrial metabolism – namely, their reactive oxygen species – control bioenergetics, gene expression, and the whole of human biology far more than genetics. Yes, the field has learned a great deal in the last twenty years. Yet thought leaders agree: the majority of discoveries in redox reactions have yet to be made. The implications of redox science are huge, as we're just beginning to realize.

5

EARTHING
(aka "grounding")

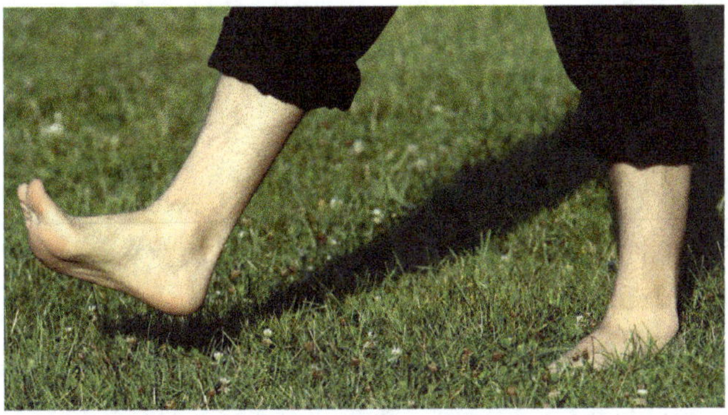

Earthing has been called one of the greatest health discoveries ever

That's because "earthing" (aka "grounding") gives you oodles of a vital resource your body needs to live, heal and resist aging: electrons. Yet it was considered pseudo-science prior to 2010. Indeed, when you learn what earthing does in and for the body, you can't help but agree with its devotees: grounding is a basic nutrient that keeps us healthy when we're well and heals us when we're sick. Call it vitamin 'G'.

As we now know, biology uses electrons to maintain itself. And the earth is our best source to get them. Unfortunately, modern society's infatuation with creature comforts such as heating, air conditioning and lighting give us every reason not to go outside and touch real earth. Amenities like these help us avoid the smallest discomfort in the moment. But they also separate us from a biological resource we can't live without.

Of course, we can get electrons from other sources. But they all come at a higher cost biologically. So, considering what it does for us, earthing has to be Nature's most under-appreciated method of acquiring electrons, redox potential and healing capacity (virtually the same thing). Indeed, earthing benefits the human body so profoundly even former skeptics now consider it a basic resource of healing and vitality.

Although used interchangeably by most people (as it is here), it is helpful to distinguish the term "grounding" from "earthing" because grounding, technically, is the removal of excess (unwanted) static charge to the earth via conductive wiring, while earthing is the addition of electrons into the body for therapeutic benefit. They are similar activities, employed for two different purposes

What on earth is earthing?

Coined by its modern researcher/developer Clint Ober (retired cable industry executive), *earthing* is the practice of physically connecting directly to the earth and receiving its infinite flow of negatively-charged electrons. Walking barefoot on a beach, or using a specially-designed conductive device, are two such ways to deliberately ground yourself. This gives you a variety of impressive health benefits, yet its mechanisms of operation are so subtle in the body you may not feel it in action unless you pay close attention to the 'before' and 'after' sensations.

That's because, like many natural healing methods, earthing is not jarring to the body. It doesn't feel artificial and forced like drugs and mainstream modalities often do (are designed to do, actually). Instead, it's 100% in alignment with the way your body is supposed to work. So you hardly notice it unless you know what to expect and go looking for it.

What's more, it's inexpensive or free. It works fast. You never develop a tolerance, or intolerance, to it. And it's simple as can be.

The value of earthing

Every living thing draws energy from the earth's electric and magnetic fields through its feet, paws or roots. As long as there's been life on earth, those life forms have charged up their "biological batteries" with electrons delivered through direct earth contact, a natural water body, or some means of conduction. But that's the key: whether plant or animal, you have to be touching the earth, or connected through an earthing device, to receive its nurturing benefits. Unfortunately, six decades ago, people started wearing rubber-soled shoes and spending most of their time indoors, disconnected from earth. Out of the same ignorance, we started insulating our pets and livestock as well. "Coincidentally," that's when chronic, degenerative diseases started taking off.

Before that, we went outside and played in the dirt. We swam in lakes and drank from streams. We wore leather-soled shoes (which are moderately conductive). Many worked outside in the fields. We even lived in homes with dirt floors, ages ago. And we were much healthier then. So, is electron deficiency and the rise of chronic disease really a coincidence? I say no. Separating ourselves from the earth's energy blocks the flow of health benefits we used to get when we lived in harmony with Nature.

What happens when we don't earth ourselves?

Being disconnected from the earth for days, months or even years doesn't necessarily make you sick or kill you right away. You just aren't as healthy and energetic as you could be. This makes you more susceptible to poor circulation, low energy, unproductive sleep, stress, slow wound healing, hormone imbalances, rapid aging and weight gain. Your mood suffers, and you aren't at your best mentally.

Earthing's biggest benefits

Electron donor. Earthing remedies an electron deficiency we all suffer from in our modern insulated culture. Earthing builds redox potential.

Antioxidant effect. Earthing is like a limitless antioxidant (reductant, actually). That's because earthing uses electrons from the earth's infinite supply to neutralize free radicals that are responsible for cellular injury and aging when they go unchecked. Alternatively, food sources, as good as they are, have their limitations.

Anti-inflammatory. Building up your supply of electrons and redox potential extinguishes inflammation.

Blood thinner. By increasing negative charge, earthing increases the zeta potential of red blood cells by 270%, which is a fancy way of saying earthing makes red blood cells repel each other with negative electrostatic charge, instead of clumping together (called "rouleaux" formation). So, instead of your blood being thick and viscous like motor oil, it's thin and easy-flowing like red wine. This makes it easier for blood to get into tiny capillaries under one blood cell-width in diameter which might otherwise clog from clumping.

Most of the time, thinning the blood reduces blood pressure as well, because the heart doesn't have to work as hard. Thinner blood also improves oxygenation, nutrient delivery and waste removal. These markers of health are virtually synonymous with life itself. If your body can't do these things well, you either have a medical condition, or you're prone to coming down with one. So it only makes sense: increasing blood flow, oxygenation and nutrient exchange improves health across the board.

For these reasons, earthing is scientifically shown to:

- thin the blood, reduce blood pressure and improve circulation
- reduce chronic inflammation and its plethora of adverse effects
- decrease pain from a variety of sources
- improve sleep in most people
- increase energy levels
- reduce stress and promote calmness by cooling down the nervous system and lowering cortisol levels
- normalize the body's biorhythms
- reduce muscle tension and headaches
- decrease hormonal and menstrual symptoms
- speed healing and prevent bedsores
- reduce or eliminate jet lag
- accelerate recovery from intense exercise
- slow the aging process.

How strong is earthing?

The effects are strong enough that most people who try earthing see obvious improvements in pain, sleep, endurance, recovery or mood. Many of these effects can be objectively verified by tests such as thermal imaging. The benefits often start with the first session and continue for as long as you earth yourself on a regular basis.

How grounding works

For a little background, electrons move from areas of high electrical charge, towards areas of lower charge.

The earth has an effectively infinite supply of electrons that it gently pushes into everything that's physically connected to it through some sort of conductive medium, whether natural or man-made. Sun exposure then increases electron flow when you're grounded. Other times, the earth *receives* electrons from sources of higher charge in brief instances. We call that static electricity. In other words, grounding moves electrons in whichever direction has a lower charge density.

Intentional grounding is thought to have originated in 9th century China, soon after the invention of gunpowder. Workers manufacturing ammunition and fireworks learned to be extremely careful to discharge static electricity through conductive grounding straps, before handling the gunpowder they were making. Otherwise, they experienced oxidation so rapid they would never forget it (aka an explosion). Same thing with oil and petroleum workers. This is where the term "grounding" comes from.

Purging surplus energies for their own reasons, the cable TV industry learned decades ago how to produce crystal-clear picture and sound by shielding their cables with grounded casing which carries unwanted EMFs away. We used to see these stray EMFs as noise in our picture years ago, but it rarely happens today because they've pretty much perfected grounding in wired communications.

Nowadays, most electrical systems are designed to deplete excess charge into the earth before this "static electricity" has a chance to damage sensitive electronics in the home, or harm patients undergoing open heart surgery, as examples. In other words, the electrical system of buildings and gadgets are "grounded" through the third prong of electrical outlets. This connects to a metallic ground rod stuck in the soil beneath the structure to dissipate static electricity through the building's wiring.

Fortuitously, this also provides a path for electrons to travel upward to therapeutic devices (and you) that are plugged into that third prong. So, today, dozens of companies make earthing products, such as bed sheets, pillow cases, mouse pads, foot pads, flooring, and lots of other configurations. However, there can be problems: Dirty electricity and nnEMFs can be transmitted through electrical wiring like an antenna. This can bring alien electric and magnetic fields right into your living space. There are ways to ground yourself safely with an earthing device through the third prong, but you need to shield your space from electosmog, and filter out dirty electricity first, which may require the assistance of an EMF remediation specialist.

So Nature's way is still the best way

- As a general rule, most materials made by Nature are good to great conductors, while most artificial materials are poor conductors, if at all.
- At one extreme, you can earth yourself Nature's way by touching bare skin to grass, sand, rock, clay or dirt.
- At the other extreme, wood, asphalt, plastic, and most man-made materials are insulators. They do not conduct electrons.

- Between the two extremes, cement and moist clay are slightly conductive when they contain moisture.
- The more moisture in or on a material, the easier it is for electrons to flow through them.
- Grounding works to some degree through cotton or wool socks – more so if there's any moisture in them.
- Specially-designed shoes that conduct electrons to the feet do an excellent job of keeping you connected. Rubber-soled shoes: not at all.
- The ultimate earthing experience is taking a dip in an ocean, lake or stream. Water can absorb massive amounts of electromagnetic energy, so it shields you from nnEMFs like a Faraday cage. That gives you full electron flow, with no foreign frequencies.
- Lastly, sunshine (the positive "anode") hitting the earth (the negative "cathode") stirs electron mobility. So physically touching earth on a sunny day gives you the very best grounding for your efforts.

What does earthing feel like?

It's funny but sad that we even have to describe what it feels like to be connected to the earth, because it should be as familiar to us as living and breathing. But that's "progress" for you. The physical sensation you get from earthing is easily deniable. You almost have to consciously pay attention in order to feel what it's doing. However, its health benefits are not so easily dismissible.

When you set your feet on slightly moist grass, or put your hand on a grounded earthing pad, most people notice a warm, tingly sensation where the flow of electrons enters the body. It's a warm, fuzzy feeling most people say is mildly stimulating. That's the nurturing effect our bodies are designed to receive several hours per day. Yet we've led ourselves astray since we started isolating ourselves from the planet.

The quick and easy way to experience what earthing feels like is to touch a metal faucet. Almost all bathroom and kitchen faucets are grounded because many plumbing pipes are metallic and run straight into the ground. That's one reason showers feel amazingly refreshing: You're getting drenched with massive quantities of negatively-charged water particles. Same thing with throwing water on your face directly out of the tap for a quick 'pick-me-up' – lots of negatively-charged particles. Unfortunately for many people, those are the only times they get to experience the benefits of earthing today.

Earthing summary

Earthing is an outstanding way to increase circulation, healing efficiency, sleep quality and energy level. It's a potent way to reduce inflammation and increase redox potential. It's among the cheapest, most beneficial practices you can do for yourself. However, as of 2025, the challenges to earthing yourself as Nature intended have grown considerably.

Risks

Over the past 5–15 years, electrosmog, jump conduction and ground current have become legitimate risks that make it harder for you to ground yourself safely. Earthing, in principle, is still great for you. But now, nnEMFs are corrupting the process – basically using you and your equipment as a low-resistance path to get to ground. To illustrate, which would you rather have: electromagnetic frequencies pass through you to get to ground, as earthing methods do? Or have them hit you and not exit? The answer is neither. Passing nnEMFs out to ground may sound a little better for you. But, ideally, you don't want any nnEMFs to be hitting you in the first place, because any electromagnetic pollution touching you *can't not* harmfully affect you.

Word to the wise: A building's electrical system picks up ambient high-frequency EMFs like an antenna, in addition to low-frequency dirty electricity riding the power lines. It sends those frequencies throughout a building's wiring to every grounding port on a circuit (the third prong). In doing so, earthing equipment can inadvertently transmit nnEMF pollution into you. In fact, Dr. Dean Bonlie actually measured earthing equipment increasing the amplitude of nnEMFs going into a person's body by 300%.

Electrosmog: (Unwanted) electromagnetic fields.

Jump conduction: Transference between conductors of static electricity from nnEMFs.

Ground current: Stray electricity in the ground underneath you (not good).

Dirty electricity: Unwanted spikes, surges, and frequencies riding the power lines (below the frequency of wireless communications).

Earthing Rx

By far, the best way to ground yourself is to get outdoors and touch the earth with bare feet and hands. In most places that's still a good idea. However, it might not be that way for long – particularly as 5G ramps up and stray electricity pollutes the land as ground current. See page 202, section "Ground current," to learn more about testing for a safe level.

The second-best way to earth yourself is through a dedicated ground rod and wiring, placed as far away from electrical service panels as possible (such as a smart meter). Much more problematic is earthing yourself through your home's ground wiring. It's almost never free of unwanted frequencies. Bottom line: Whether through dedicated earthing equipment, or through your house grounding wire, you can earth yourself indoors safely when electrical body voltage is under 0.1 volts on your skin, and when radio frequency fields are under 0.001 microwatts/cm^2 (peak). If your room exceeds these levels, earth yourself with caution.

6

WATER

Water is our largest mammalian battery pack

Water is essential to the energy needs of the body because it stores energy from our surroundings and later releases that energy to power cellular processes. That makes water a battery and the sun our best battery charger. Without water, most bodily functions don't work properly. To give you some perspective on the importance of water, biochemists have traditionally studied biology by looking at its composition. They take a cell or a tissue, break it down, empty out the water, and study it in a dehydrated state. But how much can you really know about how biology works when you remove the single most important driver of cell function?

As one pillar of the biophysics triumvirate, water is an inseparable sculptor of health or sickness because it powers, contains or otherwise facilitates the physics and chemistry of the body. Generally speaking, you can take the cell out of the sea, but you can't take the sea out of the cell. Without a doubt, there's much more to water than you might think. But supporting the body's biochemistry is only the beginning. Only recently has science come to learn most of the work in the human body is done by water's bio*physical* properties we never knew it had. So who would have thought two decades ago that water's most important contribution to human health comes from its electrical potential? Wait, water is a power source? That's crazy.

Images of the exclusion zone ("e-zone") in chapters 8 and 10 are used courtesy of Prof. Gerald Pollack.

Principle 1: Water Has Four Phases

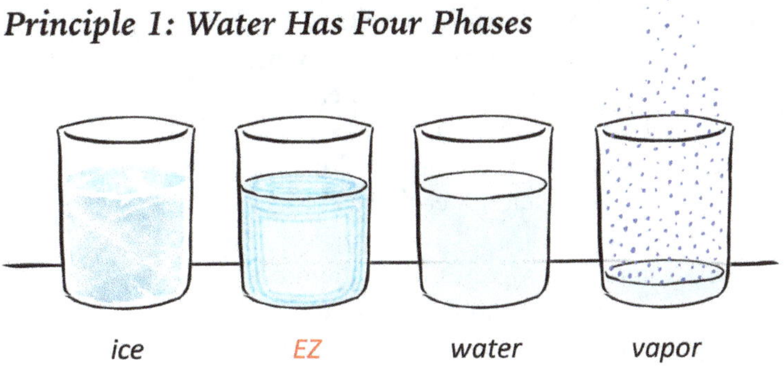

The fourth phase of water

Defined scientifically by Professor Gerald Pollack and colleagues (based on concepts developed by Gilbert Ling, Albert Szent-Györgyi, Walter Drost-Hansen and James Clegg), the fourth phase of water is one of the most intriguing scientific discoveries made in the last century, because it explains dozens of natural phenomena that happen around us every day, but science and medicine had yet to explain.

Also called (1) "exclusion zone" water, (2) "EZ," (3) structured water (not the same as vortexing to de-cluster molecules), (4) interfacial water (the interface between a surface and the water around it), (5) charge-separated water, (6) H_3O_2, (7) coherent domains, or (8) "e-zone" herein (think 'electron-zone' or area of heightened electrical charge). Its applications in science and medicine boggle the mind, because e-zone is a distinct phase of water between a liquid and a solid. It's found almost everywhere in nature that normal water is present. And it has unique structural dynamics, as well as its own remarkable properties.

Electrical charge is Nature's favorite force

To give you some background, most activity in the domains of biology, hi-tech, physics, or geology revolves around electrical charge. That's because most work in the universe is done by 'like' charges repelling each other, or unlike charges attracting each other. Gravity, in contrast, is many orders of magnitude weaker at making things move (by a factor of 10^{38}). Positive and negative charges are the driving force that moves things in biological systems, man-made devices and geology.

Nature uses electrical charge in so many ways because electrostatic attraction and repulsion (aka Coulomb force) is astonishingly strong, particularly at the subatomic level. To illustrate, suppose you could collect the electrons flowing through each of two incandescent light bulbs for one second. Position those negative charges a few inches apart from each other, and how much repulsive force do you think they'd have? Try the weight of 50,000 garbage trucks. Now remove 1% of the electrons from each of two people, and those electrons placed a foot apart would repel each other with the force equal to the weight of the earth.

How water's positive charge separates from its negative charge

Regular water (H_2O) is composed of two hydrogen atoms, each of which has a +1 electrical charge, and one oxygen atom, which has a -2 electrical charge. When combined, this gives water a neutral electrical charge (1+1 of hydrogen – 2 of oxygen = 0). Hence, water's neutral electrical charge can't do any work (attract or repel) in its normal state. Yet those charges do contain tremendous energy *potential*, as we shall see.

Electron-zone water is able to power cellular processes (i.e., chemical reactions and movement) by separating positively-charged hydrogen atoms from negatively-charged OH groups, and rearranging them into hexagonal sheets, with a pool of those hydrogens beside it. Separate pools of charged particles thus create the electrical equivalent of electrolyte (the liquid with all the juice in a car's lead-acid battery).

Honeycomb Sheet

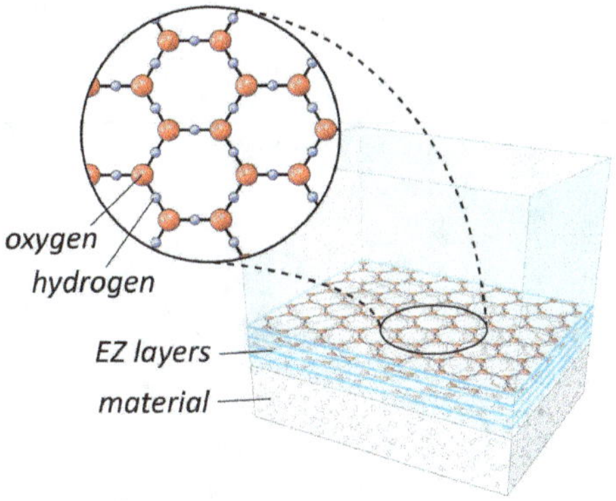

What happens is, around surfaces (especially water-loving surfaces), water naturally rearranges its atoms into layers of a honeycomb shape, which are one atom thick.

The honeycomb sheet is the EZ's unitary structure. Sheets stack parallel to the material surface to build the EZ.

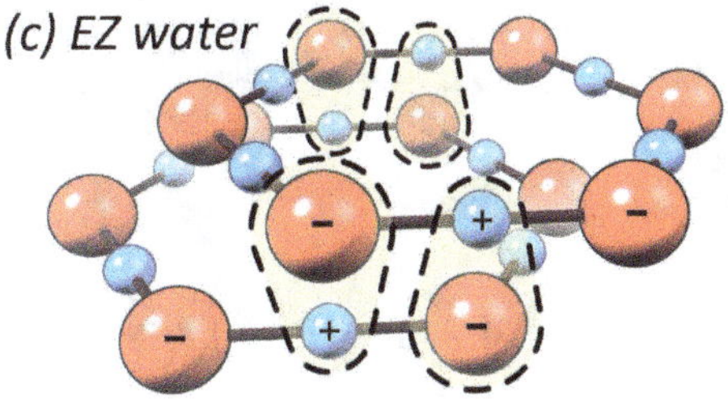

The polarized layers build up on top of each other, shifted sideways by one atom (i.e., oxygens in one layer are lined up with hydrogens from another). This forms a crystal lattice that's easy to break apart due to its weak bonds.

Interfacial Battery

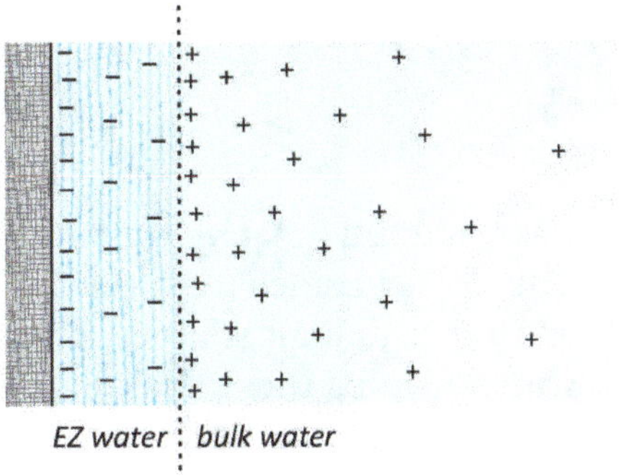

These configurations, like a game of musical chairs, don't have room for some of the hydrogen atoms in its hexagonal matrix. So the e-zone kicks the orphaned hydrogens into the surrounding space. These hydrogen atoms, and their positive charge, collect in a pool next to the e-zone layer, which is called "bulk water" or "unstructured water."

Thus, the layers of e-zone accumulate negative charge, while the water around the electron-zone concentrates protons and positive charge. *Violà*, you have a battery.

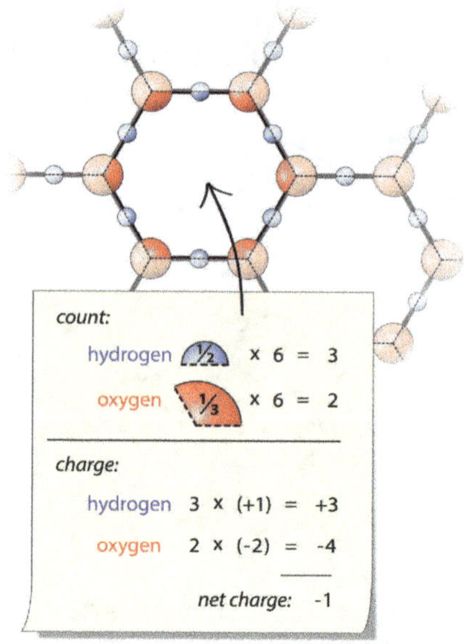

Configurations like this are called liquid crystals because of their honeycomb-ordered structure (crystal) and weak bonds (viscous fluid).

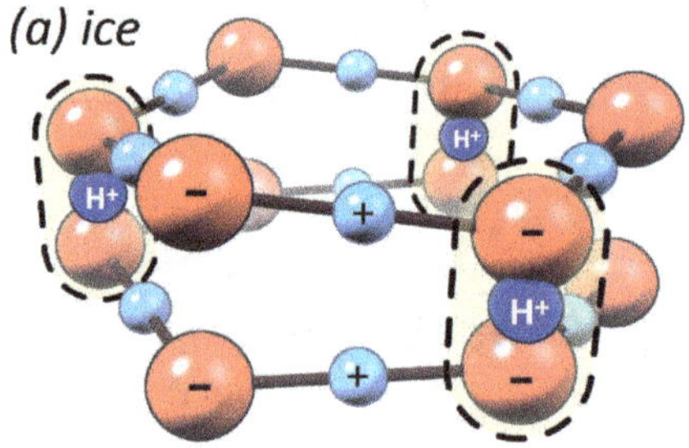

In contrast, ice uses the same hexagonal structure as e-zone, only the layers are stacked directly on top of each other. The negatively-charged oxygens in one layer are bound to oxygens in the adjacent layer through the positively-charged loner hydrogen we just talked about, thus forming a strong, solid bond. This strong electrostatic connection is what makes ice solid.

Interesting to note, water goes through the e-zone phase before it turns into ice. And when ice melts, it becomes e-zone briefly before it turns back into regular water. For these reasons, sailors have noted for centuries that water surrounding icebergs is more viscous than regular water. The liquid crystal properties of e-zone explain why ships experience friction as if sailing through Jell-O.

Radiant Energy

Radiant energy charges the battery. The energy comes from the sun and other radiant sources. The water absorbs these energies and uses them to charge the battery.

But the difference-maker for biology is that the size and strength of exclusion zone grows when it's exposed to light – particularly IR – thus separating more charge, storing more energy, and making a bigger battery. The wavelengths that are best at building e-zone are infrared, followed by visible frequencies, ultrasound and UV This is how water stores the sun's energy for our cells to use.

Light charges up your water

Water. E-zone starts with pure water – no impurities. In particular, fluoride, bromine, chlorine and deuterium spoil the size of the crystal lattice and thus e-zone's battery capacity.

Radiant energy charges up the e-zone. The best way to build e-zone in the body is by getting sunlight on the skin. Infrared wavelengths penetrate soft tissues 10–30 cm. This is how sunlight directly charges your battery. Visible frequencies also help, as do most natural energy sources – including radiant energy from people, pets (i.e., warm bodies) and things.

IR light/heat released by mitochondria. When mitochondria make heat, the IR light they release shrinks water around the respiratory proteins. This condenses the proteins, making those mitochondria more efficient. And, just as helpful at powering cell function, that infrared light charge-separates water into e-zone.

Atomic, molecular physics of e-zone. What's happening on an atomic level when water becomes e-zone is that light changes the bond angle of hydrogens so sheets are formed from the hydrogen and oxygen atoms, instead of bulk water molecules conforming to the shape of a tetrahedron (4-sided pyramid). The semiconductors inside us, notably melanin, accentuate this process by feeding water the best light with which to build e-zone. The energy (trans)formed from this charge-separation of water represents over 90% of our cells' energy requirements.

Magnesium needs water to work

Magnesium is hydrophilic. It needs water to function optimally. So when you're dehydrated due to nnEMFs and mitochondrial dysfunction, etc., magnesium deficiency makes it harder to activate a parasympathetic state of rest and digest. Magnesium is also used in over fifty enzymatic reactions. So dehydration throws a monkey wrench in dozens of key biochemical process through magnesium deficiency alone – including helping ATP get to where it needs to go in the cell.

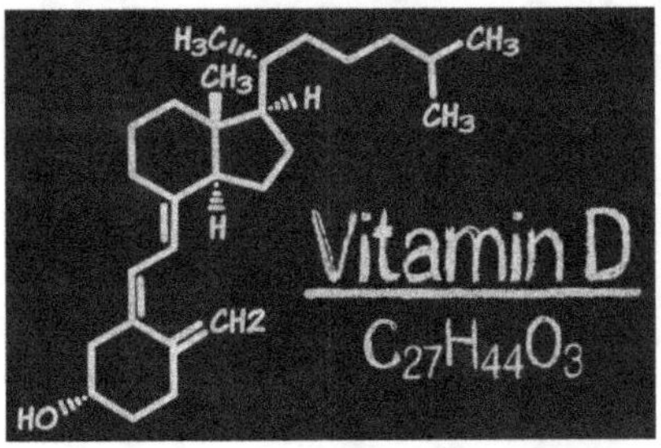

Vitamin D needs water too

Vitamin D is best-known for enhancing the immune system. It controls bone growth. And it helps us avoid obesity, heart disease, brain disorders, multiple sclerosis and cancer. Deficiency of vitamin D plays a large role in causing allergies. In fact, vitamin D is used by the body in so many ways, some even consider it to be a (master) hormone.

Most people know vitamin D is made by UV sun on the skin. Unfortunately, centralized medicine has brainwashed people into believing sun exposure is bad for you, and needs to be blocked with the latest super-duper sunscreen. Truth is, UV light is not bad for you. UV frequencies that regulate and regenerate our biology only *appear* to damage the skin because most people are dehydrated, deficient in sulfur (which blocks harmful rays), and have not developed a "solar callus" through morning sun exposure.

The body needs water to make vitamin D, but almost everyone today is at least moderately dehydrated. That impairs the body's ability to turn LDL cholesterol into vitamin D, called the "isomerization" step. How do we get dehydrated? It's mainly nnEMF exposure, weak mitochondria producing less metabolic water, inadequate intake, fluoride in water, and low moisture content in our food, along with the electrical differential that draws water into cells.

Therefore, you solve intracellular dehydration by doing the opposite: You improve your mitochondria, drink more water, avoid toxins, and eat more whole foods. You eliminate as much wireless radiation as you can, and fix those other problems. That helps you make more vitamin D. And then magically you'll be able to stay out in the sun longer, you won't get burned, and you can harvest the sun's energy like you're supposed to. Dr. Jack is a perfect example. He's fair-skinned, freckled, of Irish descent. But he's out in the sun all the time, and does those other things to help capture sunlight better. Thus, he can tolerate strong sun exposure for more than five hours a day.

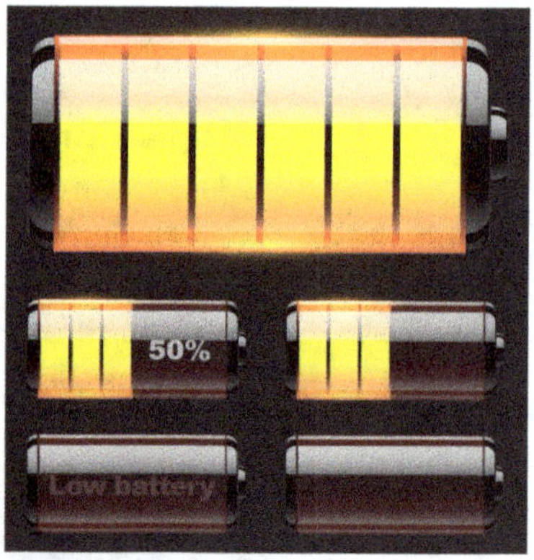

Is your mammalian battery in shape?

There's a way to test what kind of shape your mammalian battery is in. Meaning, is it low and just needs to be charged up? Or is there a fault with something in the system that has to be fixed before your biology will run at full power? (In this instance, mammalian battery refers all stores of electrical charge in the body, the largest of which is e-zone.) That biological "battery checker" is your vitamin D level. You need some combination of all three of the following for your battery to operate at full strength:

- **Sun.** What's your solar yield like? If it's low, your solar panels must be either covered up by clothing, you're not outside long enough, you're out at the wrong times, your location is not conducive to getting UV and IR, or you're not incorporating sunlight into your physiology properly.
- **Water.** Are you well-hydrated? If you're low, you could be dehydrated from being around too many people, using too many microwave devices, for too long. In other words, your environment is toxic.
- **DHA.** Got DHA? If it's low, it could be because of too much blue light, or a dietary deficiency of DHA, or poor recycling of DHA (blue light impairs the long-loop recycling DHA through the liver).

The best water is made by mitochondria

We all know it's important to drink lots of water. But did you know the water you drink is not nearly as beneficial to the energy needs of the body as the water your mitochondria make? For perspective, water found in nature averages around 145–155 parts-per-million deuterium. Unfortunately, the more deuterium in water, the more resistant that water is to forming e-zone. On the other hand, the water that mitochondria make possesses a very special quality: When all's well, it should be almost deuterium-free. This is crucial in creating optimal energy and wellness, because the production of this metabolic water helps dilute your deuterium level into a healthy range (along with breathing, urinating, sweating and other waste-removal pathways).

Bottom line: the stronger your mitochondria are, the more deuterium-depleted water they make. The e-zone layer in the cell then gets bigger, more DC electricity is available, and redox potential increases. What's more, that e-zone is where you need it most: inside the cell and mitochondria. Yet another way the road to wellness or illness goes through your mitochondria.

Hydration strategies

1. **Drink more water.** Duh. And purity matters. Make it easy on your detox systems by testing your water supply, filtering it or buying water that helps your body remove toxins, not adding to your already high toxin load.
2. **Fluoride is not your friend.** Do whatever it takes to make sure there's no fluoride in your water. Filter it, or buy natural spring water. Even reverse-osmosis water is better than fluoridated water.
3. **Chlorine is pretty bad too.** Reduce your water's chlorine content by letting it sit in an open container overnight. Since chlorine is a gas in its natural state, it off-gasses over several hours. However, chloramines (a chemical cousin of chlorine used to disinfect) evaporate much more slowly. So this trick doesn't work with them.
4. **Reduce your nnEMF exposure.** The EMF frequencies all around us are mostly microwaves. And what does everyone know about microwave ovens? They dehydrate. They bombard food with microwave radiation so strong it shakes water molecules loose from their surroundings. That's why steam pours out of food heated in a

microwave oven. And that's how cell phones, Wi-Fi and IoT devices unquestionably dehydrate *you* in their presence. The solution? Stop using hand-held microwave devices, put more distance between you and the devices, drink more water, or all of the above.

5. **Reduce beverages and foods that dehydrate.** Avoid diuretic beverages like coffee, soft drinks and energy drinks because caffeine dehydrates (i.e., makes you pee out more than you drink). Stay away from fruit juices and sweetened beverages because sugar dehydrates as well. Their concentrated sweeteners are a burden on your system. Avoid alcohol for the same reasons. And cut down on salty foods when you can. They soak up water.
6. **Barbara O'Neill's sea salt "hydration hack."** Dissolve a crystal of Celtic sea salt under the tongue as you drink some water, or before. It's absorbed quickly through the oral mucosa and beats the water into the bloodstream. The salt then gets into cells first, which pulls water inside from the magnesium in the salt (it's hydrophilic). Just make sure not to put the salt directly in the water you drink. It will just absorb water before it enters the cell and not help you hydrate.
7. **Spread out your intake.** Drink small amounts of water throughout the day if you can, instead of all at once. That makes it easier for cells to absorb. You may want to add a mild natural flavoring like lemon juice to pure water so the digestive tract takes longer to process it.
8. **Get more water from moisture-rich foods.** Unsweetened tea, unprocessed broths, and homemade soups are good ways to hydrate, as are fruits and vegetables with a high water content. This is how we used to get much of our water.
9. **Drink water sparingly with meals.** Water dilutes stomach acid, which neutralizes its acidity and impairs digestion. Instead, drink most of your water up to a half an hour before, or 1½–2 hours after meals.
10. **See also Chapter 11: Deuterium** for more information on drinking deuterium-depleted water to bring your level down.

7

MAGNETISM

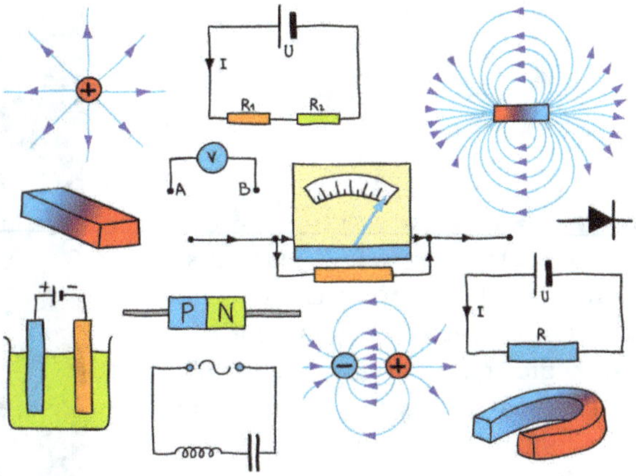

Magnetism is essential for life and health

All life forms need to be in a magnetic field to build structure, move materials and power their physiology. We must have unidirectional magnetism to live, because it supports our biology in ways we never imagined. There would be no life on earth without it. In fact, planets like Mars that don't have a magnetic field can't support any life at all.

Reference: "The Infinite Mind," Valerie Hunt, pp. 30–38.

To illustrate what happens in the absence of magnetism, Dr. Valerie Hunt, Professor of Physiological Science at UCLA, hooked human subjects up to monitoring equipment such as EKGs and EEGs, and put them in a 7' x 7' enclosure completely shielding them from all magnetic fields. Within minutes, the subjects began to sob uncontrollably. "We feel like we're falling apart," they said. A loss of sensation and muscle control started at the extremities and worked its way inward and upward. After two hours, the heart and brain were in such great distress the experiment had to be stopped. Dr. Hunt was certain they would have died had the experiment been allowed to continue another hour or two. That's how dependent we are on magnetism. Let me repeat that so it sinks in. **Without magnetism, you'd be dead in about three hours.**

MuMetal: Magnetic shielding material made of nickel, iron, copper, chromium and molybdenum.

Shocked and amazed by Dr. Hunt's results, Dr. Dean Bonlie, DDS, did a similar experiment years later on six healthy, adolescent mice. He wanted to know what the physiologic effects of a *partially-reduced* magnetic field were. Where is the threshold between dysfunction and lethality? He put six littermates in a MuMETAL box that reduced earth's magnetic field from 0.5 gauss (normal) to 0.1. Within 15 minutes, the mice went into slow motion. They were barely able to move, or even get up, after being flipped over. After 24 hours, one mouse died. The others that survived ate twice as much food to compensate for the loss in energy. Left in a 0.1 gauss field, they quickly became so morbidly obese they were actually round.

The earth's magnetic field has been declining for millions of years

Some say it's due to a cooling and slowing of the earth's molten iron core that makes magnetism, or energetic changes in the sun, or the area of the universe we're travelling through. Some suspect it's caused or worsened by proliferation of man-made EMFs ionizing our skies (i.e., electrically

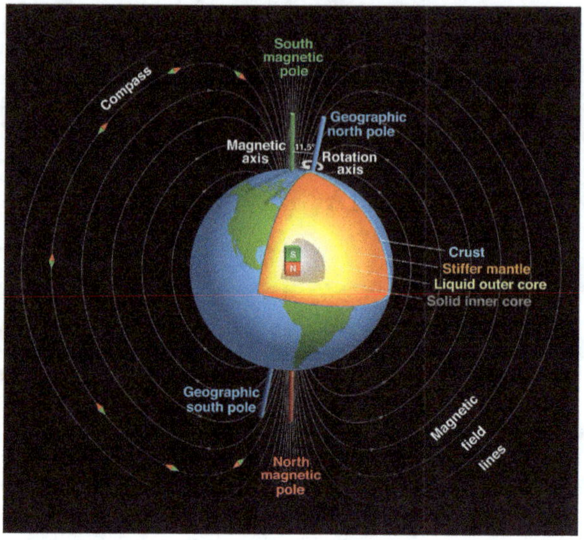

charge so it's more reactive). Whatever it is, the magnetic field has decreased from a whopping 300 gauss in dinosaur days (as measured by alignment of magnetite crystals in geologic formations), to 2.8 gauss 4000 years ago in ancient Babylonian days, to an average of 0.5 gauss today, depending on location.

Some experts say this is barely sufficient to support life. Still, the earth's magnetic field is dropping about 5–7% per century. At this rate, NASA projects it will go to zero within 500–800 years. Unfortunately, the decline appears to be accelerating, and we're left to deal with the health effects in the meantime. So, at the high end, magnetic flux supports life on a massive scale. Indeed, science suggests prehistoric life forms were way bigger back then simply because stronger magnetism gave them more energy to support their great size and no doubt longer lifespans. Whereas, down low, when the gauss level bottoms out, the magnetic field reverses, the poles swap places (called a "pole shift"), and a mass die-off occurs from lack of energy. In fact, these pole shifts and mass extinctions are shown in the geologic records to have happened 183 times in the last 83 million years. That's about 450,000 years apart. And we're due for another one soon.

Devil's Tower, Wyoming.

How long can you stay disconnected?

Most people in good health can tolerate being disconnected from the full strength of earth's magnetic field for a short time, like flying at 35,000 feet. Others with poor mitochondria and redox potential can't tolerate being disconnected even for a few hours, like those who suffer a heart attack or acute psychotic episode in the air, not to mention a blood clot or stroke. But, ultimately, no one can stay disconnected from earth's magnetic field forever, because **magnetism energizes the matter within us, regenerates us in sleep, reduces free radical damage, and helps regulate our biorhythms.** The body and mind pretty much fall apart at every level from the micro to the macro, when you don't get your daily dose of magnetism.

Magnetism energizes the matter within us

Magnetism increases the velocity that valence electrons spin around an atom's nucleus. This phenomenon is described in physics by the Larmor Frequency Formula ($W_{Larmor} = \frac{e}{2m}\vec{B}$), where

- W_{Larmor} = angular velocity of electrons.
- e = charge of electrons.
- m = mass of electron.
- $\vec{B}$ = magnetic field vector.

Valence electrons are the outer orbiting electrons involved in the body's chemical reactions. They are the electromagnetic glue that makes atoms cling together to form molecules, or it keeps molecules apart from one another. The negative charge on electrons makes molecules sticky or repulsive, as the case may be. This has enormous implications in biology.

In the same way that heat affects a liquid, increasing the speed/energy of valence electrons improves the bioactivity and reactivity of molecules in basic processes such as the ETC, electron transfer of oxidation and reduction, enzyme function, detoxification, and digestion of food. Simply put, magnetism enhances the physics and chemistry of the body by making electrons faster and friendlier (more open to forming relationships).

For comparison, *stirring* a solution speeds up chemical reactions by moving molecules by meters (macro), which gives them greater opportunity to interact with others. *Heating* a solution can speed up chemical reactions even more by inducing motion at an atomic, molecular and macro level. While *magnetism* increases molecular action by making electrons dance faster at a sub-atomic level.

More active electrons not only move more. But a stronger magnetic field also causes electrons to cut through more lines of flux as they dance around their nucleus. That raises electrical charge just like an alternator would, given stronger magnets. That means if you increase a magnetic field by 10,000 times (5000 gauss), electrons orbit 10,000 times as fast. *And* they cut through a magnetic field 10,000 times as dense. That amounts to 1 million times the charge in those electrons (10,000 x 10,000 = 1 million). Now that's hot.

Crucial to health and life, electrical charge aids energy production, blood flow (by making RBCs repel each other), healing and detoxification. And just as light turns into DC electricity via the Photoelectric Effect, magnetism can also turn into DC electricity via the Inverse Spin Hall Effect. All of which **makes magnetism the mother of all catalysts** – even more essential than enzymes, more foundational than grounding, and more basic to our biology than sunlight. These are some of the reasons why we can't live without magnetism for more than a couple of hours… and why the low magnetic flux today is causing us to get fat, sick and tired.

Inverse Spin Hall Effect: Discovered around 2008, the ISHE explains that current spin (in this case from magnetic flux) creates an electrical current at a 90° angle.

Image used under Creative Commons 4.0 license. Author: Eleanor Holmes. I modified: aspect ratio, colors and fonts.

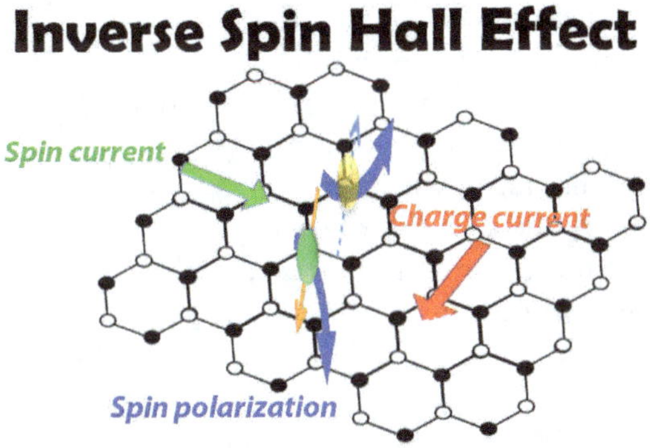

The brain recharges one organ/tissue at a time while you sleep

The brain conducts this renewal effort by sending pulsed DC electricity to one organ or tissue at a time, while you sleep, in order to charge up their electrons – similar to charging up your smartphone. Recharging of electrons also happens during the daytime to a lesser extent. Brain cells called "astrocytes" put tissues and organs in a better state of health by sending them voltage along nerve pathways so their electrons move faster.

When electrons are more energetic, the physics and chemistry of the body works better. Dr. Dean Bonlie calls this recharging of tissues by the brain's electromagnetic pulses "resonance." Still almost completely unknown in 2025, the theory of electromagnetic resonance is mostly the work of Dr. Robert O. Becker, refined by Dr. Bonlie using acupuncture principles. In other words, this is a breakthrough concept, should the theory be proven true.

The strength of resonance, which is simply voltage, greatly influences how well cells and mitochondria function. Voltage from astrocytes equals potency of regeneration. Unfortunately, when stress, heavy metals, sleep deficiency or foreign frequencies reduce electrical output from the astrocytes, then organs and tissues don't get their fair share of reenergizing resonance. So that, right there, is one of our biggest opportunities to bring more healing and vitality to our lives: supporting our astrocytes in producing DC electricity to revitalize cells. That can consist of removing barriers to resonance, and/or it can mean supplementing your sleep with more earth-type magnetism so brain cells make more electricity.

Astrocytes create voltage

One of the primary functions of astrocyte cells in the brain is to convert chemical energy into electrical energy in order to heal and regenerate the organs and tissues of the body. Loosely interpreted, the Chinese call the body's capacity to rejuvenate tissues with electricity and blood flow *chi, vitality* or *life force*. (To them, the act of restoration, and resulting vitality, are considered one and the same.) The electricity that astrocytes have available to effect healing is determined primarily by the strength of their mitochondria, because mitochondria turn the chemical energy in glucose, etc. into electrical energy in the form of electrons and protons.

Chi: Vital life force, or essential energy, running through all living beings that makes us alive.

Both production and consumption affect availability of chi.

- One-way magnetism makes more ATP and voltage via "hotter" electrons feeding *through* the electron transport chain.
- Stress uses up electricity/chi faster.
- While heavy metals and lack of sleep prevent chi from being made.

By supporting your chi with concepts like these, you can tilt supply and demand in your favor. Astrocytes will then be able to send sufficient electrical flow to the tissues that need it in order to charge up your cells and mitochondria with resonance.

Natural vibrational frequency (aka fundamental frequency)

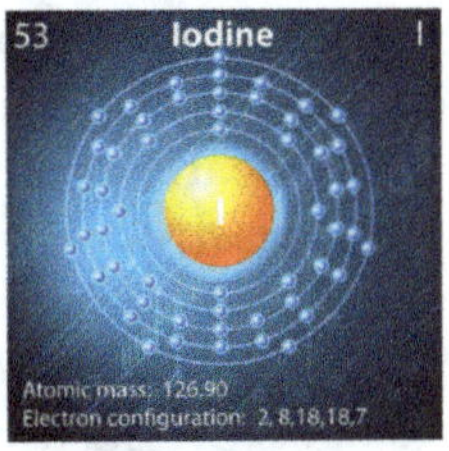

Despite the way they're often depicted, electrons don't actually spin around an atom's nucleus in perfect circular orbits. Electrons are only presented like planets circling a sun in order to show discrete energy levels they can occupy, called "orbital shells," which are basically stair-stepped levels of energy that electrons can reside at. More accurately, quantum mechanics says electrons hang out in predictable, but not precisely definable, *patterns of probability* around the nucleus, which we call "orbitals."

In the graphic below, each box represents a different orbital which can host *a pair* of electrons, or *one* in the case of H^+ hydrogen or a free radical. Each orbital has a certain energy level and corresponding shape. Eastward is a higher energy state, as is southward. Each line (designated by the numbers 1–7) is an orbital shell (the above graphic represents orbital shells as concentric circles). The lowest-energy orbitals in each orbital shell – the "ground state" – are spherical in shape. Higher-energy orbitals are shaped like a dumbbell, or a looping sine wave. So the more electrons an atom has, the more complex its combined orbital geometry is, because electrons try to stay away from each other due to their 'like' charges repelling each other.

The horizontal axis depicts number of electrons an atom has from fewest (left), to most (right). The vertical axis depicts orbital shells – each possessing more energy – from lowest (top), to highest (bottom). Blue regions are positive, the lighter ones are negative.

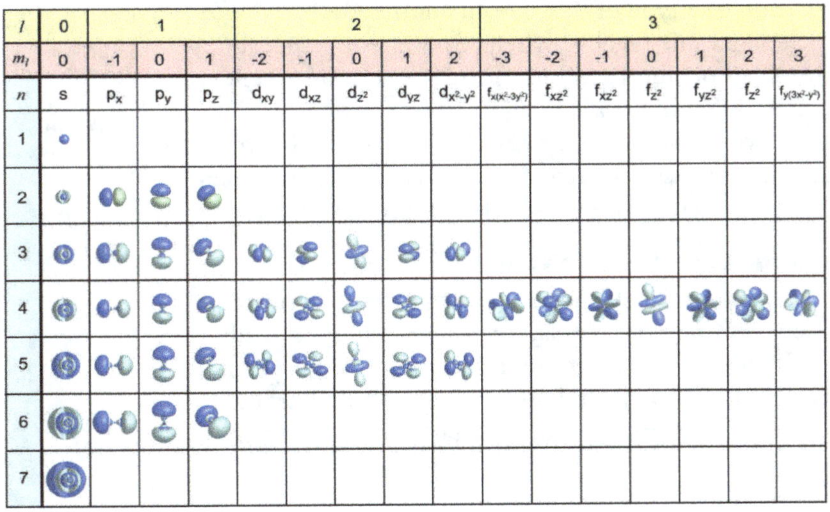

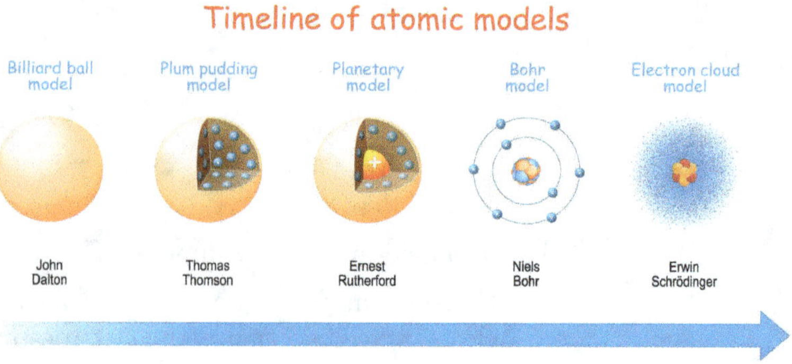

But the funny thing is, through Heisenberg's Uncertainty Principle, it's impossible to know exactly where an electron is, or where it will be next, until you observe it. We can only predict with 90% accuracy where an electron will show up at any given moment within these *regions of probability*. The closer you get to an electron's orbital shape – generally a sphere or hour-glass shape – the greater the chance you'll find it where you expect it. But it literally could appear anywhere in the universe. That's quantum physics for you. It's actually a lot more complicated than that, especially in compound molecules. But, to keep it simple, we'll envision circular orbits for the rest of this discussion.

Now, the majority of the time, the weight and electrical charge of electrons are somewhat evenly balanced across the atom. But, every so often, electrons cluster more on one side of an atom than the other, which over-weights that side with polarity and mass. Like a washing machine spinning drying when out-of-balance, the atom vibrates as one side gets overloaded again and again. In the case of body tissues, this happens about 1–100 times per second.

Important for biology, when electrons cluster together, an atom's attractive and repulsive force on that side increases briefly because its charge and momentum bunch up. An imbalance of charge and momentum from this momentary grouping creates wobble and vibration, called "precession." And the rate at which an atom, molecule or organ oscillates in this fashion is its natural vibrational frequency (which anyone could detect, given proper equipment and training). Consequently, the frequency at which a material vibrates is one surefire way scientists can distinguish one material from another, and one type of tissue from another.

See electron orbital shapes/ energy level in YouTube video entitled "Orbitals, the Basics: Atomic Orbital Tutorial" by Crash Chemistry Academy, or "A Better Way To Picture Atoms" by minutephysics. Or visit webpage: chemguide.co.uk/atoms/properties/atomorbs.html

What's happening at an atomic level is, when atoms are together for some time, as they are in an organ, the orbits of their outermost valence electrons sync up with each another so they all spin around their nuclei at regular intervals. Hence, you get signature rates of vibration the brain can use to target just that organ or tissue for regeneration. Astrocytes can then address each tissue individually, based on need and supply.

How "resonance" energizes and heals tissue

Like any pure, unadulterated material, each tissue in the body oscillates at its own distinct frequency. When the oscillation is strong enough in an *inorganic* material, and within the range of human hearing, we hear the frequency vibrating air molecules as sound. For example, we all know the sound a wine glass or tuning fork makes when it's tapped. Similarly, if a strong oscillation is below the range of human hearing, we might feel the vibration with our sense of touch.

Now resonance is when you expose an atom, molecule or material to the exact frequency at which it naturally vibrates – whether that application of energy is a sound wave, magnetic force or electrical charge. You see, atoms and pure materials are constantly vibrating at a certain frequency – whether you can hear it, or feel it, or not. Increasing the intensity of that vibration can turn imperceptible frequencies into perceivable ones. **So resonance is simply adding more of an atom/material's natural frequency to make it vibrate more intensely** – meaning, stronger as in amplitude, not faster as in higher-pitched. Best example is an opera singer breaking a wine glass with sound. By singing at the exact frequency at which crystal naturally oscillates, she introduces more energy into a wine glass. The glass continues to absorb this energy until the resulting vibration exceeds the material's fracture strength and the glass shatters.

Normally, resonance in the body can't be felt because its amplitude is too low. But when super-accelerated healing is taking place under the colossal magnetism of the MME machine (more on that shortly), resonance can be so intense that broken bones, degenerated discs or diseased livers can feel as if they're vibrating so strongly they want to leap right out of the body. Both patient and clinician can feel it. What's happening at a subatomic level is that resonance gives an atom's electrons a tiny push each time they hit the same spot in their orbit (i.e., when they're most lined up). This gets those electrons going faster and faster with each push, which *supercharges* the cells, tissues and organs they reside in.

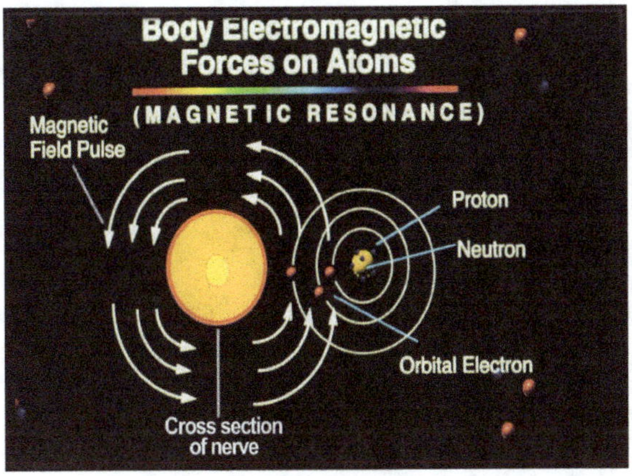

Image used courtesy of Dr. Dean Bonlie, DDS.

Foundational to biophysics and chemistry, orbital velocity makes electrons more enthusiastic dance partners in electron exchanges such as the ETC, enzymatic reactions like digestion, and in protein production. Vital to maintenance and repair, electron transfer builds proteins to make new cells, tissues and biochemicals. Or it breaks them down in processes such as apoptosis.

To help you understand how resonance supercharges electrons in the body, picture this: **Astrocytes broadcast a pulsed DC current to the entire body** along the outer layer of motor and sensory nerves. These electrical pulses travelling through nerve casings produce a pulsed *magnetic field* at a right angle to the nerve pathway, which gives nearby electrons a push at just the right moment in their orbit. This targets tissues by their frequency, not by "connecting with tissues on a private line," so to speak. Targeted tissues receive the effect by resonating, while nonparticipating atoms are unaffected because they're vibrating at a different frequency.

To illustrate this effect, how do you get a kid on a swing to swing higher with minimal effort? All you have to do is give them a tiny push each time they stop moving backward and start moving forward – not a moment sooner or later. Same thing with charging up electrons in living tissue: perfectly timed pushes. Except, in the case of biology, orbital speed of electrons translates into more energy for growth and maintenance.

So timing is crucial, because if you push electrons at the wrong moment (slightly before or after they hit their "apex"), electrons don't continue to gather speed. Rather, some pulses add velocity, while others subtract it. The result being, overall electron speed does not change, vibration of the atom stays the same, and you get no resonance. This is what happens when astrocytes and cells around the body aren't communicating properly: lack of resonance equals lost opportunity to renew.

Similarly, if you give electrons a push in the *wrong direction*, instead of at the wrong time, you unequivocally slow down their orbital speed, decrease the atom's vibration, and reduce its chemical activity. Not good for life. You get this effect in a reverse-polarity magnetic field. It's a loss of energy and health, which is why you've got to watch out for unstable, non-uniform magnetic fields: You speed up some electrons, while slowing others down, thus upsetting communication between controller glands in the brain and organs in the body, as well as organ function itself.

Without a doubt, bipolar magnetic fields and chaotic fields (like those found in a modern home or workplace) are not good for the body on a continual basis. Applied repetitively, weaker (but still uniform) *bipolar* magnetic fields activate a stress response, deplete the adrenals, and drain your tissues and organs of the energy they need to renew, and even maintain, themselves. This is what small magnets and PEMF devices do when used too often.

Pulsed electro-magnetic field therapy devices (PEMF) stimulate nerves, muscles, and blood flow using on-off electricity to heal an area.

But even worse, *highly irregular* magnetic fields – meaning intensely disordered – can easily cause the worst diseases imaginable, such as brain cancer. Guess why. A portion of that exposure slows down valence electrons, reduces ATP production, inhibits enzyme function, and cuts protein production to repair cells and make biochemicals. This disastrous situation can happen when your home is constructed with wiring errors.

How, then, does the brain know which frequencies to use, and how much to apply to each tissue? That's the job of **biofeedback.** Through biofeedback, the astrocytes know which tissues need to be resonated with electricity and which don't. Through this feedback-control system the brain targets one tissue at a time with its favorite frequency to bring it up to speed, instead of wasting resources resonating tissues that don't need it.

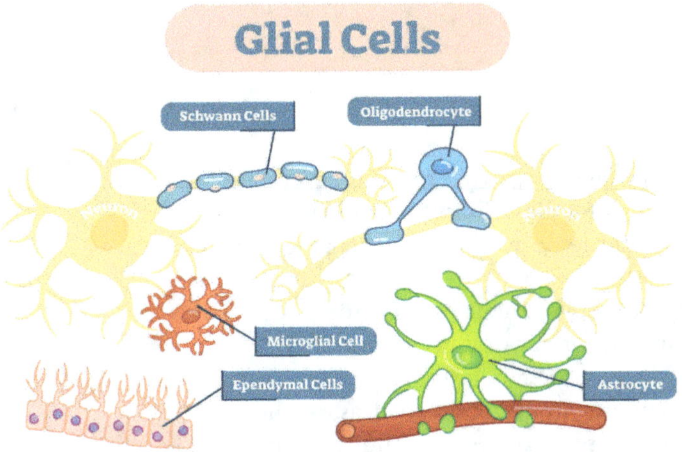

How astrocytes resonate tissue

Not well known, but instrumental in healing, the body's electrical regeneration system is a closed circuit, with energy and information going out to the body to charge up cells and tissues, and coming back in through return pathways of acupuncture meridians to tell the brain what cells need.

As far as I know, this is the only place in which resonance is described. However, you only need to know how it works in order to explain magnetic therapy with the MME, or in reversing disease conditions with the Magnetico.

Here's a basic overview of the healing and communication circuit:

1. **Astrocytes convert chemical energy into electrical energy.** Mitochondria in the brain's astrocytes turn glucose into electrical energy. This pulsed DC current is broadcast to tissues around the body. Astrocytes use up their electricity during the day when you're active. And you charge them back up at night when you sleep.
2. **Carrier frequencies employ macro-frequencies to heal.** Slow-wave pulsed DC electricity (negative only) from the astrocytes flows out to distant organs and tissues to resonate them. This energizes and regenerates tissue by charging up their electrons.
3. **Message frequencies employ micro-frequencies to communicate.** Encoded on those outbound healing frequencies (by piggybacking small waveforms onto the larger carrier wave) are special instructions for stem cells to tell them where they're needed, and what to mature into when they grow up (e.g., nerve cells, blood vessel cells, bone cells). These directions are message frequencies. When these messages are not received properly for some reason (most often because of nnEMF interference, weak signal or both), stem cells don't proliferate and differentiate into new tissue like they should. That means full healing fails to take place. Mitochondria are given their own set of instructions on message frequencies to respond to the ATP needs of tissues.
4. **Astrocytes collect data.** Information is collected about the health status of organs and tissues based on the frequency of their vibration, which is then encoded onto the large carrier wave for its return trip. Similar to 120 volt power returning to the power station, the carrier frequency loops back around and returns through the covering of the spinal cord, now loaded with information for astrocytes to decipher about the condition of tissues.

Fascia: Thin, filmy casing surrounding muscles, nerves, organs, blood vessels, and bones that holds each tissue in place, separates it, protects it, and relays information.

On a technical note, electrical resonance frequencies are transmitted to organs and tissues through the semi-conductive outer layer of nerves, made up of Schwann cells. And, on the return trip, the fascia and spinal cord covering are the semi-conductive layers that transmit the combined signal back to the astrocytes. From the intel gleaned in the resonance phase, astrocytes find out which cells in the body have lost their mojo *vs.* which ones are happy campers.

Astrocytes then direct cell repair and replacement in real time as the carrier frequency is boosted and sent out for another go-round. The lesson for the day is, the electrical system of regenerative resonance and cellular communication forms a complete circuit. It is broadcast to the entire body on the outside of motor and sensory nerves. And it returns through the acupuncture meridian system. This is the body's electrical regeneration system of resonance.

Why we sleep in 90- to 120-minute cycles

While we sleep, our astrocytes cycle through a range of frequencies at which tissues resonate, once every 90 minutes to two hours. They manage this not by matching frequencies exactly, but by hitting each tissue with a *harmonic* of its natural frequency once per cycle. The cycle is repeated until something makes us wake up, or we've gotten sufficient sleep. Resonating each tissue repeatedly, and adding energy into their electrons, is how the brain recharges, renews and re-synchs organs back to being their best selves. This includes the replenishment of neurotransmitters and hormones, DNA making more RNA to build proteins and cells, as well as mitochondria making more ATP due to freer flowing electrons.

Now, very important in hydration and heavy metal removal, ATP actuates sodium-potassium pumps to raise electrical differential at the cell wall. Positive charge from sodium and potassium helps bring more oxygen, water and nutrients into the cell, while it pushes out trash more easily – notably mercury, lead, aluminum and other positively-charged toxins. On the other hand, when organs or tissues are not resonating properly, these processes break down and disorder sets in. That's a state of incoherent operation we call chronic inflammation, degeneration and/or disease.

Cancer is the ultimate expression of communication breakdown between the brain and a group of cells. No matter how hard the brain tries to talk to cancer cells and get them on the same page as the rest of the body, the rogue cells just keep on doing their own thing. In this situation, astrocytes aren't resonating the cells properly. Instructions from astrocytes aren't getting through to the cancerous cells, telling them to sacrifice themselves with apoptosis. And cancer cells aren't communicating back to the astrocytes on message frequencies. It's a mess.

What throws the body out of resonance? What can we do about it?

Stress, in all its forms, is the #1 cause of tissues not being able to resonate properly – for example, a high-pressure job, sleep deficiency, too much blue light, stimulant drinks, toxins, oxygen deficiency, poor nutrition, hectic lifestyle, past traumas, and of course foreign frequencies on top of it all. Stress depletes the body's biochemical and biophysical reserves. What happens is, stressors place extra wear and tear on an organ/tissue. Through biofeedback, those cells tell astrocytes to send extra electricity and blood flow to help them recover from the extra demand. However, there's only so much of that to go around. If these coping mechanisms aren't sufficient – in capacity or duration – stress management chemicals such as adrenaline, cortisol, calcium and ATP are called in for support.

On a temporary basis, that's fine. The bad news is, these compensatory chemicals contribute to chronic disease by launching processes that bring **heavy metals** into the cell. Heavy metals interfere with the mitochondria's ability to make energy and avoid disease. But the thing is, without stress, heavy metals would not be able to build up in cells because, when all's well, they get pushed out of the cell by resonance and high electrical charge before they can ever build up.

Bottom line: Heavy metals, particularly mercury, are a villain that "pulls the trigger" in Alzheimer's, Parkinson's, multiple sclerosis, chronic fatigue, fibromyalgia, cardiovascular disease and general failure to renew. Stress starts the cascade by bringing mercury into the cell. It's mercury, as well as lead and aluminum, that cause most of the damage by reducing mitochondria's ability to make ATP, electricity, magnetism and resonance from astrocytes. See also chapter 14, section titled "Biochemical stress responses" (page 218), to learn how stress brings calcium into the cell to suck in more oxygen for respiration. When that happens, mercury gets pulled in by accident. Calcium later leaves easily, but the mercury is much harder to get out.

Scar tissue through the fascia also blocks healing and message frequencies, so avoid surgery whenever possible.

nnEMFs: Very important in today's wireless world, when astrocytes are able to, they turn up the power on the carrier/healing frequency to overpower man-made frequencies. However, their capacity to do this is limited by the amount of electrical power they're able to generate. So as foreign frequencies around us get denser, we're seeing more and more extreme sensitivity to nnEMFs in the form of electro-hypersensitivity (EHS). It's a vulnerability to foreign frequencies so severe they scramble the more powerful carrier/healing signal.

The person's system then malfunctions in all sorts of unpredictable ways that look like hormone, electrical or signaling problems. More commonly seen, communication on message frequencies is far more susceptible to interference from weaker nnEMFs because it's a much fainter signal than resonant healing frequencies. Three consequences of this failure to communicate are chronic inflammation, stem cells that can't differentiate and slower healing.

Luckily, fortifying the brain's electrical output with a Magnetico Sleep Pad may fix EHS, because exposure to enhanced earth-type magnetism improves electron transfer of ATP production, charge on the cell wall, and excretion of heavy metals out of astrocytes. Astrocytes then have the chi/vitality to emit stronger resonance frequencies and message frequencies than competing nnEMFs. Conclusion: Magnetism is the main biophysical force that helps the brain's astrocytes produce enough electricity to resonate organs and tissues at the proper voltage, frequency and timing. Hence, the Magnetico Sleep Pad is one of the fastest, most powerful ways to reverse many chronic conditions caused by mercury mayhem. It can dramatically increase your body's ability to recharge its chi/vitality.

Magnetico Sleep Pads are made of thousands of tiny magnets that act as one giant magnet, amplifying the earth's magnetic field by 10, 20, or 40 times. See the end of this chapter, and the Recommended Resources section at the end of the book, for more information.

Exercise strengthens the body's electrical system

Dr. Robert O. Becker's books, *The Body Electric* and *Cross Currents*, taught us that bones, muscles and the ends of the ligaments are piezoelectric. That means muscles are like rechargeable batteries. Each time you contract or relax a muscle, the trio (bones, ligaments and muscles) emit static electric charges of electrons. By adding piezoelectricity into the resonance system with *movement*, astrocytes then don't have to work as hard converting chemical energy into electricity for regeneration.

For this reason, physical activity enhances electrical messaging of the brain and body through donation of electrons into the system – particularly when that movement is strenuous enough to be considered *exercise*. Thus, the brain's got more power to heal – and be heard above competing frequencies – which is a big deal when you're immersed in electrosmog every minute of every day. Piezoelectricity of bones, ligaments and muscles is another reason exercise offers the body biophysical benefits we're just beginning to understand after Dr. Becker wrote about it in the 1980s.

How magnetism affects us all

Magnetism makes the ATP synthase spin faster – all by itself

When we're closer to the earth's magnetic field, the fifth cytochrome (the ATP synthase) in the ETC spins faster. This makes mitochondria more efficient at producing ATP. And that means more energy is produced from the food you eat, fewer calories consumed, and less waste. It could translate into easier weight loss.

One place you can get more magnetic flux, and thus make more energy, is in the Gulf South of the United States. When the Chicxulub asteroid hit the earth 65 million years ago near modern-day Mexico, it left a crater 12 miles deep and 93 miles in diameter. That means the earth's crust is thinner from the coast of Louisiana to Florida, which puts earth's magma chamber closer to the surface. Hence, the entire region is in a stronger magnetic field, which makes mitochondria pump out more energy.

To illustrate what this energy boost means to life in the area, the Gulf of Mexico is closer to the equator. It should not be as fertile as colder water. It should be lower in seafood because warm water carries less oxygen and iron. However, the gulf has something special other equatorial waters don't: superior magnetism. It's this magnetism that gives sea creatures more energy to thrive. The perfect place to see this phenomenon for yourself is in New Orleans. Seafood is so plentiful there it's not just another food category to residents; it's a cultural cornerstone.

The earth's magnetic field drops during the day and increases at night

At night, the magnetic field of earth and sun add to each other. So when our biology is supposed to be recharging and renewing with magnetism, that's when the earth's magnetic field is strongest. On the other hand, during the day, the sun's magnetic field – which is ⅔ as strong as earth's – competes with earth's magnetic field, pushing it into the planet. As a result, the side facing the sun experiences a lower magnetic field.

Bad for biology then, the magnetic field emanating from some areas of the planet are so weak the sun's magnetic field completely overpowers them, making them reverse polarity. The South Atlantic Ocean is one such area. Large stretches of the Atlantic Ocean now have a reversed magnetic field during the day, and a normal (reduced) magnetic field at night. This is a big deal because the earth's magnetic field (magnetosphere) also keeps the ozone layer in place. So if the magnetic field goes away, so does the ozone layer. That means we lose our protection against UV rays, x-rays and gamma rays. And that's exactly what is happening in the area: Low magnetic flux in the South Atlantic destroys ozone layer over the North Atlantic, where the magnetic flux returns to earth.

Few people live in the affected areas. But, for the first time, scientists have reported seals in *Atlantic* polar regions getting sunburned. In fact, UV rays have been measured there to be as intense as those at the equator. Unfiltered light is melting ice in Greenland, Canada and the North Pole dramatically faster than the North *Pacific,* where the ozone layer is intact. There, the icecap is actually growing. That's primarily what creates the *appearance* of higher temperatures and more ice melt. So-called "global warming" in the North Atlantic is caused mostly by diminished magnetic field in the South Atlantic. Waning magnetism breaks down ozone layer in the region, allowing more UV light to hit icecaps, seas and earth. Current CO_2 levels do not warm the planet significantly.

For us, this means weakening magnetic fields reduce healing at night when we sleep. It means by day, the ozone layer is broken down in certain parts of the world, so unfiltered light is wreaking havoc on the biology beneath it. It means serious polar ice problems for those ecosystems. And that also means, not too long from now, the earth's magnetic field will reverse, as it's done many times before, according to the geologic record.

The earth's magnetic field helps regulate our circadian biorhythms

Proteins go through a daily/nightly cycle of "breathing" energy in and "exhaling" energy out – which is their version of energy respiration in a daily cycle. Our proteins literally expand by day to accept energy into them by unfolding. And they contract by night, when we sleep, to recover from a day's exertion. The molecule that expands them is ATP. And the fundamental force that brings our proteins back together after a day's work is magnetism (aided by gravity).

On the other hand, without magnetism recondensing proteins, the collagen and DNA in your body slowly unwinds – making your bones, cartilage and muscles break down more quickly. These are mysterious, but meaningful, reasons humans can't live without magnetism for long: The earth's magnetic force acts like a tuning fork, bringing you back in-line when you're connected. Conversely, the farther away from the earth you get, the lower the magnetic field, and the sicker you're going to be. Unfortunately, the earth's magnetic field appears to be bottoming out as we talked about. And magnetic flux weakens with distance, according to the Inverse Square Law.

Rebar in concrete walls conducts magnetism

Dr. Dean Bonlie studied this effect. He found the high iron content of rebar conducts magnetism toward it. So the outer load-bearing walls, elevator shafts and corners of concrete buildings (where executives tend to have their offices) have heightened magnetic field – as much as twice the norm. On the other hand, magnetism is conducted away from the center of tall buildings. In fact, the middle of these rooms had about half the magnetism compared to outer walls and elevator shafts. That encourages employees in common areas to eat more in order to keep up their energy level. So they tend to get fatter working in a reduced magnetic field. On the other hand, executives with the best offices had a health advantage due to a stronger magnetic field and more energy. Hear that kids? Stay in school ;-)

Feeling low on magnetism? Take a trip to the North Pole

Lines of magnetic flux around the earth are denser at the poles than the equator (picture iron filings around a bar magnet). The magnetic field around the equator averages about 0.3 gauss, while the magnetic field at the poles measures 0.6 gauss. So, once again, we see how clever Nature is at supporting life all over the planet. The tropics are blessed with abundant heat and sun, resulting in a more prolific food chain. But

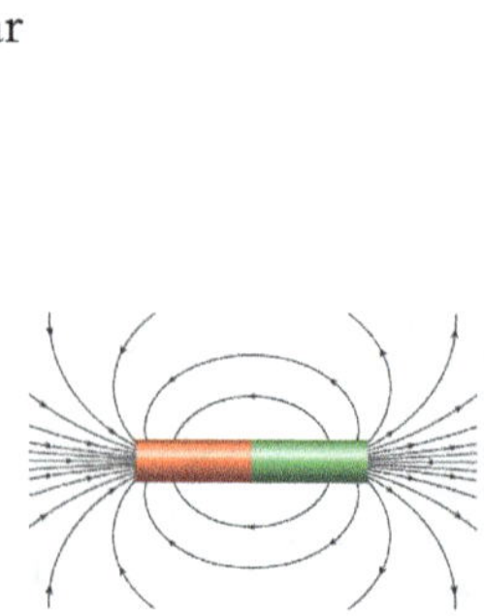

Stock image of a random guy in space.

colder climates of the North and South poles have more magnetism, lots of water, iron and higher O2 saturation in water, which makes for more plankton and fatty fish in the area. Without biological balancing mechanisms like these, extreme latitudes would be barren wastelands like some other planets in our solar system that shall remain nameless.

Want to know what THE worst mitophysical environment will do to a person? Spend some time in high earth orbit

The space program doesn't like talking about it, but living in space is THE worst environment you can experience in biophysical influences. Life in high earth orbit is a virtual hell hole of inappropriate light exposure, low magnetism, dehydration, low oxygen, less weight-bearing and zero gravity. All of that decimates human health, much of which is irreparable. When astronauts stay in space for long periods, the alien environment around them uncouples their biology from earth's natural rhythms. All their chronobiology programs that run sleep/awake cycles, metabolism, regeneration and hormone signaling are thrown out of synch. In a word, their biological programming becomes *incoherent.* And that makes organ systems malfunction like an engine management computer misreading its sensors. Erroneous signaling spells trouble for both people and computer-controlled systems.

The International Space Station, for example, travels at 17,150 miles per hour. At that speed, astronauts/cosmonauts onboard see the sun rise and set every 90 minutes – 16 of each per day. Plus, the electromagnetic frequencies they receive are radically different, and more intense, than those at ground level because the sun's rays in orbit are not filtered by earth's atmosphere. That means astronauts are exposed to the full power density of the entire electromagnetic spectrum without any attenuation from earth's ozone layer and atmosphere. Imagine what all that ionizing radiation does to a person's cells and circadian rhythms.

Just as upsetting to human biology, there's no gravity. There's no natural grounding effect in space or Schumann resonance. They're constantly working under blue light. And they're discouraged from drinking a lot of water due to lack of conventional toilets. Worst of all, the vast majority of magnetism is from the sun, which is ⅔ that of earth's, at that distance. What's more, the frequent positional changes of being in a space station subject the occupants to unstable magnetic fields – in contrast to being upright or lying flat most of your hours on earth. This circadian nightmare is so radically different than what our biology is built to tolerate that astronauts experience decades' worth of aging in months.

For example, Russian cosmonaut Valeri Polyakov spent 437 days aboard the Mir space station and was found to have lost 80% of his bone density when he returned to earth. American astronauts have become sterile permanently after long space flights. What basically happens in space is human biorhythms are sped up, deranged and decoupled from one another, while regeneration programs are impaired from lack of native EMFs regulating these processes. So more damage is done faster to their bodies, while regeneration almost completely collapses.

But the most important idea to incorporate into your consciousness is **environmental exposures that astronauts receive in space are just a more concentrated version of what we all get living in a world that worships its wireless devices.** Astronauts just get their exposures confined in space and time, whereas we get the same exposures spread out in location and duration. To illustrate, Uncle Jack flatly states that every single person living in a populated area today has some degree of bone loss from nnEMFs infiltrating our atmosphere. It's as if we're all living in a giant MRI machine no one bothered to turn off.

Magnetism helps make the world's best wine

Through centuries of observation and refinement, the French have mastered the art of winemaking. Their most famous is Champagne. But French wine isn't in a class of its own simply because of richer soil, better grapes and more passionate people. Its superiority comes from minding the details. And the way the French store their wine while it's aging is an important part of that process. The way it's transported contributes to the quality. You see, down through the centuries, the French learned that aging their wine underground, where it's closer to the earth's magnetic field, yielded a better product. Storing it in cellars prevents photo-oxidation. Shipping it in colder months protects it from temperature oxidation.

So whether they knew it or not, the higher magnetism underground, and the consistently cool and dark conditions, gives wine the ideal environment for their microbes to ferment sugar. The little critters in wine benefit from being closer to the earth's magnetic field, just like our own cells and microbiota do. Once again, we find physics to be fundamental, with chemistry bringing up the rear.

Some physiological effects of magnets

To see what magnets can do on a small scale, you can put a tiny shower curtain magnet on an acupuncture point on the wrist and get these results:

- 45% increase in beta-endorphin
- 24% increase in serotonin

- 15% increase in ACTH (an adrenaline precursor)
- 12% reduction in cortisol.

Even more impressive, Dr. Gumiel (PhD of science, researcher with Project Genesis, The World Development Organization) – his placebo-controlled study showed that 23 different insects live five times as long when put in a magnetic field 10 times stronger than earth's. Likewise, human tissue cultures and mice live 2½ times as long in that field. However, there is a difference between tiny magnets and Dr. Gumiel's results. It comes down to the type of magnetism used and how each operates.

In the graphic below, the red magnetic flux lines (positive) flow upward. The green magnetic field lines (negative) flow downward, which is the same as the earth's magnetic field in the Northern hemisphere.

Crucial to understand, bipolar magnetism operates on a different principle of action from unipolar magnetism. Unipolar magnetism is an amplification of earth's own magnetic field, giving only positive benefit. Whereas bipolar magnetism (two-way field) redirects the body's limited resources to an area of need, which is essentially an emergency healing response. Failure to understand this critical difference has led to a lot of skepticism and confusion about magnetic therapy. Prime example: People have gotten mixed results applying hand-held magnets to painful backs and arthritic joints. For this reason, the medical profession, and public in general, have not taken magnetism seriously as a healing modality.

Here are more reasons why magnetism can deliver benefits that seem miraculous to casual observers:

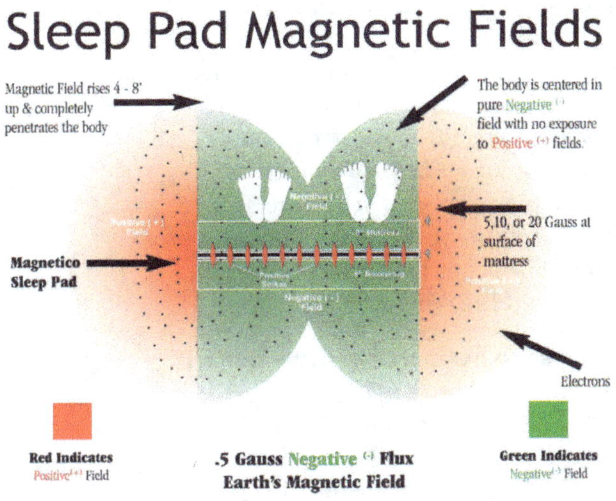

The Magnetico Sleep Pad

A reliable way to give yourself more energy and resiliency (as our toxic world tries to rob you of yours) is by sleeping on a properly-designed magnetic bed pad. As long as your EMF environment is not too bad, it's one of the few products or practices in the world that deposits vitality into the health bank account of everyone using it – regardless of age, condition, lifestyle choices or genetics. And sleep is the time when we need magnetism the most, because that's when the body goes into active repair, replenishment and regeneration mode.

For these reasons and more, Dr. Dean Bonlie invented a unidirectional magnetic sleep pad for home use, and an industrial-strength device for clinical use. Both give you the benefits of the earth's magnetic field, every night times 10, 20, 40 or 12,000. He calls the sleep pad for home use the

Magnetico. It delivers 5, 10 or 20 gauss, depending on model, compared to today's 0.5 gauss. And he calls the machine designed for clinical use the *Magnetic Molecular Energizer* (MME). It looks something like an MRI machine, and delivers 3,000–6,000 gauss.

How the Magnetico Sleep Pad improves your health and resiliency

Dr. Bonlie believes much of the Magnetico's disease-fighting, anti-aging magic happens primarily as a result of increasing the orbital velocity of valence electrons. That makes atoms wobble more (or vibrate, called "precession"), which makes them more chemically active. Free radical production is also decreased. In fact, he's seen free radical production reduced by 80% at 3pm, after sleeping on a Magnetico the night before... 5–9% after just 20 minutes. Blood cells also pick up more oxygen from increased polarity.

The Magnetico basically energizes matter, enhances chemical reactions, oxygenates more binding sites on hemoglobin, and protects cells from oxidative stress. It's even been shown to repair damaged DNA, diseased livers, and malignant tumors. For his part, Dr. Jack Kruse believes the Magnetico improves electrical and magnetic charge from the mitochondria, water chemistry (structuring), and the Photoelectric Effect, as well as enhancing DHA, melatonin and cortisol cycling. It also maximizes the benefit you get from Cold Thermogenesis, as well as detoxification by enhancing redox potential.

In particular, a major benefit of sleeping on a Magnetico is it improves sleep quality, so each sleep cycle is more effective. In other words, **you can think of the Magnetico as a *sleep accelerator*. It is the #1 most powerful non-natural intervention I know of to improve sleep quality in healthy people.** Indeed, many users report that they're able to sleep fewer hours and wake feeling just as refreshed. What it's doing is speeding up chemical reactions happening in sleep, including recycling neurotransmitters, digestion/metabolism, and breaking down toxins.

Conditions for which the MME or Magnetico have shown huge benefit

Of course, Magnetico makes no claims about treating the following diseases directly. But enhanced, earth-type magnetism has helped alleviate the *root causes* of fibromyalgia, chronic fatigue, arterial plaque, low back pain, congestive heart failure, Parkinson's, Alzheimer's, neuropathy, cerebral palsy, multiple sclerosis, stroke, PMS, arthritis, allergies, migraines, sleep problems, cancer, autism, ADD and heavy metal toxicity.

Wow, a miracle cure for everything! I know it sounds that way. But when you act on foundational levels of biology – which are subatomic energy, electron exchange, ATP, redox potential and biochemistry – those symptoms of energy deficiency listed above disappear to the best of the body's healing ability. Over-supplied with the basic natural resource of magnetism, inconceivable healing becomes possible, even commonplace.

Specifically, the MME is shown to grow new blood vessels and nerves. It has repaired degenerated discs in the back, and dead heart muscle after heart attack. Even the Magnetico, making "just" 10, 20 or 40 times the earth's magnetic field, has been shown to unclog plaque from arteries, change malignant tumors back into healthy tissue, and improve function in autistics and Alzheimer's, as well as grow new brain tissue after parasites had literally eaten huge holes in it. Consider it canon: 9 times out of 10, disease is nothing more than a shortage of energy to recondition tissues. That's is. And the MME delivers that in extraordinary quantities. The Magnetico, for its part, does similar things in the body, albeit its results are more like respectable to impressive, but more within the realm of believability.

You need maximal ATP to push heavy metals out of cells

Non-toxic metals leave the cell at around a 40–50 millivolt differential at the cell wall (58 millivolts is normal), and even easier when the cell divides. But brain cells don't normally divide. So to get toxic metals out of the brain, you need maximal electrical charge of 80–110 millivolt differential. That high a positive voltage is needed to push positively-charged heavy metals out of the cell with electrostatic repulsion, while negative charge outside the cell helps suck the metals into the blood, lymph and interstitial space. To reach about twice the normal voltage, mitochondria must be running at full blast. Mitochondria need to oversupply the sodium-potassium pumps with ATP (their fuel) in order to raise voltage at the cell wall from 58 millivolts to around a hundred.

Some things that crank up ATP production:

- magnetism
- IR and UV sun exposure
- low deuterium level
- low heteroplasmy rate
- a ketogenic diet (periodically)
- exercise
- and Cold Thermogenesis.

The Magnetico is masterful at getting heavy metals out of the brain

There is nothing more powerful at mobilizing heavy metals out of hard-to-reach areas like the brain than a strong one-way magnetic field. Chelating agents alone aren't able to access heavy metals in the brain. But Dr. Bonlie's 1.3 tesla MME machine has the power to do it. So does the Magnetico (less aggressively). They're able to increase ATP production, and voltage at the cell wall, so heavy metals are released from the brain.

The challenge then, once mercury, lead and cadmium are mobile is to capture and remove them from the bloodstream before they recirculate and cause a Herxheimer reaction. Meaning, people with a high heavy metal load (particularly MS and Parkinson's) can get sick if you don't bind up the heavy metals as they're released from cells. DMSA and chlorella are Dr. Bonlie's chelators of choice.

Dimercapto succinic acid (DMSA) is a potent clinical-strength chelator of heavy metals that is over-the-counter and gentle enough to use at home. It grabs hold of metals such as lead and mercury and escorts them out of the body through detox pathways.

Conclusion: Standard chelation methods can get heavy metals out of the body. Good going. But if you want to get heavy metals out of the brain, where they cause a plethora of chronic diseases, you need special help. You first need more earth-type magnetism to help push heavy metals out of brain cells. Then you need a chelating agent such as DMSA or chlorella to bind the metals so detox organs can pick them up and escort them out.

Most impressive, is how fast magnetism has healed bone fractures

A machinist who did some work for Dr. Bonlie broke the radius and ulnar bones in his wrist, in places that are fairly hard to heal. He was treated conventionally by an orthopedic doctor on a Saturday, and had his arm put in a cast. On Monday morning, he called Dr. Bonlie to see if he could do something about the horrible pain he was in. The swelling was making his cast too tight for comfort. He went to Dr. Bonlie's clinic and put his forearm in the MME's 10,000 pound electromagnet. His wrist started to resonate so intensely he asked the staff if the vibration was normal. They assured him it was. Three-and-a-half hours later, the bones were completely healed.

They coaxed him into doing a pull-up on a door to prove it to himself. Needless to say, he was impressed more than words can convey. He revisited his orthopedist to mess with him. He jiggled his arm in front of the doctor and asked if he should have the cast redone because it was too loose to be doing any good. His doctor panicked at the prospect of disturbing his work. But, unable to argue with the machinist's logic, the doctor ordered a new x-ray to take a look. The three x-rays were placed next each other – the *before* picture, *after* setting the bones, and the one just taken. His orthopedist was in utter disbelief at what he was seeing. Totally mystified, his doctor could not find the break he himself set just days earlier. He asked the machinist what kind of voodoo he'd been into.

Another man, a 34-year-old, broke the fibula bone in his lower leg falling out of a tree. He had it professionally set. Three-and-a-half hours in the MME and he could walk on the leg again. Another two hours in the machine and the bone was completely healed. Meaning, doctors could not find the break when x-rayed. But the funny thing is, results like this are typical when astrocytes are able to resonate tissues in a magnetic field 12,000 times that of earth's. These cases were resolved quicker than most, but not unheard of, when you understand how magnetism gives the body thousands of times the resources they normally have to heal at incredible rates.

Limitations of the MME and Magnetico

Unfortunately, the MME and Magnetico do not shield you from non-native EMFs themselves (darn it). EMFs go right through magnetic fields, bending their direction only slightly. But the MME and Magnetico do help protect you from adverse effects of nnEMFs by replenishing chi/vitality faster. This is particularly helpful in relieving electro-hypersensitivity. And the Magnetico does deflect alien magnetic fields fairly well. The strongest magnetic field wins, basically. The MME and Magnetico do not kill viruses, parasites, bacteria or other microorganisms directly. But unidirectional magnetism does strengthen the immune system so your defenses are fortified against microbial pathogens.

Until a few years ago, Dr. Bonlie had used the ultra-powerful MME in his clinics for many conditions suspected to be caused or worsened by heavy metals in the nervous system such as Alzheimer's, dementia, Parkinson's, multiple sclerosis, arterial plaque, cerebral palsy, thyroid dysfunction, autism and ADD. **But, due to FDA regulations, the MME is no longer in use** (a good sign it works, in my opinion, because it was shown to be extremely safe and effective, as claimed). Magnetico Sleep Pads are a good way to do most of the same things at home, albeit more slowly and with a lot less power.

In summary, the Magnetico improves energy production and healing capacity so you can tolerate more nnEMF exposure before they harm you. Its one-way magnetism boosts your constitution. But shielding your home against nnEMFs may be necessary if you live in a densely-populated area and want the best health and resiliency possible. Man-made frequencies are basically the MME's kryptonite (i.e., disables its superpowers) and a tax on the Magnetico's performance that varies from person to person.

The difference between a one-way magnetic field and a two-way

The earth's magnetic field is so big it's effectively unidirectional to the life forms living within it. Meaning, the magnetic field travels in one direction through you because the lines of flux come out of the planet in the Southern hemisphere, and they go back in in the Northern hemisphere.

An interesting implication here is that exposing yourself to a unidirectional field that's directed against the earth's own field will indeed slow down your valence electrons and deplete you of energy and healing capacity. This actually happened to Dr. Bonlie's customers early on when he first started selling Magneticos south of the equator. They stopped using it after one night because they felt it was killing them. Meaning, they were getting the wrong polarity, at 20 times the intensity. He put on his thinking cap to figure out why this was happening, and once he understood the problem the solution was simple: He told them to turn the sleep pad over. That fixed the problem. With the polarity now headed in the right direction, the Magnetico was then adding energy to the user's valence electrons, instead of taking it away. The lesson learned for everyone: *polarity matters;* direction matters.

In contrast, a magnet you could hold in your hand is two-directional. The magnetic field it produces is small enough so both its positive and negative poles radiate through you at the same time. Generally, you get 60/40 exposure, because the field must loop around and return to its other side as the opposite polarity. This is what we call a bi-directional, or bipolar, magnetic field. At first, the body is stressed by the disharmony of a bipolar field. A "wrong way" field upsets the body's own delicate electrical system by reducing ATP and charge on the cell wall. So, if they're able to, astrocytes (brain cells) send more voltage to an area to regain control and give it extra resources. Blood flow also increases. However, the effect is limited because it robs from Peter to pay Paul. The way bipolar fields act on the body are not natural. But that's not to say they're without benefit. They just act differently and temporarily.

Bipolar magnetic fields can work well initially. But they deplete the body's electrical resources when you use them too much. The benefit goes away the more often you do it, and the longer you do it. For instance, adrenal fatigue happens a lot when you overuse bipolar magnetism. Keep pushing the stress button over and over, and pretty soon, your adrenals do an on-the-job slowdown, if not a strike. Reason being, bipolar magnetic fields operate on the same principle of action as acupuncture: they aggravate the electrical system, and cells notice the disturbance. The body then responds by sending more electrical current and blood flow to overcome foreign frequencies and revitalize the area.

It's basically a call for help that reroutes resources to an area of need. For this reason, acupuncturists are trained to treat patients no more than once a week, or else the benefit goes away. But, even worse, are oscillating magnetic fields and chaotic fields, like those produced by electricity running through your walls, wires and devices. Oscillating magnetic fields from AC power and wireless sources can devastate a person's health like few other exposures in the modern world – in both totality and timeline.

Changing hemispheres

When you're introduced to a magnetic field going the opposite direction, your cells need time to reorient their electrons' direction of orbit. That means if you go from the Northern hemisphere to the Southern, you change the polarity of magnetism your cells and atoms are subjected to. This temporarily makes you feel depleted because the new magnetic field depresses your energy production, as well as hormones and biochemistry.

But as cells divide, electron orbits on the new cells intrinsically match the magnetic field they're born into. So, for the first month or so you're in a new hemisphere, you lose energy and vitality. You feel run down and then clobbered. But from that point on, your energy starts to pick back up, until you fully acclimate to the new magnetic field. After about 10–12 weeks, you're back to where you started. Unfortunately, if you change hemispheres again, you have to repeat the process.

There are good magnetic fields and bad

To save money, Magnetico copycats contain far fewer magnets. By constructing the pads with fewer magnets, they have to space them farther apart. And that makes each small magnet create its own bipolar magnet field on the body, as opposed to acting like one big (single) magnet. In other words, imitators expose you to bipolar fields that may give you benefits when you first start using them. But the benefits wear off over time, until you're back to where you started 6–12 weeks later. Indeed, you can even go backward after that. Most bipolar magnetic bed pads then end up gathering dust in the user's garage because of this blunder of basic premise.

Conversely, the Magnetico's patented design acts like one giant magnet, because it's made of an array of small magnets placed an inch apart. Its field is so large it acts like a stronger version of earth's own magnetic field. For this reason, you place the Magnetico underneath your mattress, on top of your box spring, and not on top of your mattress like cheaper knock-offs. Their field is too small and weak to reach you through a mattress. When you (try to) pick up a Magnetico Sleep Pad, you'll feel what I mean. It's extremely heavy. Their king size version, for example, is made of four pieces, weighing 99 pounds each. The sheer mass and strength of magnets is why the Magnetico affects the body in subtle to stunning ways other brands can't touch. It's just that powerful at enhancing the earth's own essential magnetic field. Think of it as an *earth's magnetic field, enhanced.*

The ultimate benefit any therapy can give you: Time

Precious but rare, the Magnetico has the distinct advantage of putting time on your side. Just think: Most therapies (1) require you to carve out time in your schedule, (2) there are ongoing costs for treatment or consumables, and/or (3) their effectiveness wanes over time – think yoga, red light therapy, massage, hyperbaric oxygen, Cold Thermogenesis, sunbathing, acupuncture or exercise. Even taking supplements, meditation or preparing healthy meals takes time (to hopefully gain time back). These are big expenses of resources you may not have considered.

But sleeping on a Magnetico is entirely effortless and automatic. You get the benefits whether you're actively pursuing them or not. Plus, it's always present when you are. The hardware never wears out. And the effect never wears off (though it may be beneficial to take a three-four week break every so often). Consider shielding your sleeping space for the very same reasons: *passive benefits*. Conclusion: Everyone might need to sleep on a Magnetico in the coming years just to maintain their health and fend off disease. The earth's own magnetic field is getting to be that feeble, while non-uniform magnetic fields are that pervasive and offensive to your person.

Update: I just did a challenge-rechallenge test for the third time with my Magnetico Sleep Pad to confirm my beliefs. After sleeping many months on 30-gauss, earth-type magnetism, I tried sleeping without it. Result: I woke up repeatedly through the night for no reason, which is unusual for me. I'm now 100% positive that the Magnetico, as claimed, is one of the very best modalities you can use to improve your sleep quality, recovery and biochemistry.

Note: See Recommended Resources section at the very end for Magnetico contact info and special promo code.

Take aways

When mitochondriacs talk about the importance of magnetism, they mean both external and internally-produced magnetism. This chapter emphasized external magnetism from earth's molten iron core and magnetism produced by resonance from astrocytes. But the magnetic force from the ATP synthase, charged particles in blood, and streams of DC electricity are equally imperative at moving materials, creating chemical reactions and informing regulatory systems of the brain. As examples, ROS and RNS are key messengers that operate on magnetic flux. Size of magnetic field around nerves controls tissue development (morphology) and regeneration. In short, DC electricity, magnetism and electrostatics are corollaries of one another. They're inseparable elements of one system.

8

CIRCULATION AND PARAMAGNETISM

Paramagnetism: weakly attracted to magnetic fields

The heart doesn't *pump* blood the way we think

We've been told the heart manually pumps blood through the body's many miles of arteries, capillaries and veins. Biologists say this pumping action is the driving force that circulates blood around the body. But that's a false assumption first published in 1628 by William Harvey and petrified into doctrine ever since. Before that, a Greek physician for Roman soldiers, Galen, said blood is made in the liver and sent to organs, which consume that blood – never to be reused! So, to this very day, not only is our belief about the heart and blood wrong; it's absolutely impossible, as you soon must admit to yourself or at least question.

William Harvey's theory of blood flow (1628): Exercitatio Anatomica de Motu Cordis et Sanguinis in Animalibus, *commonly referred to as* "de Motu Cordis."

The mystery that medical science can't adequately explain: How can the heart push individual blood cells through the body's many miles of tiny capillaries, considering the fact red blood cells (RBCs), 6–7 micrometers (μm) in diameter, are actually wider than the smallest capillaries (3–5 μm)? Does centralized science forget RBCs need to be deformed into an elliptical shape to fit through capillaries? Those two factors – distance and deformation – would create unimaginable resistance to fluid flow. Can you imagine the friction? Can you imagine the hydraulic pressure needed to push a viscous liquid through miles of microscopic tubes AND distort the RBCs at the far reaches of their circulation? Not to mention the pressure needed for the return trip.

It would require hydraulic pressure orders of magnitude greater than the heart can possibly produce and the vessels can withstand. So that line of thinking clearly does not hold water. Just try blowing air, which produces far less friction than water, through a straw more than ten feet long. You'll see what I mean. Now try pushing water through rubber tubing a millimeter in diameter, 3000 feet long, and clogged at the end. Which leads us to only one conclusion: There has to be more to blood flow than meets the eye. The heart can't possibly operate the way we think it does. Instead, blood flow must be driven by other forces in order to avoid unimaginable resistance which would theoretically be present if blood flow worked the way people thought.

Viktor Schauberger. *© Schauberger-Archive. Used with permission.*

How blood circulates through the body

What follows is my take on how the cardiovascular system operates, based on the work of Prof. Gerald Pollack (dynamics of charged particles), Viktor Schauberger (implosive energy flows of vortexing), my own analysis (zeta potential unifies the blood), Dr. Jack Kruse (paramagnetism of blood), and Rudolf Steiner/Dr. Thomas Cowan (Cowan said Steiner recognized the heart acts like a hydraulic ram pump).

The circulatory system uses multiple mechanisms of action to operate efficiently. Healthy blood flow is a combined effort of: (1) the electrostatic repulsion and attraction of charged particles that cancel friction and interconnect the blood; (2) the heart making blood flow pulsate and vortex; (3) the suction effect of the heart's magnetism; and (4) electrical charge expanding vessels to then pump blood upon relaxation. When the circulatory system receives contribution from each of its elements, blood flows around the body with minimal effort/maximum conservation of energy. But when these forces weaken due to disease, or when adjusting to exertion or temperature, the body compensates so the right amount of oxygen is delivered and waste evacuated.

The blood behaves as a single organ, with lots of roving parts

It's tempting to think of the blood as being little more than a collection of cells in water. But Nature designed it to function more like a liquid-solid system, with countless components distributed throughout the body – all adapting to their local terrain, but working as one, courtesy of polarity. Blood, to put it simply, is not just a variety of cells floating in liquid. It is exquisitely engineered bio-machinery whose movement is more ingeniously-designed to negotiate radical changes in topography than its material or chemical properties suggest. Indeed, the technology contained in the blood and circulatory system is so advanced, yet subtle, it's taken till 2022 for someone to come along and break down the brilliance of its design into laymen's terms. Specifically, we have the blood's electrostatic properties, combined with the heart's structuring of blood flow, to thank for the blood's ability to circulate as easily through microscopic vessels as it does through tubes the size of a garden hose.

Speed of blood flow varies, but total momentum stays much the same

Counter-intuitively, total momentum of blood in the circulatory system does not change much from one point to the next. Its toughest challenge is probably to maintain *some* momentum as vessels branch into 3–5 micrometer capillaries… not just because the vessels are smaller than RBCs (certainly part of it) but, rather, the capillary bed is so spread out that individual cells each carry negligible momentum when they need it most: through the tiniest capillaries.

That's why the body often resorts to raising blood pressure when the blood loses polarity and cohesiveness, and/or when the heart is not pulsing and vortexing the blood properly: cells need a minimal amount of momentum to keep flowing in the right direction and not stall out somewhere. Constricting the arteries to increase blood pressure is how the body usually hikes circulation to meet this 'force of flow' requirement. However, this has consequences.

We need to rethink our understanding of blood flow

Key to knowing how the circulatory systems works, blood is not pumped around the body by mechanical forces of high pressure pushing it into areas of low pressure. Rather, electrostatic repulsion and attraction do most of the work turning many individual cells into a marching band with many moves. Polarity makes the blood's journey around the body virtually frictionless. Here's how I came to this incredible realization:

Watch micro blood flow in YouTube videos entitled "CapillaryBloodFlow.wmv", link: https://youtu.be/QuWhKN1bHLA, and "Blood Thur Veins2 Microscopy blood circulating flowing capillaries corpuscles vessel FVSS master Ser", link: https://youtu.be/wu01vlf4ORM.

In watching real footage of how red blood cells move through small vessels, it was clear to me that areas of high pressure and low pressure could not possibly make cells move the way they do. Blood moves through vessels too freely and homogenously (as-one) for that to be the case. First conundrum: How do red blood cells get into the smallest capillaries? Why do capillaries less than one cell width in diameter get any RBCs to flow through them at all, when nearby vessels are far bigger, and offer far less resistance? You'd think blood cells would avoid the hassle of having to contort themselves when larger vessels offer a speedier, more comfortable ride, with no distortion needed. Doesn't make any sense.

Second problem: How does the heart generate enough pressure to push red blood cells through cramped capillaries? The heart is too far away, and blood flow too widely distributed, for this passage to happen against the friction we would expect. Third, how do red blood cells exit the smallest capillaries and rejoin their co-workers in line? RBCs with a normal zeta potential are supposed to repel each other with negative charge. But I saw very slow-moving RBCs leaving a capillary and entering a perpendicular, extremely fast-moving larger vessel that was completely packed with cells. There seemed to be no room to enter the fray, so they should have been blocked from entering by the repulsion of traffic flying by.

But the exact opposite was happening. After an instant of hesitation, there appeared to be some mysterious force violently yanking the RBCs into the smallest gaps in the flow. Something is grabbing them and pulling them in, which isn't simply the force of suction. Judging by the way they get stretched out like a spaceship entering warp speed, the force couldn't be mechanical suction; it has to be electrostatic.

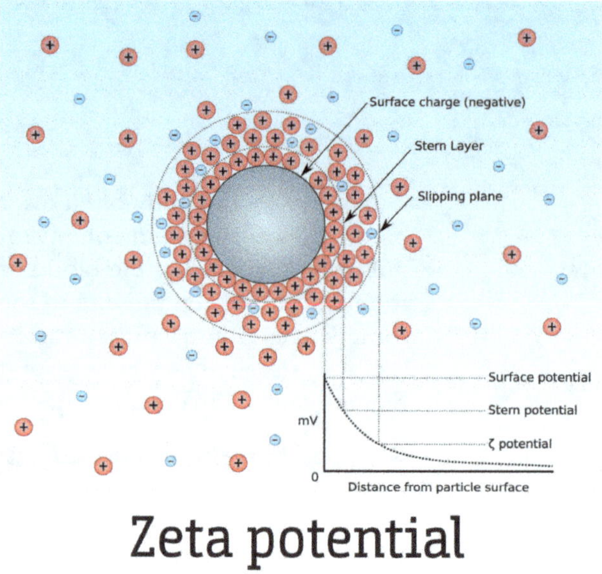

Image used under Creative Commons 3.0 license. Author: Larryisgood, modified by Mjones1984.

For blood to flow effortlessly, first you need to cancel friction

This is accomplished with zeta potential. Zeta potential is often described as making red blood cells slippery so they don't clump together. That's good enough for newbies, but a more complete understanding will greatly increase our ability to diagnose and treat cardiovascular disease. From a layman's perspective, zeta potential is probably the first thing you want to fix when your cardiovascular system malfunctions.

Zeta potential is actually a net-negative cloud of charged particles around red blood cells that encourage them to stay uniformly spaced from each another – meaning their round sides are very close together or touching, but their surfaces are not smushed together or very far apart. Blood cells are neither intimate partners nor strangers; they're good friends with all their cell-mates. Conversely, when RBCs are low on zeta potential their concave faces stick together like a stack of coins, because RBCs are made of sticky glycoproteins. That's incredibly bad for your circulation, oxygenation and organ function.

Basically, charged particles form an invisible buffer zone, or lubricating layer, around RBCs that give them freedom to move independently, or help them catch up when they fall behind their squadron. *Close*, but not clustered, is where RBCs feel most comfortable, because that's the best distance for maximal flow rate and nutrient exchange.

Zeta potential: Electrical charge of particles attached to an object (larger dotted circle in above graphic, composed of two zones), compared to ambient solution. In biology, zeta potential describes the net negative charge around red blood cells, which makes them repel each other and slippery to their surroundings.

Zeta potential keeps red blood cells from sticking to vessel walls

Cells of the endothelium (inner vessel wall) normally have a negative charge. So repulsive force is what propels net negative RBCs through negatively-charged blood vessels like a maglev train's levitation system. The net negative charge of zeta potential cancels friction between blood cells and vessel walls – especially important in one-cell capillaries. Not so coincidentally, like charges repelling each other is Nature's favorite way of making materials slippery such as ice. By definition, zeta potential is the net charge of particles around RBCs, compared to that of the surrounding fluid. In biology, zeta potential is mostly the measurement of electrons around RBCs, compared to protons in the surrounding serum.

Serum: The blood's liquid.

At the same time, protons in the serum give blood cohesiveness

Essential to blood's interconnectedness, particles of positive charge in the serum (derived from charge-separation of water) are the attractive force that gives whole blood its urge for solidarity. Protons pull RBCs along when they need to accelerate – much more than mechanical force of suction (i.e., low pressure). Wherever the serum goes, it sucks the RBCs along, using positive-negative attraction. Imagine tiny magnets in the serum pulling RBCs through the smallest passages, or merely into gaps in arteriole (small artery) flow.

Electrostatic *attraction* is as important as *repulsion* to blood flow:

- This explains a more plausible means by which RBCs get pulled through the narrowest capillaries, other than pressure differentials.
- This explains why small gaps in a large, fast-flowing arteriole violently suck stationary RBCs into the flow from a single-file capillary. When red cells are spaced out in a small capillary (i.e., more positive fluid, fewer negative cells), the net negative charge in a crowded arteriole (lots of RBCs) pulls the positive charge of the serum into the flow and the red cells follow. RBCs in the capillary are pulled into the current like yanking a big water balloon out of someone's hand.
- An absence of attraction explains the occurrence of low "ejection fraction." Attraction between the liquid and the cells is low, so blood flow doesn't have its usual oneness. Each contraction then leaves more blood in the chamber.

Ejection fraction: Percentage of blood ejected from the heart with each contraction, compared to how much remains in the chamber. EF represents the heart's efficiency.

- Tissue polarity is another way that blood manages to get through the capillary bed without losing much momentum. Oxygen-depleted tissue in the capillary bed (i.e., positive) literally pushes proton-rich serum out of the way to get to oxygen-rich (negative) RBCs. RBCs are connected to the serum like a chain of rubber bands.
- And this explains why, when you actually watch it, blood flow appears incredibly agile and energetic, as if red blood cells enthusiastically move as a unit, rather than a series of independent billiard balls smacking into each other.

No doubt, conventional mechanics helps mobility of the blood a little, such as fluid molecules crashing into the back of RBCs, and the pressure differentials we always assumed were the blood's sole motive for movement. But it's clear: mechanical forces are completely inadequate to explain how a one-pound heart can push a viscous fluid through 60,000 miles of small-to-microscopic tubing. Now can you appreciate how RBCs repel each other, while at the same time the positive charge in the serum keeps whole blood moving in a train-like fashion? Can you see why, except for when the heart interrupts the flow to create pulsation, all the blood in the entire circulatory system does its best to move in unison?

Of course, blood flow must adapt to differences in vessel size, shape, pressure, occlusions and temperature-related constriction/dilation. But these are understandable reasons why the design of the cardiovascular system has tricked people into believing the *heart is a pump* paradigm for thousands of years. It's because the blood's most important properties are not mechanical or chemical in nature. Instead, they're centered around electrical charge, physics, and the flow of energy.

Look closely at how blood really moves through medium-sized vessels and you'll see groups of cells flowing as a unit. Then, when a breakaway group enters a single-file capillary, those cells act more independently (i.e., spacing becomes more random) as friction goes up and the influence of mechanical pressure goes down. Finally, as cells re-enter bigger vessels, they go back to behaving as a group.

Gerald Pollack described this cohesion action in his book *The Fourth Phase of Water*. He wrote that objects of like charge in a solution are attracted to each other because their clouds of opposite charge surrounding each object comingle, pulling the objects together due to a doubling of electrostatic attraction between them (see image top right). When (or if) the objects touch, the particles of opposite charge then glue the objects together with "like-opposite-like" charges.

This explains why "like likes like" in physics. Wet sand clings together using this phenomenon (unlike dry sand) to make sand castles, as do water droplets to form clouds, among many more scenarios. Cells in whole blood exhibit the very same dynamics, only the negative charge around RBCs is not a single layer of charged particles; it's a zone composed of two layers. And blood is not fixed; it's fluid.

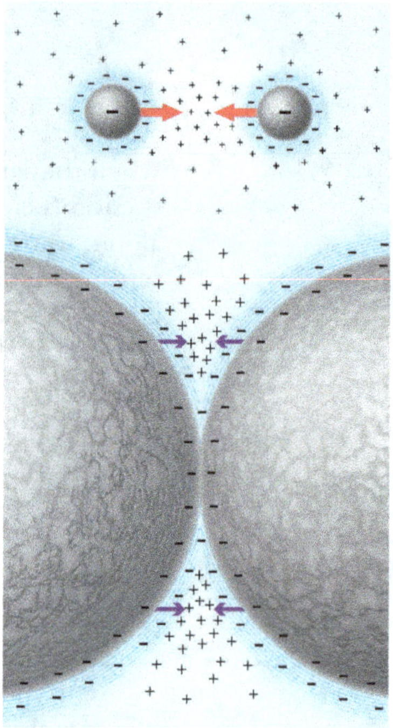

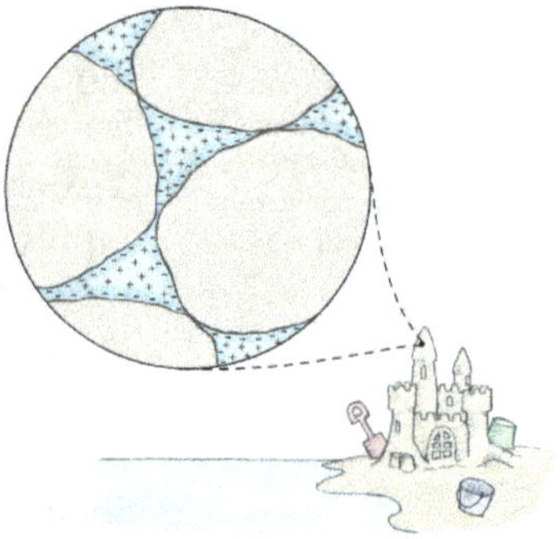

Above images are used courtesy of Gerald Pollack.

The heart has 6 main jobs (none of which are to pump the blood)

1. **Makes blood flow pulsate.** The heart stops blood flow for a split second to build a pressure differential between the supply side of the tricuspid valve and the delivery side. When the valve opens, a pressure wave is launched into the preceding one.
2. **Vortexes.** It converts a laminar (flat) flow into a vortex (swirling).
3. **One-way valve.** It keeps blood flowing in the right direction.
4. **Pulse rate**. When you need more oxygen, the autonomic nervous system makes the heart beat faster to increase the frequency of its pressure waves. This is how the heart regulates *quantity* of blood flow.
5. **Adds propulsion with the blood's paramagnetism.**
6. **Structures water molecules in the blood.** The heart's magnetic field structures the blood's water molecules into smaller clusters. This enhances charge-separation and erases energetic residues.

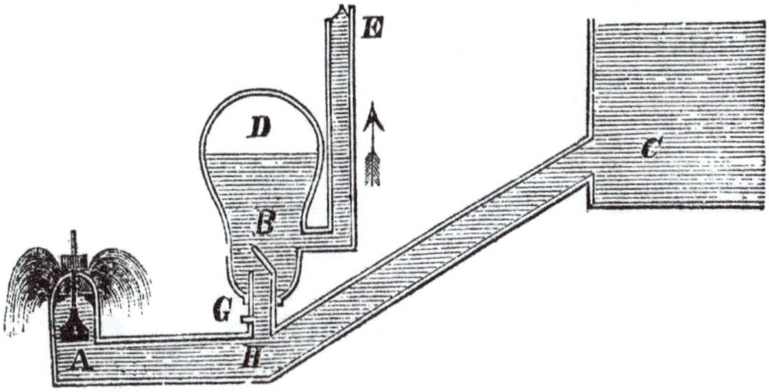

How a hydraulic ram pump operates: Water from source 'C' causes one-way valve 'A' to close. Water pressure increases around 'H', which pushes water through valve 'G' into vessel 'B', while the air contained in 'D' gets compressed. Soon, valve 'G' closes from the high pressure in 'D', which pushes water up pipe 'E'. Meanwhile, valve 'A' has opened from the low pressure around 'H'. That's one full cycle.

See a hydraulic ram pump operating in the YouTube video entitled "How the ram pump works": https://youtu.be/i31hGJ93OTg.

The heart doesn't pump blood the way we're told

It does not create an area of high pressure, which then drives blood out to the tissues and back. That's laughable. Instead, the heart controls the *quality* of blood flow first and its *quantity* second. The heart shapes the way the blood flows, which is a large reason why blood is able to circulate around the body at all. If the circulatory system were not designed around Nature's principles of implosion, acceleration, anti-friction, polarity and conservation of energy – then the blood would not get very far, and complex life forms would not exist.

The heart operates similar to a hydraulic ram pump

Hydraulic ram pumps (aka "water hammer" pumps) have been used for centuries to move water uphill, using pressure differentials created by running water. Ram pumps take in water from a fast-flowing stream (C) into a pipe (H) with a pressure accumulator (D). Water pressure slams a one-way valve shut (A), which causes positive pressure to build up on the supply side (H) of the one-way valve (G), which sucks water into the delivery side (B) of the valve (G). When valve (G) closes in the second half of the cycle, water is pushed through the system by the air pressure in accumulator (D). In this way, the kinetic energy of flowing water (C) is converted into pressure, which is used to move materials against gravity or resistance (E).

The cardiovascular system uses a similar concept to help regulate blood flow, except in biology it's a completely closed system, the components are in-line, and there's no waste overflow. The heart is essentially a modified hydraulic ram pump. In fact, hydraulic ram pumps are 70% efficient at best, while a healthy ejection fraction (a measurement of heart efficiency) for a fit person is also around 70%. Coincidence? I think not. Furthermore, the blood is already moving at full speed when it returns to the heart. It leaves at that same speed, slows down each time the system branches, then picks up steam on its way back to the heart. But, crucial to our understanding, blood is motivated to move through the system by other means, which are described below. On the other hand, if the heart did all of the work pumping blood around the body, its pressure and velocity would be lowest before entering the heart, and highest when it leaves. But it's not.

The heart makes blood flow pulsate and spiral

The tricuspid valve in the heart stops blood flow for a fraction of a second. This causes positive pressure to build up on the supply side of the valve, which expands the vessels (superior and inferior vena cava) with blood, while a vacuum builds up on the delivery side of the tricuspid (the

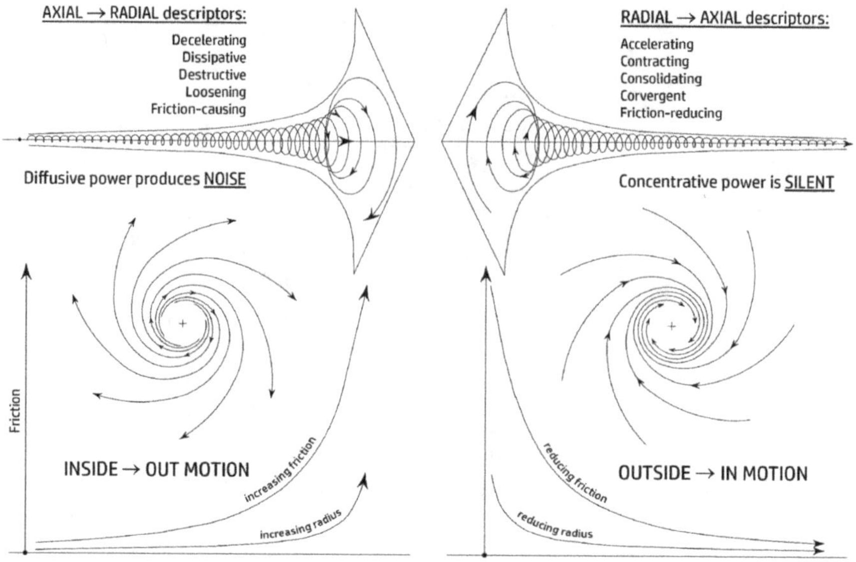

Centrifugal vs. Centripetal motion

right ventricle). When the tricuspid opens, the blood is both pushed and pulled out of the heart, which vortexes, cools, condenses and structures its constituents. The left side of the heart structures blood flow the same way, along with oxygen from the lungs.

Vortexing employs the secret science of implosion, acceleration, energy conservation and friction negation

Johns Hopkins University and Leonardo da Vinci both showed that the heart turns laminar blood flow into a vortex pattern (picture water swirling down a drain). The centripetal (outside-in) acceleration of vortexing takes the flow's momentum and concentrates that energy through the center of vessels. The coldest, densest portions of blood accelerate faster through the middle, sucking the warmer, less-dense fractions after them. The warmest fractions along the vessel wall form counter-rotating eddies, which act as ball bearings for the colder, core fractions. This flow pattern allows the warmer components of blood greater opportunity to form e-zone on the vessel walls (from IR light/heat), as well as protons, which go into the flow.

Indeed, Viktor Schauberger showed in testing of his specially-designed plumbing pipes that when water is vortexed, friction is dramatically reduced – briefly dropping into the negative range at times, thereby creating a supernatural effect that a critic of Schauberger's, professor Franz Pöpel (Director of the Institute of Hygiene), called "negative friction," reminiscent of levitation. Schauberger believed the implosive motion of a vortex adds extra-dimensional energy into the system which has been called orgone energy, scalar waves, zero-point energy, torsion fields, dark

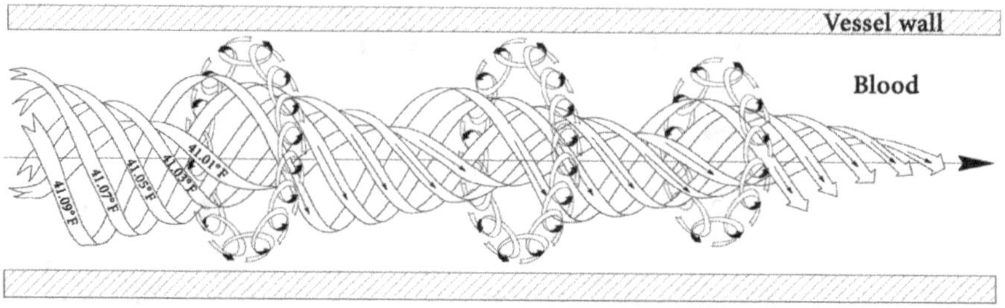

Note: temperatures for illustrative purposes only.

energy, tachyon energy or source field energy. Vortexing implodes matter to manifest small amounts of energy, which is the opposite dynamic to that of an explosion.

In summary, vortexing the blood does a spectacular job of conserving energy, and it vitalizes the water in the blood by structuring its molecules. But, most weird and wonderful, vortexing generates energy in the circulatory system that ancient traditions have described as "filling the body with life force," for it is at this time that energy enters the material world to breathe life into the body. It's even been said that the moment the heart pauses and restarts circulation to vortex the blood is when God enters the human body. Vortexing imbues water with life-giving properties that modern humans have done their best not to notice.

Blood is pulled into the heart with paramagnetism

Paramagnetism: Substance that is weakly attracted to magnetic fields because of its unpaired electron(s).

The heart, by virtue of hosting more mitochondria than any other organ, creates a magnetic field around itself. The more prolific your heart's mitochondria, the stronger the electron flow through their ETCs. A livelier stream of electrons creates a more powerful magnetic field with which to draw blood into the heart. Paramagnetism, as this attractive force in the blood is called, is a primary means (probably not the strongest) by which blood is propelled through the body. Paramagnetic materials like the blood are drawn to induced magnetic fields because their unpaired electrons have an electrical charge that is attracted to other electromagnetic charges.

In this case, the heart is an electromagnet from its mitochondria. And the blood is paramagnetic from its oxygen, cells, (iron-based) hemoglobin, and zeta potential. Hence, the heart sucks blood into its chambers using this attraction we call paramagnetism. The attractive force is much weaker than conventional magnetism, but significant in systems where conservation of energy is essential to efficiency of operation.

The heart's magnetic field structures water molecules in the blood

For decades, health enthusiasts have vortexed their water in a magnetic field to make it healthier to drink. This is done by attaching magnets to the outside of a blender, or between water bottles spun mouth-to-mouth, to break apart large clusters into smaller ones so they're easier to absorb and energetically purified. Vortexing water in a magnetic field mimics the turbulent movement of a mountain stream – the tumbling over rocks, swirling around bends, and descending from falls. In nature, these asymmetrical movements energize water with increased oxygenation, more mineralization, improved clustering and enhanced electrical properties, while at the same time it kills pathogenic microbes. In a home-made vortex, you get all that minus the minerals.

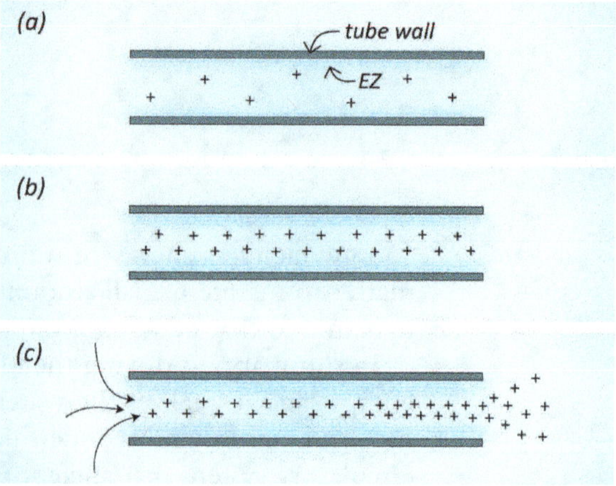

Image used courtesy of Gerald Pollack.

Vortexing with magnetism also erases the energetic imprints of chemical toxins, drug residues, mistreatment through city water systems, and people's negative emotions embedded in the water. You see, water records exposures both good and bad in its hydrogen bonds. It retains these imprints until it freezes as snow and thaws (the natural way), or its molecular arrangement is reset through vortexing – sometimes with magnetism to break apart its clusters (the beneficial, man-made way). Water then transmits these "memories" onto its consumers as a health benefit or hazard. Homeopathy uses water memory for therapeutic benefit. In the same way, the magnetic field around the heart structures water in the blood as the heart ejects it. This breaks up larger clusters of water into simpler ones, which condition the molecules with maximum ability to form e-zone, zeta potential, and optimal blood viscosity.

Watch the documentary "Secret of Water" on Gaia, or other streaming platforms, to learn more about water's memory and restructuring it.

E-zone adds protons, propulsion and cohesiveness to the blood

Gerald Pollack showed that when you place a hydrophilic tube in a container of water, the formation of e-zone on the inner surface of the tube releases protons into the tube. Positive polarity pushes the water out of the tube as protons try to distance themselves from each other. Water will then continue to flow into and out of the tube on its own in whichever direction the flow was initiated (or toward a wider side), powered only by light or a native EMF.

Similarly, Prof. Pollack theorized in *The Fourth Phase of Water* that e-zone is formed on the inside of capillaries, which would propel blood through the same mechanism as a hydrophilic tube: protons in the fluid. Blood vessels are hydrophilic tubes, after all. Investigations by other researchers appear to prove Prof. Pollack correct: The interior of all blood vessels are lined with a fuzzy surface that holds e-zone in its hairs. Looking somewhat like microscopic fur, the "glycocalyx," as this layer is called, holds gel-state water which propels blood and protects endothelial cells from friction. At the same time, the charge-separated water adds protons to the serum, thereby adding polarity, propulsion and cohesiveness to the blood. Plus, propulsion from narrow to wide veins helps returning blood reach its top speed, right before it enters the heart.

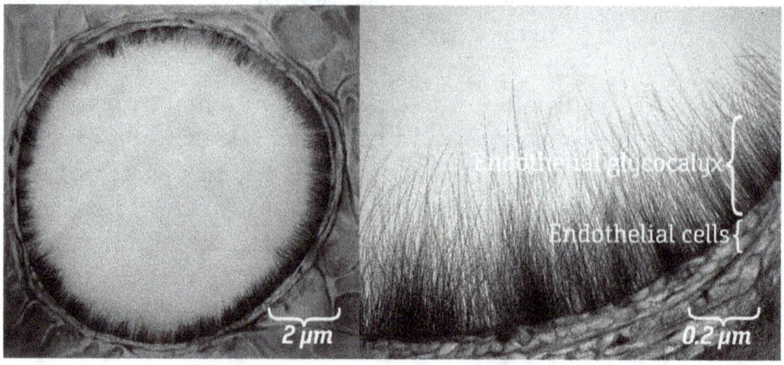

Arteries employ a peristaltic "squeezing" action

Major arteries expand and contract in rhythm with the beating action of the heart. The question is, whether expansion and contraction of major vessels is a passive reaction to pressure produced by the heart, or an active driver of blood locomotion by design. My theory is, it's the movement of blood's charged particles into arteries that causes their smooth muscle fibers to expand in order to pre-tension the vessels with pressure, thus storing potential energy. The rationale: ATP, a storage molecule for electrical energy, relaxes muscles, enabling them to stretch to their maximum length. When the artery is full, its muscles contract and the vessel passively constricts, thereby creating a pumping action.

States of vessels' smooth muscle cells: Activated is shortened, thereby constricted. Neutral is medium length/diameter. Relaxed is expanded.

Think about it: Passively stretching vessels would consume blood's momentum leaving the heart. That would waste energy. That's not efficient. Nature would not design blood flow in biology to be anything less than 100% efficient. Inconceivable. On the other hand, programmed (electrical) *expansion*, then automatic (mechanical) *contraction*, of vessels is way more efficient at sucking blood into vessels, then pushing it out. Therefore, I think vessels do more of the pumping than commonly believed, while the heart does far less. It just appears as if the heart is doing all the pumping because it's obviously doing something. So we just assumed it was doing everything, based on our biases toward classical Newtonian mechanics. We mistook the expansion and contraction of blood vessels to be incidental to cardiac ejection of blood when, in reality, constriction of vessels may be one of the blood's main motives of movement.

Observational evidence: Professor Kurt Bergel of Berlin witnessed this vascular pumping action when he opened the top of bird eggs after a few days of incubation. He noticed that blood vessels surrounding the yolk-sac pulsated before the heart had even been formed. How it that possible? Other researchers have seen blood flow continue in mice, rats and dogs from 15 minutes to several hours postmortem (i.e., no heartbeat). The most likely reason: Charged particles in the blood are not only negating friction, but also expanding vessels to add energy into the system.

Where does blood get the energy to return to the heart?

After slowing down to swap oxygen for carbon dioxide, energy for the blood's return trip comes from propulsion away from the capillary bed and suction toward the heart (both electrostatic and mechanical). As we learned from Prof. Pollack, electrostatic expansion propels fluids away from areas of high proton concentration to low concentration (smaller vessels to larger ones). Like squeezing a wet bar of soap, blood flows away from the capillary bed as its protons repel each another.

As blood travels toward the heart, suction forces unite. The current progressively picks up speed as its kinetic energy aggregates into fewer vessels of larger size – similar to the way a slow wetland current turns into raging whitewater as more current passes through a narrower channel. To add to that, Schauberger informed us that oxygenated blood has greater volume, so the propulsive effect is greatest from the lungs to the tissues. As that oxygen is dropped off to tissues, pressure decreases. This creates higher pressure and propulsion before the capillary bed, and a slight vacuum on the way back. This helps blood get over the "hump" of the capillary bed, far from the heart. Plus, paramagnetism of the blood, combined with electrostatic cohesion, contribute a suction effect of their own. These factors make the blood's return trip virtually effortless, as blood is pulled back up to its highest velocity by suction, polarity and momentum.

How does the heart know when to close its valves and contract?

Schauberger theorized that closure of the valve and contraction of the heart are triggered by a difference in electrical charge between oxygenated blood and deoxygenated/carbon-rich blood. My (loose) understanding of this mechanism: The two states carry opposite charges, so the heart uses that electrical differential to know when to contract (perhaps playing a role in powering its contractions). Each contraction is activated when opposing chambers of the heart fill with blood and reach a certain voltage. The upper chambers work together, as do the lower chambers (four in-total), to create a double-segmented heartbeat ("lub-dub").

When the right and left chambers reach full capacity, the negative and positive charges held in their blood merge and equalize to zero. This is the signal for the heart to contract. This is why Schauberger said the heart does not "pump" the blood, so much as it "is pumped" by the blood. Mind-boggling to imagine we've had it backwards this whole time: the heart does not pump the blood the way we're told. Rather, it is blood flow, combined with gas exchange of the lungs, that causes the heart to beat.

Circulation summary

Our classic conception of blood flow conflicts with science, reason and reality. Just imagine: a high-pressure area making physical contact with a long succession of molecules, one after another like billiard balls, to push blood through 60,000 miles of tubing. There's no way on earth that could be correct, based on logic and what blood flow actually looks like. Far too much energy would be absorbed and wasted by the very squishy RBCs, plasma and vessel walls. To illustrate the scenario differently, imagine if RBCs were negative-only magnets in dry tubing. Flow properties would be chaotic and extremely inefficient because electrostatic repulsion would make cells want to do their own thing *independently*, while the entire circulatory system absorbed and wasted the kinetic energy of cell movement.

On the other hand, it's easy to understand how the momentum from each surge of cells and fluid *electrostatically* drives preceding groups forward. To start the show, charged particles set the stage for blood flow to occur. Their job is successful when friction is close to zero, and pulses of pressure are causing a ripple effect. Once the blood is moving effortlessly and homogeneously with the polarity of zeta potential, the physics behind pulsation and implosion raise the second act of the circulation story to a crescendo.

Third, structuring water molecules with magnetism, forming e-zone with IR, triggering the valves with voltage, and pumping the blood with vessel contractions bring this story to its triumphant conclusion. Each element is a piece to the puzzle, a subplot in the story, of how the blood circulates around the body without breaking a sweat. Conversely, when an actor is unable to perform its duties, the show doesn't usually end there. Others can pick up some of the slack. Adaptation protocols are called into action and the show goes on.

Here are some of those corrective measures:

What causes blood flow to break down? What can we do about it?

Let's deconstruct some potential causes of poor blood flow, so we can understand them and find out what might be required to fix them.

Excessive friction. Negation of friction is the first fundamental of blood that flows freely. For this to be accomplished, red blood cells have to be surrounded by a cloud of charged particles that contain more electrons than protons. The polarity of charged particles keeps blood from sticking to vessel walls and each other. Equally integral to the blood's hydrodynamic efficiency, protons from the formation of e-zone unify the blood so cells and fluid move with an affinity for each other.

So what would happen if zeta potential drops and friction rises? Poor blood flow, oxygenation, metabolism, hormone function and detoxification would be the obvious effects. The body would then raise blood pressure by constricting the vessels in order to increase flow rate at the expense of volume. When blood is not moving well, the vessels take whatever momentum is available in the blood and concentrate their investment into fewer cells.

Colloid: A solution with free-floating particles mixed in (e.g., milk).

FYI: Before its use in medical research, zeta potential was used in manufacturing and formulation to measure the stability of colloidal solutions such as cosmetics, inks, dyes, coatings, detergents, electronics and pharmaceuticals – mainly to optimize flow and absorption by avoiding aggregation. Thus far, zeta potential is not commonly used in routine office visits to analyze properties of the blood such as clustering and clotting. Rather, it has been used mostly by researchers to study polarity's effects on a wide variety of other bodily functions.

Zeta potential is calculated from how fast RBCs move in an electric field, thereby inferring electrical charge around RBCs. It is not a direct measurement. When blood cells move freely, you have a healthy zeta potential (minus 9.3 mV to minus 15 mV is normal). When they stick together, you have low a zeta potential (under 9 mV). The three best ways to improve your zeta potential are earthing, sleeping on a Magnetico and getting sunlight on your skin. Supporting strategies are drinking better water and more of it, cutting nnEMF exposure, exercising regularly, eating an alkalizing diet, and avoiding the usual stuff such as smoking, stress and drinking too much alcohol.

To learn more about how zeta potential effects the blood, read this research paper: Chevalier, G., Sinatra, S., (2013). "Earthing (Grounding) the Human Body Reduces Blood Viscosity – a Major Factor in Cardiovascular Disease." *The Journal of Alternative and Complementary Medicine, Vol. 19, No. 2, 2013, pp. 102–110.*

Inelastic blood vessels. The smooth muscle fibers of blood vessels are encased interiorly and exteriorly by an elastic membrane, called the "internal and external elastic lamina," respectively. These smooth muscles contract and relax to control vessel diameter and blood pressure. But what happens, according to Dr. Jerry Tennant's working theory, is that intrusion of fluoride breaks down the collagen and elastin fibers of the *elastic lamina* that are supposed to limit the movement of smooth muscles like a netting.

Without the restriction of this netting, the smooth muscles are then free to overstretch themselves. Dr. Tennant believes this creates wounds in the vessel, which the body tries to repair with cholesterol, which becomes an arterial plaque. My guess is, Dr. Bonlie's theory of plaque buildup (page 324) is more accurate (or more prominent), in which heavy metals and their positive charge break down the glycocalyx and falsely attract repair processes, including cholesterol.

In either case, the point to consider here is that breakdown of the *elastic lamina* would impair the ability of blood vessels to expand and contract, thus hindering the pumping action of the vessels – whether the vessels' elasticity is designed to actively do the work of pumping the blood or is just a passive reaction. If fluoride weakens the *elastic lamina* to the point of incompetence, elasticity of the vessels is certain to be compromised and, along with it, blood flow.

What would the body do then? Unable to expand and contract fully, the body might keep the arteries constricted permanently (and thus unresponsive). We would see this situation expressed as high blood pressure and/or narrow arteries. Calcified arteries also make the arteries inelastic. So whether fluoride damages vessels on the way to being repaired with cholesterol, or lead and mercury cause a calcified plaque, both scenarios result in a loss of vessel elasticity that cuts down that input of energy (propulsion) into the system, thereby making blood flow less efficient.

Possible solutions: The Magnetico Sleep Pad helps remove heavy metals by improving mitochondria function, thereby clearing up the source of clogged arteries. The chelating agent EDTA has been shown to unclog arteries – I believe because of its ability to detox heavy metals – but it's not FDA-approved for such use. Best combo: The Magnetico, with DMSA, plus iodine, minus fluoride.

The heart can't pull the blood into it. Two situations can cause this: Frailty of the heart's mitochondria, or the blood has lost paramagnetism from low zeta potential. Compensation mechanisms the body would then employ to make up for the energy loss might include: increase blood pressure, narrow the veins/arteries, or alter the heart's anatomy such as enlargement, muscle thickening, or distorting its shape. Is it making sense how mitochondrial toxins such as smoking, drinking, stress, nnEMFs, prescription drugs, lead and mercury connect to the cardiovascular outcomes listed above? Hint: hypertension and changes to the heart and vasculature are symptoms, not causes. How then should a person remedy these root causes? A mitochondriac would reread the list of mitochondrial toxins on page 57 and start by fixing those issues.

Poorly structured blood flow. Blood flow must pulsate, and it must spiral, to give blood the boost it needs to avoid losing too much momentum through the capillary bed. Both stem from the hydraulic ram effect centered around the closing and opening of the heart valves. Breaks in the flow generate pulses of pressure, which leave the valves in a helical pattern that reduces friction and consolidates energy. The centripetal acceleration of vortexing preserves kinetic energy present in the blood. And it is thought to bring more energy into our 3-D space at the moment blood flow is halted by the closing of the valves – like squeezing small amounts of energy into this realm by condensing matter.

It's not a lot, but it is enough to keep total momentum of the bloodstream at breakeven (resting, sustainable) or better (athletic, resilient). That's how I would describe the roles of pulsation and vortexing in the efficiency of blood flow. Unfortunately, when body parts such as vessels, valves or heart muscles deteriorate, you get corrective measures like high blood pressure, congestive heart failure or inflammation. Hopefully, now you recognize how most anatomical changes are preceded by a decay in your biophysical state – meaning electrical charge, magnetism, light, water and mitochondria.

Misshapen heart. The heart is shaped like an inflated seven-sided "chestahedron" (as its discoverer/inventor, Frank Chester, named it). This expression of sacred geometry gives the heart special energetic properties. To give you some idea, Nature uses the "golden ratio" of 1:1.618 (aka the

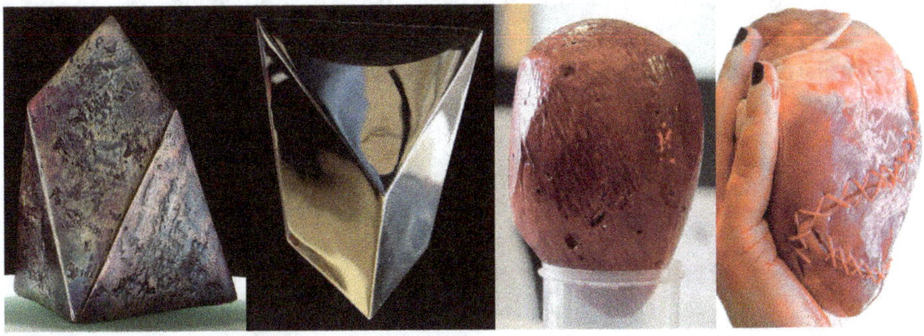

First three images from the left are used courtesy of Frank Chester.

"golden mean" or "divine ratio," symbolized by the Greek letter phi) in everything from joints of the arm and hand, to tree branches, to snail shells, to the spiral shape of galaxies, because those proportions are where etheric energies coalesce into material form.

Similarly, the heart uses sacred geometry to maximize energy flows entering the bloodstream. The heart is significantly less efficient without this chestahedron shape. For example, in congestive heart failure, the heart becomes more spherical in shape. This upsets the flow pattern through the lower chambers. The heart then can't eject as much blood with each beat, and its vortices are malformed. A woo way to describe it is, when you lose the electrical, implosive and geometric properties of Nature's design, you dip under the over-unity arrangement that makes the cardiovascular system capable of running problem-free for decades. When that happens, the body needs to jerry-rig the way your anatomy functions in a variety of sub-optimal ways to make up for the loss. However, you can never match, let alone beat, the efficiency of your original equipment and the way it's designed to work.

"They" say congestive heart failure (CHF) is due to a breakdown of the heart muscle. I say CHF and other forms of heart inflammation can be traced back to irregularities with your electrical charge, mitochondria, (para)magnetism, lead and mercury in vessels, fluoride and seed oils disrupting repair processes. Something's not right with one or more of the factors we're examining, and the heart is forced to change its muscle thickness, composition or architecture as a result. I say, focus on your blood and your vessels first, because they probably caused the heart to change its shape.

Impotent e-zone. When it's hot out, your circulatory system needs less help delivering a decent amount of blood because you're collecting more energy from your environment. IR light builds more e-zone in blood vessels which adds polarity and propulsion to the blood, at a time when UV light is increasing flow volume by dilating vessels with nitric oxide. On the other hand, when you get cold, or when your internal biophysics are abnormal, the circulatory system goes into adaptation mode because you don't have the zeta potential to delete friction with electrons, and help propel the blood with protons. What does the body do then to compensate for a low-flow situation? It constricts the vessels to keep the blood's momentum up.

Similarly, when your extremities get cold, those vessels shrink to minimize heat loss. When IR exposure drops low enough – both exogenous and endogenous – vessels close and blood flow ceases entirely. But did the blood stop flowing because the vessels were closed off? Or did the vessels shut down because there was no IR-induced e-zone to keep them open? My answer would be the former, while the latter seems

Exogenous: From outside the body.

Endogenous: Made in/by the body.

more likely to be a coincidence of complimentary factors. But it is an interesting whodunnit.

Consider a similar condition, significant to millions: A shortage of charged particles has got to play a major role in poor circulation in diabetics (aka "peripheral artery disease"). Is low polarity, propulsion and cohesiveness to blame for poor circulation in our diabetic population? I would be shocked if the factors I'm laying out here were not a whole lot closer to the truth than the cooked-up stories about bad fats blocking up arteries and causing neuropathy (nerve damage), nephropathy (kidney disease) and retinopathy (eye damage). Even blood-sugar probably doesn't directly damage nerves as much as it weakens zeta potential and e-zone.

Arterial plaque. I can't prove it but, based on the way Nature operates, I wouldn't be surprised if intentional deposition of cholesterol narrowed the arteries on purpose in an effort to maintain blood velocity.

Cholesterol and saturated fats do not cause heart disease. Awake and aware observers know that cholesterol and saturated fats, such as animal fats, have nothing to do with clogged arteries. Eating natural fats simply does not cause heart disease. In fact, people who have low blood-cholesterol are at increased risk of diseases involving: the heart, brain and nervous system, mood and behavior (from inability to make hormones and neurotransmitters), metabolism, infertility and libido, as well as the maintenance and repair of cells. And the whole story about "good cholesterol and bad cholesterol" is nothing more than a fictional narrative – an excuse to sell more statins.

Paramagnetism

Once blood is flowing properly due to phenomena we just examined, the body is infused with life by transferring substances from the bloodstream to cells. Here are some of the methods by which nutrients are transported around the body, and organ function is coordinated via hormones.

Oxygen is paramagnetic

High O_2 in the lungs/low O_2 in its capillaries make it easy for O_2 to enter the bloodstream and hop aboard the hemoglobin molecule in RBCs. Downstream, in the capillary bed, high CO_2 in the tissues and low CO_2 in the RBCs make it easier for RBCs to give up their oxygen for CO_2. These are conventional pressure gradients that promote the gas exchange of respiration. But the force seldom mentioned is paramagnetism. A force that helps oxygen make it to its final destination in the mitochondria is paramagnetism (weakly attracted to magnetic fields). Paramagnetism helps oxygen find its way through multiple membranes from the RBCs, to the interstitial space, into the cell, and to the end of the electron transport chain, where it can be used for respiration.

For context, ferromagnetism (conventional magnetism) is when the direction of magnetic charges in a material such as iron are easily aligned with a magnetic field. When far more of a material's atoms are attracted to a magnetic field than repelled, you have conventional magnetism. But paramagnetism is when a *small majority* of a material's atomic charges align with an induced magnetic field (due to its unpaired electrons). This is what gives oxygen paramagnetic properties: its unpaired electrons give it slightly more attraction to the magnetic field of productive mitochondria than repulsion. This is what pulls oxygen into the respiratory process.

Unfortunately, when cells' magnetic field is weak due to poor mitochondria function (i.e., low electron flow), oxygen is not sucked into the cells and mitochondria very well. That's one of the biggest causes of hypoxia (low oxygen), poor metabolism and free radical damage. It's an inability to transfer oxygen all the way from the red blood cells to the ETC – both caused, at least in part, by weak electric and magnetic fields in mitochondria.

Diabetes is primarily a disease involving weakness of mitochondria, oxygenation and magnetism

Diabetic degeneration is also caused by crashed mitochondrial function. Sluggish mitochondria produce a weak magnetic field, which compromises blood flow, oxygenation, DHA delivery and hormone function. Many diabetics know from experience what can happen symptomatically when you have less magnetic field – though most of them have never been told that weak mitochondria cause their poor circulation, low oxygenation and resulting nerve damage. Those breakdowns are responsible for causing complications such as peripheral neuropathy, retinopathy, impaired metabolism (from insulin dysfunction), and a host of other problems.

Alzheimer's is a similar story

Functional medicine educators have called Alzheimer's "diabetes of the brain" because, in this condition, brain cells lack the energy they need to support themselves. Mental health experts say faulty metabolism in the brain impairs its function. But it's more accurate to say broken-down mitochondria in brain cells weaken the magnetic field around them, which cuts blood flow, oxygenation, nutrient exchange, DHA delivery, detoxification and energy availability in the area. These are some of the under-the-radar causes of the neuro-degeneration we call Alzheimer's. You see, the brain should be a powerful electromagnet like the heart. But with Alzheimer's, mitochondria in the brain are absolutely exhausted. Brain cells are low in redox potential. That deforms their proteins, which causes long-term memory circuits to fail.

Many modern diseases are caused by "hormonal disconnection"

The reason so many people have hormonal issues today is not because of hormone production or availability. What's happening is, hormones operate on paramagnetism, and the magnetic force is not present in the organs to get those hormones where they need to go. Dispensing them appropriately is also a common problem (as a result of circadian disruption). We'll discuss that later.

Crucial controller organs like the thyroid, pituitary and adrenals are making the hormones, and may be releasing them okay. But worker glands such as the pancreas, testes, ovaries and pineal gland don't have sufficient magnetic force to pull them inside, so hormones go undelivered. This breaks down the connection between the chiefs and the workers. Organs then don't perform the way they're supposed to. Hormones are being made and shipped out okay. They're just not being received the way they're supposed to. That's one of the primary causes of so many modern diseases of metabolism, sleep, digestion and sexual function/fertility. The first cause is unproductive mitochondria in organs, which reduces each organ's magnetic field. Secondary causes are either lack of polarity in the blood to induce paramagnetism or low flow.

How a magnetic field is generated around endocrine glands

Influential endocrine organs like the pituitary have dense networks of blood vessels encircling them. Charged particles in the blood travel through these arteriole and capillary networks, creating a magnetic field around the organ and making its hormones more responsive to magnetism. Physics tells us that the movement of charged particles creates a magnetic field at a 90° angle. Unfortunately, when hormones aren't properly "magnetized" (from the *cause* side of the equation) they can't get off-loaded into glands and tissues when they arrive at their destination.

This drop in magnetism that begins as mitochondria weakness then contributes to hormone problems that precipitate into diabetes, weight gain, infertility and many more chronic conditions. It's one of the single biggest drivers of chronic disease that stumps modern medicine as to both source and solution… but biophysics explains perfectly. Simply put, electromagnetism is essential to the operation of many endocrine functions, while biochemicals such as insulin and testosterone are just middlemen that get blamed for breakdowns that make our lives miserable. In reality, the transportation and communication systems are to blame, most of the time, not production.

PART 2

Energy Management/ Mis-management

9

WEIGHT GAIN, WEIGHT LOSS

The Sneaky 6 reasons people gain weight (not in order)

Here are six *covert* reasons people gain weight and *gain it back* after losing it. However, they don't fit neatly into marketable paradigms of diet, exercise and supplements, so they may come as news to you.

1. Microbiome lacking diversity.
2. Leakage of light.
3. Electron deficiency.
4. Weak mitochondria.
5. Leptin resistance.
6. High deuterium level.

Artist: Mike Bocianowski.

The remainder of this chapter covers the first four subjects. The last two, leptin and deuterium, are individual breakout chapters. More widely-recognized reasons for weight gain, such as poor food quality and toxins, are covered at length in this book's predecessor, *Gut-Brain Secrets*.

1. Microbiome lacking diversity

Research into the microbiome tells us there's a strong correlation between obesity and narrowing of microbial species in the gastrointestinal tract. So most gut-health experts now agree that when certain bacteria strains dominate the gut, they promote weight gain, while diversity does the opposite: it normalizes weight.

However, the very latest research by Jeff Leach strongly suggests that classic risk factors, such as diet and antibiotics, don't narrow microbial diversity in the gut by themselves. Instead, it's chronic exposure to an altered light spectrum that leaves the door open for poor diet and antimicrobial threats to corrupt the microbiome. It's both non-native frequencies *and* ingesting toxic substances that damages the microbiome.

How did Jeff Leach figured that out? He took an indigenous tribe of hunter-gatherers that live near the equator, the Hadza. He fed them a standard American diet – with soft drinks, candy bars and antibiotics. He examined their feces for microbial content. And you know what he found? Their microbiome didn't change at all. **Good sun exposure protected their microbiomes from corruption, despite a crappy diet.**

So Dr. Jack and Mr. Leach now firmly believe natural sunlight protects the diversity of the microbiome. Native light exposure on the eyes and abdomen are the microbiome's best defense against functional decay. On the flip side, they believe our microbiomes in the West lose diversity mostly because of artificial blue light, lack of real, full-spectrum sunlight on our eyes and skin, as well as too much wireless radiation. Lack of diversity in the microbiome then sets the stage for obesity to occur.

2. Leakage of light

Every cell of every organism emits ultraweak UV light while it's alive. But when cells are stressed or diseased, they release more biophotons than they would normally. Obesity is one such condition that's associated with biophoton loss. That means when your internal electric and magnetic fields are weak, your cells can't retain light as well as they should, and you leak energy out of your light "gas tank" – energy that would have, should have, and could have been used for productive purposes.

Those with excess fat leak more light – all kinds of light, including biophotons, IR/heat, and frequencies released from food electrons, as they hop through the ETC. If the loss of light is happening below the neck, the circadian system motivates you to eat more to try and compensate for the energy loss. Unfortunately, if you don't fix the sources of the problem – which are bad mitochondria, poor redox potential, chronic inflammation and a toxic environment – your internal programming leaves your appetite in the 'on' position, which undermines your best efforts to lose weight.

3. Electron deficiency

Low electron intake. We've already talked about electron deficiency, so we'll just recap here: Whatever your diet may be, all food ultimately breaks down into electrons, protons and light. So that's what you need to focus on: the quantity and quality of photosynthetic products that food can give you. That determines the totality of resources your body can extract from calories you consume.

In other words, focus on how many electrons the macronutrient can send through the electron transport chain, relative to calorie count. Whole foods and natural foods are naturally higher in electrons, whereas processing depletes electrons, especially in carbs. Electron density tells you how many units of ATP your mitochondria can make in relation to calories consumed. That, right there – ATP per calorie – is much more on-target as a formula for weight loss than the theory of *'calories in, calories out'*.

Electron loss. An inability to retain electrons is another way you can find yourself electron-deficient. You can lose electrons through blue light, non-native frequencies, high heteroplasmy rate (the first three usually go together) and low DHA. And what happens when you're electron-deficient? You can't capture and use as much light.

The less connected you are to earth and sun, the more food you need to eat to make ATP

Radical concept: One-third of the electrons needed to make ATP are supposed to come from the food you eat, and two-thirds are supposed to come from grounding and sun exposure. That's right: we're biologically designed to have 66% of our electrons supplied by grounding and sunlight (along with DHA and being properly-hydrated). The simple act of touching the earth (directly or through earthing equipment) gently pushes electrons into your body, because the earth is an electron donor.

Grounding is even more effective when your eyes and skin are getting sun exposure (i.e., without sun-blockers like eyeglasses, sunglasses, sunscreen, window glass or clothing). Like a solar panel, skin turns sunlight into electrons and electric charge. Yet how many people today ground themselves daily, or get full-body exposure to the sun? Hence, we need to make up the deficit by eating more calories. And that's reason #3 that people are fat: We're making up for a lack of sun and grounding by eating more food – particularly carbs – to get our fill of electrons.

4. Weak mitochondria

We've discussed mitochondrial productivity in previous chapters. So, to summarize, the strength of your mitochondria is the single biggest factor in when and where disease strikes you, and the speed at which you age. When your mitochondria are strong, you make a lot of ATP relative to calorie consumption, with little waste. But when your mitochondria are struggling to make ATP, you take in more calories than necessary to make ATP. Hence, you need to eat more.

So it's not so much an excess of calories consumed that makes you gain weight. It's inefficient conversion of one form of stored energy to others the body can use that's largely responsible for weight gain. You're basically not getting enough sunshine and grounding on the front end, while electrons and protons are not being converting efficiently into ATP, light, water and magnetism. This lack of energy into, and out of, mitochondria is a primary cause of weight gain. It's this conversion inefficiency, combined with signaling breakdowns involving leptin, hormones and infradian rhythms that trigger weight gain. In other words, weight gain is the result of poor "metabolism management." And fixing metabolism mismanagement is the future of weight loss, as you will soon learn.

Introduction to Dr. Jack's Leptin Prescription Reset

The Leptin Rx Reset: Dr. Jack Kruse's protocol to improve the body's response to leptin in order to achieve a healthy weight and fix hormone/ endocrine dysfunctions such as hypothyroidism. For more information, visit "jackkruse.com" and his book *Epi-Paleo Rx* to learn more about resetting your metabolism management. The following is just an overview.

First, check to see if you are leptin resistant

- If you're overweight, you're probably leptin resistant. The very presence of excess body mass is a strong indicator that leptin receptors in your brain are not responding properly to leptin in the blood by decreasing appetite and increasing your resting calorie consumption.
- On the other end, if you're excessively thin, that can mean your leptin receptors have lost sensitivity to leptin as well.
- Large appetite and carb craving, especially at night, are two more signs that you're leptin resistant.
- It is possible for those at a normal weight to be leptin resistant. You can test for leptin resistance by checking your *Reverse T3* (an inhibitor of the thyroid hormones T3 and T4) level on a blood test. Elevated *Reverse T3* is a sign of leptin resistance in fit people. Scoring high on a salivary cortisol test is another indicator of leptin resistance as cortisol is thought to regulate leptin secretion.

Second, know *what* to eat. A big part of the Leptin Rx Reset is following Dr. Jack's Epi-Paleo Diet

There's a lot more to the Epi-Paleo Diet than can be explained here. Here's just a taste:

- **Produce.** Eat vegetables and fruits that are in season near you. If it's sold at a local farmer's market, it's likely to be conducive to the diet.
- **Protein.** Eat lots of good-quality proteins. Seafood is best. Grass-fed, pasture-raised organ meats are next best. Grass-fed muscle meats are third best.
- **Fats.** Cook and flavor with fats according to season. In spring and summer, favor these fats: coconut oil, pastured butter, olive oil, raw cream, avocado oil, ghee, palm oil, duck fat, beef tallow, bacon fat

and duck fat. In fall and winter, look for animal fats such as ghee, pastured butter, duck fat, beef tallow, bacon fat, non-hydrogenated lard and raw cream.
- **Oils.** Avoid nut or seed oils when starting the Epi-Paleo Diet.
- **Broth.** Eat bone broths made from grass-fed animals (to heal leaky gut) and seafood broths.
- **Fermented veggies.** Eat all the naturally-fermented vegetables you can.
- **Carbs.** If you're overweight, limit carbs to 25 grams with breakfast. If you're of average weight, limit carbs to 50 grams. If you're fit, limit carbs to 100 grams.

Dr. Jack Kruse's food pyramid

1. The base of the pyramid (most plentiful) is shellfish like oysters (other than crustaceans). They're best for brain function.
2. Next up are crustaceans.
3. Followed by fish.
4. Then organ meat of pasture-raised animals – especially liver.
5. The fifth level is grass-fed muscle meats.
6. Pastured eggs (unless you have allergies to them).
7. At the top (least plentiful in your diet) are nuts and seeds.

Foods to avoid

1. All grains.
2. All pasteurized and homogenized dairy.
3. Nightshade vegetables – if you have chronic inflammation or are low in vitamin D.

Third, know *how* to eat

- Eat as soon as you can after waking – preferably within 30 minutes of rising. Breakfast may include pastured/organic eggs, grass-fed meat, poultry, fish, protein shakes (less ideal).
- Eat three meals a day, to start. As your hunger/cravings go away, eat two times a day.
- Don't snack between meals. Snacking activates the liver more frequently than necessary, and upsets the timing of your circadian rhythms that work with leptin. You need to retrain your liver to make glucose on its own (aka ketosis).
- Do not count calories.
- Allow 4–5 hours between dinner and bedtime.

Most will notice a change in cravings within 4–6 weeks.

Supporting strategies

- Don't exercise before or right after breakfast. Exercise after 5pm.

- Trouble sleeping? Do 3–5 minutes of light exercises before bedtime (i.e., pushups or squats). Avoid activities that elevate cortisol.
- Within an hour of sunset, make your surroundings as dark as possible. Most important, reduce your exposure to artificial blue light (less insulin). This helps turn off cortisol, turn on melatonin, and increase detection of leptin by the hypothalamus.

What to expect when the Leptin Rx is working for you (i.e., increased leptin sensitivity)

- Rapid weight loss for men.
- Women will notice mood changes (calmer, more centered), improved sleep (huge clue). Clothes may fit differently, but weight may not change much initially.
- For both men and women: change in sweating pattern.
- Better recovery from exercise.
- Higher energy level.
- Diminished hunger and cravings.
- Feeling refreshed after sleep.
- Hair, nails and skin will look healthier.
- Carb cravings subside.
- You'll feel warmer, but body temperature will actually go down.
- Thoughts, mood and personality normalize.
- Mental acuity, insight, intuition and libido improve.

Cold Thermogenesis Protocol

Dr. Jack Kruse's Cold Thermogenesis Protocol is the practice of purposely exposing your body to cold temperature in an effort to raise mitochondrial efficiency and, with it, magnetism throughout the body. Here's how it works: When you soak your body in cold water, your thermal-regulatory functions tell your mitochondria you're now living in a cold climate. Your mitochondria respond by making infrared heat from stored energy such as brown fat, and by increasing heat production from the food you eat.

Cold Thermogenesis Protocol: Dr. Jack Kruse's name for using cold exposure to increase mitochondrial efficiency and magnetism throughout the body.

Heat shrinks the water around the respiratory proteins of your mitochondria (whereas cold temperature expands water). Condensing the respiratory proteins improves electron flow through the ETC due to a massive drop in electrical resistance. In fact, each angstrom that respiratory proteins move closer together makes it ten times easier for electrons to jump between their redox centers (structures that shuttle electrons between them). Increasing electron flow then makes the rotating head of the ATP synthase spin faster. And we know from physics anything that spins really fast makes a magnetic field perpendicular to the flow of current. So cold exposure speeds electron flow, which increases the magnetic field around mitochondria, cells and organs. More magnetism then improves the movement of blood, oxygen, hormones and DHA. The cold also makes inflammation levels plummet, as we talked about.

Acclimating your body to the cold

It's easier to get used to being cold when you stair-step the process, instead of starting "whole hog." First try dunking your face in cold water once or twice at a time, for a full breath. Most people don't mind their face being cold, while others find it exhilarating. Face dunks tell your temperature receptors you're in a cold environment, so your body turns on its thermal adaptations. This activates mammalian programming all animals have, but modern humans avoid through clothing and indoor heat. You don't have to freeze your face off; 50–55°F is fine. Any exposure below 98°F makes mitochondria release more heat. This speeds up electrons, making more magnetism and ATP, as mentioned. Done regularly, it also decreases heteroplasmy rate long-term, as if you were tuning up your mitochondria, which is exactly what you're doing.

After you get used to face dunks, soak yourself in a tub of cold water as long as you can stand it, as many times per week as you're able. You're not trying to give yourself hypothermia, just build up to 15 minutes, 30 minutes or more. The more you do it, the easier it gets. The lower the temperature you can tolerate, the better. But even small temperature challenges help. So if you can't stand the temperature straight out of the tap (about 50–55°F in most places), start with just five or ten degrees colder than your comfort zone. Remember, it's not a contest. It's a practice with impressive long-term benefits to your energy production, disease resistance, longevity and weight management.

For example, I bike, run or swim wearing a minimal amount of insulation. Works great when it's around 50°F out. By doing Cold Thermogenesis while I'm exerting myself, it's not nearly as disagreeable as it might be just soaking in a tub of ice water. So lately, I've noticed a rising sense of exhilaration the more I cold-challenge my body – kind of like an out-of-body experience the brain uses to protect you against trauma.

A word of caution though: Use caution with cold exposure when trying to heal a sports injury because coldness shuts down inflammation. It suppresses the healing response. I learned this the hard way when I hurt my back exercising on two separate occasions. It was healing slower than usual. I'm rarely hurt to begin with, and never take pain killers, so I always heal completely. This last time, it got worse and worse for two-and-a-half weeks as I continued to work out on my usual schedule. I put on my thinking cap and realized that cold exposure had been decreasing inflammation, blood supply and muscle repair. When I figured this out, I stopped the cold exposure, gave it more heat, and reduced my workouts to every other day. That allowed the healing to catch up. The pain and spasms got better… while I continued to work out.

10

LEPTIN

Leptin: The weight- and energy-balance hormone

The discovery of leptin is a monumental breakthrough for weight loss and energy regulation

Leptin is the master hormone that influences energy production and usage in the body, including psychological programming affecting motivation and activity. It's the once-mysterious force that controls programs, processes and perplexing conditions even the most renowned healers in history could not explain before. Like a symphony conductor, leptin controls the activity of hormones used to regulate one's weight, resting "idle speed," immune function, fertility and emotional state. Yet most people know nothing about it, having just been discovered in 1994.

How the leptin system regulates body weight

The endocrine system uses blood-leptin level to adjust your appetite up or down so your weight stays as stable as possible. The way it works is simple: Leptin is made by subcutaneous fat cells in proportion to a person's total fat mass. The more leptin that leptin receptors in the hypothalamus find circulating in the bloodstream at night, the more it suppresses appetite. The less they find, the more your appetite increases. So through this feedback and control mechanism, fat mass is designed to regulate not just how much you eat, but also how much energy you expend.

More specifically, the presence of body fat and leptin should give you an earlier, stronger sensation of fullness when you eat in order to encourage you to lose weight. Conversely, the absence of body fat and leptin should make you more hungry, more often, to help you gain weight. At least, that's how it's supposed to work. Through our new understand of how leptin works, we now realize this feedback-control mechanism is malfunctioning in millions of people, breeding rampant obesity, adrenal dysfunction, diabetes and fertility problems, among other conditions so common we consider them normal.

To help you understand the biological programming behind these breakdowns, when leptin receptors in your hypothalamus are unable to see how much leptin is in your system – because of leptin resistance – your body thinks it's starving. So, like a car with a broken gas gauge, it makes you fill up more often than necessary by increasing your appetite prematurely… just to be on the safe side.

Equally problematic, the endocrine system lowers thyroid function, the immune system, insulin signaling and reproductive function in order to conserve energy for higher priorities. Your mental and emotional states are depressed to discourage activities that burn energy. Leptin resistance (i.e., when the system is under-performing) basically creates a starvation response in people by chronically activating all sorts of adaptive mechanisms to cope with the perceived energy shortage. It down-regulates organs, processes and behaviors to conserve energy. That spells disorder all over the body. Bottom line: Leptin is commander-in-chief of weight maintenance. And it operates with impressive influence over bodily functions and programming. So the other systems and processes we thought were responsible for weight gain/loss act subordinately to leptin's direction. Insulin and thyroid function are but two of leptin's subjects.

Nature planned for us to get leptin first in breast milk

Those that weren't breastfed missed out on a valuable step in the metabolic priming process. Research now shows that leptin in colostrum helps establish methylation early in life. Methylation controls the replication of genes (transcription) and wide range of essential processes. So its absence can alter gene expression through epigenetics. This is one way obesity and other family traits are passed on to future generations.

Methylation is the transfer of one carbon atom and three hydrogens (CH_3) – called a "methyl group" – to another molecule. Methyl groups control: detoxification through glutathione, immunity, inflammation, gene expression, repair of free radical damage, neuro-transmitter production for brain function, energy production, the stress response, and more. Methylation defects are thought to contribute to autism, slow detoxification, and many disorders.

When leptin reception tanks

When the hypothalamus is unable to see how much leptin is in the system – that is, when you've lost leptin sensitivity – it can't tell how much fat mass the body has. This makes you lose the ability to regulate appetite, energy expenditure and ultimately body mass. Therefore, obesity is primarily a failure of the leptin system to respond to high leptin levels in the blood (>10 ng/ml), while anorexia can be a failure of leptin receptors to recognize low leptin levels (<4 ng/ml). Both are caused by dysfunctional leptin reception spoiling your appetite regulation.

The good news is, when you "wake up" your leptin system, the body realizes its fat mass is dramatically different than what it thought. It then puts enormous pressure on your energy management devices to get you to your ideal weight. This causes the body to make all kinds of adjustments large and small to get you to eat less, store less and burn more. And that's how people are losing staggering amounts of weight in record time using the Leptin Prescription Reset and Cold Thermogenesis protocols. Internalize this like a mantra: The new, improved road to substantial weight loss is the journey to *leptin sensitivity*. On the other hand, weight loss without regaining leptin sensitivity is a big reason why the pounds so frequently come back with after a successful diet.

Many people think they have a thyroid problem, when it's actually leptin resistance

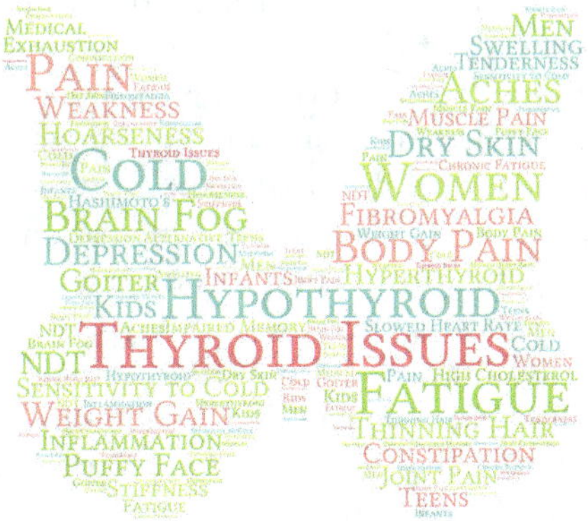

Surprising common today, *leptin resistance* has a nasty habit of turning thyroid function way down. "Hypothyroidism," as depressed thyroid function is called, can cause symptoms such as extreme fatigue; weight gain; weakness; cold hands, feet and body; hair loss; brittle nails; dry skin; constipation; poor memory and depression. In other words, leptin resistance can slow your metabolism, give you the appearance of thyroid dysfunction, and make weight loss extremely difficult because your resting calorie consumption is too low. It can even prevent people from losing weight when they exercise, because the metabolic benefits of exercise are not carrying over fully into the resting state. As a result, leptin resistance can cause practitioners to misinterpret thyroid function tests, because thyroid hormones may not be malfunctioning. Leptin *insensitivity* may be the real problem – predominately, or as a contributing factor.

Women are more sensitive to leptin, and their environment

Women are built to be more sensitive to biophysical forces around them – temperature, food and EMFs, for instance – than men. They have to be so that their mitochondria can adjust their metabolism to suit their surroundings. They then pass those survival-enhancing adaptations on to their children in the form of mitochondrial haplotype and heteroplasmy rate, as well as epigenetic changes to nuclear DNA.

Women also average more fat mass than men, on a percentage basis. That means more leptin. The thing is, leptin is pro-inflammatory, having thought to have evolved from interleukin-6 (IL-6), a pro-inflammatory chemical made by the body. That's how leptin works: leptin from fat creates an inflammatory state (and signal) that the leptin receptor reads, then tells endocrine organs to either gain weight or lose it.

Those two factors – chronically higher leptin levels, and greater sensitivity to the environment – **put women closer to the edge of having chronic inflammation all the time.** It's in their nature to have a finer line between *inflamed* and *not inflamed*. **This is a huge reason why women typically gain weight more easily, and have a harder time losing it.**

Leptin overload contributes to type 2 and type 1.5 diabetes

High leptin levels in the blood destroy "amylin," a protein made by the beta cells of the pancreas. This should come as no surprise since leptin is thought to have evolved from the pro-inflammatory chemical IL-6. When beta cells can't make amylin, the pancreas can't make insulin. And that's a foundational disturbance upon which type 2 diabetes is built. It's leptin continually flooding the pancreas and destroying insulin production. At the same time, eating lots of sugary foods that spike insulin levels stresses the pancreas even more. So type 2 diabetes is caused, in large degree, by a condition of excessive leptin wiping out insulin production, exacerbated by increased demand for insulin due to high blood-sugar level.

Interesting to note for prediction and prevention, this process of high leptin levels beating up beta cells typically precedes insulin resistance by 5–7 years. That makes leptin resistance a good predictor of type 2 diabetes. However, if leptin insensitivity develops quickly, as it can with pregnancy or leaky gut, we could be talking about type 1.5 diabetes, which is basically autoimmune-type diabetes. In type 1.5 diabetes, the autoimmune attack on pancreatic beta cells is more intense. So pancreas and insulin function fail faster than in type 2 diabetes.

Unfortunately for those affected, the medical system focuses on treating insulin-resistant diabetes after it occurs. But wouldn't it make more sense to treat the underlying problem of leptin resistance, before it turns into type 2 diabetes? To land the plane here, leptin resistance precedes insulin resistance. And, when both occur long enough, you get adrenal resistance, which means cortisol is elevated chronically so the adrenals become unresponsive. Most people call it adrenal fatigue. Not good, because high insulin and cortisol at the same time can turn into cancer and many chronic diseases.

The difference between *leptin* resistance and *insulin* resistance

"When people talk about insulin resistance, they automatically think about insulin resistance from a pathologic standpoint – meaning type 2 diabetes or metabolic syndrome. But what they don't realize is that insulin resistance also has a physiologic role [a benefit]… in ketosis, in starvation, and also in brain development for infants. They're designed to be insulin resistant up until they're almost 25 years old to myelinate their brain. So when you understand that very fact, you start to realize that insulin resistance has a key role [in evolutionary biology].

Here's an interesting conundrum that people don't realize: Insulin resistance has very little to do with fertility. What's the goal of evolution? It's to have another generation. Well guess what, when you're leptin resistant, you can't have a baby. But when you're insulin resistant, you can still have a child… So which one is really more important from an evolutionary standpoint?

...What is [leptin's real purpose?] It's an electron accountant. It counts the electrons that are in our system. It also counts how many protons are in our system. So when you have too many protons, and not enough electrons, you're leptin resistant. Leptin controls selection of eggs from the ovaries. And it controls sperm production in males. That means leptin directly controls fertility."

—*Dr. Jack Kruse.*

Leptin controls fertility (fecundity)

The leptin system is Nature's way of picking a good time to have a baby. It does this by monitoring both energy reserves, and energy in circulation, to make sure the body has enough resources to gestate a baby. If the leptin system senses a woman has enough energy to safely carry a baby to term, it picks an immature egg to develop, and Nature takes its course. When it doesn't, the leptin system makes a woman's inner terrain inhospitable to pregnancy by delivering eggs of poor quality, irregular menstrual cycles and/or reduced receptivity of the uterine lining. Leptin level is supposed to control this process. However, when leptin receptors aren't responding properly to leptin in the system, the endocrine system blocks reproductive processes. Couples having trouble conceiving experience this as infertility. A major cause, if not the single biggest reason (among several), is leptin resistance.

A major drug company cancelled its synthetic leptin trial (planned for weight loss) because it worked too well

After spending many millions of dollars, and getting well into the FDA approval process, a major US drug company halted its clinical trial of synthetic leptin because releasing the drug would have negated the need for other, more profitable drugs they, and the rest of Big Pharma, sell. The company arrived at the stunning realization that once you fix a person's broken leptin system, a plethora of the body's most vexing problems simply vanish. They realized how influential the leptin system is at controlling biological programming throughout the body, and how adept it is at doing its job.

Therefore, if they were to allow synthetic leptin onto the market, they would be teaching the entire healthcare industry what really causes dozens of chronic, degenerative diseases... and why Big Pharma is doing a horrible job at fixing them with their synthetic drugs. By releasing the cure to so many profit centers... I mean diseases... they would, in effect, be digging their own grave. It simply wouldn't do. As with all solutions that work too well, have too little markup, or are not patent-protectable, they had to stop it before it threatened their monetization and control over many disease processes.

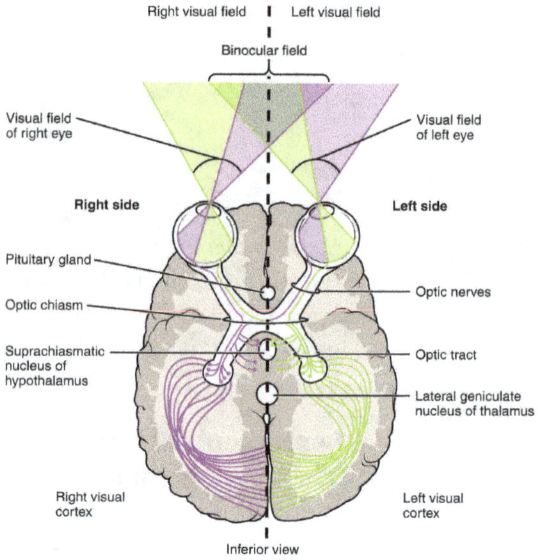

Image used under Creative Commons Attribution 3.0 Unported license. Source: Anatomy & Physiology, Connexions Web site: http://cnx.org/content/col11496/1.6/ Author: OpenStax College.

What causes leptin resistance?

Leptin receptors start measuring how much leptin is in the system four hours after your last meal of the day, or four hours after your last light exposure. Spikes of insulin block leptin in the blood from being read by leptin receptors in the hypothalamus, which is why Dr. Jack tells us not to snack after dinner. Of the many factors at play, junk light in the eye and on the skin is most responsible for creating leptin resistance.

In the eye, blue light breaks the bond between vitamin A and melanopsin, which releases a thread of positively-charged protons. In a perfect world, DHA's massive cloud of electrons is able to offset the positive charge from dissociated melanopsin. But modern living overextends our electron supply from DHA, due to blue in our manufactured light. Hence, positive charge and inflammation in the area go up, while healing and communication go down.

Structurally, free retinal breaks the aromatic (benzene) rings in leptin, melanopsin and melatonin. As we'll discuss shortly, benzene rings are photon traps (see page 309). Break these and you can't carry light to the hypothalamus so it can tell proper time and energy balance. The crush of blue light also slows the ETC locally by stretching out respiratory proteins (they're photoreceptors too), which lowers redox further. Those two factors – increased positive charge and decreased mitochondria function – combine to increase inflammation in the retina, SCN and leptin receptor. The three live along a semiconducting circuit called the "central retinal pathway" (CRP) that allows the brain to tell night from day.

Note: The central retinal pathway (CRP) runs from the retinas at the back of the eyes to the SCN and leptin receptor in the hypothalamus. The CRP helps tell the brain what time of day or night it is.

Melanopsin: photoreceptor involved in blue light injury.

Through this process, positive charge and inflammation disturb communication to the leptin receptor in the hypothalamus, while free vitamin A and melanopsin attack leptin itself in subcutaneous fat around the body. That's how blue light exposure helps fat form almost anywhere it touches the body. And it's how near-sightedness, macular degeneration, and even neuro-degeneration can develop over time: Blue light beats up photoreceptors around the CRP, anatomy gets damaged, and inflammation prevents repair programs from doing their job. Other factors then do their part to reduce leptin sensitivity, such as simply having too much leptin in the system, chronically. In the same way that insulin overload promotes type 2 diabetes, having chronically high leptin wears out leptin receptors that have already been desensitized by chronic inflammation.

How to tell when you're leptin resistant (and test for it)

The easiest way to tell if you're leptin resistant is to look at yourself in a mirror naked. The very presence, or acute absence, of fat mass tells you if your leptin receptors are reading your leptin levels correctly, and responding appropriately. That means you can tell if your leptin reception is broken based on results. Too heavy, or too thin – both mean you're leptin-resistant. It's possible for those at a normal weight to be leptin-resistant too (as other factors bring you into balance).

There are simple, straightforward tests you can do to give your healthcare providers objective measurements to evaluate. However, they're rarely used by clinicians because leptin resistance is not widely understood or practiced yet. And the tests are not covered by insurance. Low vitamin D is a good indicator of leptin resistance to start with. But even better are a "Highly sensitive C-reactive protein" test (hsCRP) and a "Reverse T3" test. Practitioners have long-used the common "C-reactive protein test" to measure inflammation. So scoring high on the *standard-sensitivity* C-reactive protein test reveals that your leptin receptors may be responding poorly to leptin due to C-reactive protein binding to leptin and interfering with its function, and/or a systemic inflammation.

C-reactive protein test is a common measure of inflammation. A Highly Sensitive C-Reactive Protein test (hsCRP) measures low levels of C-reactive protein in the blood.

The other good indicator of leptin resistance is when you score high on a *Reverse T3 test.* The hormone *reverse T3* suppresses thyroid function because it acts in opposition to the hormones T3 and T4 that run the thyroid. For this reason, leptin resistance and thyroid dysfunction usually go together. You can surmise how leptin-resistant you are by seeing how inappropriate the leptin system's instructions for the thyroid are (i.e., slowing down the thyroid when the leptin system should speed it up).

This also explains why standard thyroid tests can be misleading when you have leptin resistance: Your thyroid probably isn't broken. That's the least likely, most severe, condition. Leptin resistance is more likely to be causing slow metabolism (along with fluoride decreasing T3 and T4 levels). But leptin dysfunction is making it appear as if your thyroid needs repair (via hormone replacement).

Consequently, anything you do to try to fix your thyroid, such as hormone supplementation, is misguided and potentially counter-productive because thyroid glands are not normally the weakest link: toxicity and inflammation are (from fluoride, blue light, C-reactive protein and other pro-inflammatories). As a result, supplementation can uncouple your thyroid system from the feedback mechanism that's supposed to control it, eventually weakening your endogenous production.

Restoring leptin function

DHA. To start with, you can protect your leptin receptors from retinal and inflammation damage by having more DHA (and its electrons) available to neutralize the positive charge resulting from broken vitamin A-to-photoreceptor bonds (the blue light "beat down" we talked about earlier). Eat more seafood with DHA in it, such as oysters and cold-water fish. Liver from grass-fed animals has more DHA than regular beef, lamb, etc. But seafood is still the best.

Cold exposure. Next, you can reduce your leptin level through cold exposure. Cold exposure turns on ancient temperature regulation pathways Nature installed in mammals to help them survive winter weather. As part of this seasonal programming, cold exposure increases production of "uncoupling protein 1," which stretches out the respiratory proteins on purpose to make the ETC less efficient – meaning more heat.

Brown fat. Cold exposure also activates fat-burning pathways that turn white fat into brown fat. Brown fat's enviable ability is all of it goes into heat production, rather than ATP. These two mechanisms – *uncoupling protein 1* and brown fat usage – turn up the furnace function of your mitochondria, thus dissipating calories as heat, instead of retaining them.

Reducing inflammation everywhere. It's the side benefits of activating fat-burning pathways that will really blow your mind. Researchers believe leptin evolved from a pro-inflammatory cytokine the body uses widely to promote inflammation, called IL-6 (interleukin-6). As discussed, inflammation is beneficial when it's acute and temporary. But inflammation becomes destructive when it stays switched on too long. The point to ponder deeply here is this: As temperature on the skin goes down, so does inflammation. They're a coupled system.

Our ancient seasonal programming lowers leptin and IL-6 levels as it burns fat and generates heat. That's how cold exposure works. So when you're exposed to the cold, not only do you burn more fat, but it also stomps out a variety of inflammatory conditions such as diabetes, heart disease and cancer as a consequence of suppressing inflammation. This is why Dr. Jack calls obesity an inflammatory brain condition, and why the Wim Hof Method is renowned for reversing dozens of chronic conditions using cold exposure.

Dr. Jack's Leptin Prescription. Finally, the way to restore leptin sensitivity is to retrain your brain to see leptin at the appropriate times of day, and avoid seeing it at the wrong times. Leptin signaling is supposed to be synced to daily cycles of daylight, darkness, wakefulness and sleep. Fat-burning and the release of growth hormone, for example, work best when they happen on schedule with circadian rhythms. Hence, Dr. Jack's protocol not only tells you *what* to eat, but *when* to eat, and what other lifestyle choices contribute to your results.

II

DEUTERIUM

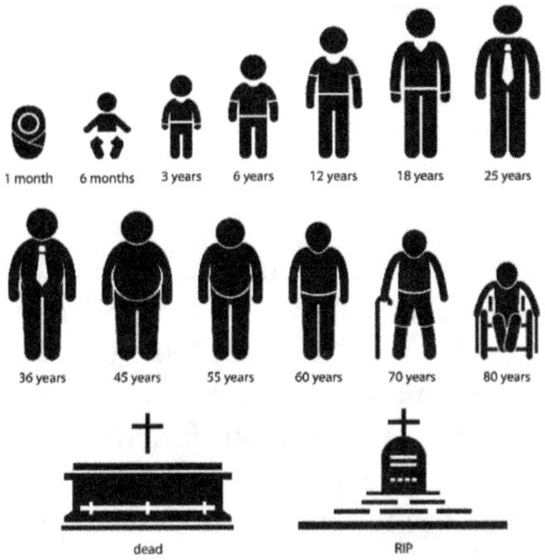

Deuterium controls survival, growth and maturation when you're younger... metabolism, disease and aging when you're older

Deuterium is a special form of hydrogen Nature uses to help regulate human life cycles. Present all around us to varying degrees, deuterium is nothing more than a common hydrogen atom with an extra neutron in its nucleus. In non-living systems, that extra neutron doesn't do anything; deuterium behaves the same as regular H^+ hydrogen (protium) in chemical reactions. But in living systems, Nature uses the extra neutron to regulate mitochondrial output, seasonal biorhythms and much more.

For example, deuterium helps babies grow in the womb. It helps myelinate the brain through your childhood and teen years. And it's used by the body to regulate your infradian biology as an adult. Problem is, after early childhood, excessive deuterium in your system turns a good actor into a bad actor when it comes to weight control, disease resistance and anti-aging efforts. That's because deuterium speeds up the aging process by keeping maturation and weight gain programs turned on after their biologic purpose is over and done with. It does this mostly by subduing mitochondrial function and causing a cascade of consequences, including fat production.

To make it simple, deuterium reduces energy output of mitochondria. And we know from Dr. Doug Wallace's work that mitochondrial dysfunction is the driving force behind modern disease as we know it. So whatever organs or systems accumulate the most deuterium in adulthood will be hardest hit by dysfunction and disease as the individual ages, because that organ's mitochondria won't be able to make as much energy to maintain and restore cells.

Deuterium serves many roles in survival, maturation and metabolism

Deuterium is a control factor the body uses to start or stop growth and metabolism programs, based on its presence or absence. For example, in the third trimester of pregnancy, mom's body offloads her deuterium into baby to fatten it up to help it survive after birth. This reduces the mother's deuterium load, which invigorates her energy production, while at the same time baby gets a growth aid. For these reasons, babies are like the fruit of a human.

Deuterium also prevents myelin from forming around the baby's brain cells while it's in the womb. This keeps babies' brains as small as possible so they can fit through the birth canal (which in humans is pretty tight). But it also delays maturation of brain function until well after the baby is outside the womb. In fact, brain function continues to mature through this myelination process until a person's mid-20s (one more reason children are especially vulnerable to wireless radiation: less insulation to protect the brain). This is Nature's way of making humans as smart as possible, given constrains of the birthing process.

Now what's been happening in recent decades is more and more toddlers to adolescents look like they were born with an "obesity gene" turned on full blast. Their weight regulation appears to be broken from inception – even more then you'd expect, given their family's genes and lifestyle choices. But why would this be happening? A: **Excessive deuterium gets into the egg before it's even been fertilized, damaging its 100,000 mitochondria.** The germ line itself (the egg) is damaged before conception, which is passed on to that generation, and their progeny, in a compromised state.

By inheriting defective mitochondria, young children then become obese far in excess of normal baby fat. So instead of young children eating anything they want and not getting fat, they actually gain weight regardless of what they eat or how much they exercise. And those frail mitochondria predispose the individual to diseases we used to associate with "getting old." Therefore, weak mitochondria are the single biggest factor in childhood obesity and disease. High levels of deuterium are causing kids to lose ATP, eat more, and get fat as a result. All three are not good. But it's actually the drop in ATP that causes organs to malfunction and disease to kick in due to compromised maintenance and repair systems, not excess calorie consumption or obesity by themselves.

Deuterium promotes growth and maturation

After early childhood, elevated deuterium from high exposure and/or slow elimination can cause precocious development and early-onset puberty. That produces rapid physical development in adolescents to early teens. But, to their detriment, it also means early development of diseases associated with aging because deuterium overload makes you live fast and die young, mitophysically speaking.

This is exactly what we're seeing with so many childhood diseases showing up decades earlier than before. For instance, we're seeing diseases such as arthritis, osteoporosis and cancer that used to show up at age 50 and 60 now showing up in 20-somethings, teens or even younger. Now you know why. It's because deuterium is Nature's special way of getting animals and plants to grow and mature by regulating their metabolism. Unfortunately, that effect does not extend to regeneration and vitality. Instead, excess deuterium continues to promote aging, even when you don't want it or need it. To put it simply, deuterium makes you age faster the more of it you have in your body.

But deuterium isn't always bad

There are some places in the body where deuterium serves a beneficial role in adulthood. Meaning, it's useful and welcome in some places and not others. It just becomes problematic when/where you have too much of it in organs and tissues that depend on mitochondria for energy. So not only does deuterium not harm you in blood plasma, red blood cells, sperm, eggs or uterine lining. But it's actually beneficial, because those pathways use it to their advantage and/or help deplete it from the body.

To illustrate, deuterium is beneficial in the blood because sunlight on the skin squeezes deuterium. That squeeze produces UV-A through V-UV and hydrogen peroxide (H_2O_2: free radical that kills pathogens). White blood cells need biophotons from deuterium to activate POMC products so they can serve their immune functions, H_2O_2 helping in the effort. On the other hand, deuterium is bad in the mitochondrial matrix because when you squeeze it, the ultraweak UV it releases aids mitosis of cancer cells. Here, biophotons help cancer grow and spread, aka metastasis.

In a nutshell, deuterium is not toxic for all people, all the time. Instead, it's more a matter of it being in the wrong place, in the wrong amount, at the wrong time that makes it detrimental to human health. In fact, our systems are designed around its presence. So it's only when society's ways concentrate deuterium in our food, water and physiology that it becomes an enemy to optimal health.

How deuterium harms the body

Deuterium behaves the same as regular hydrogen in most chemical reactions outside the body. But, in humans, deuterium becomes a villain after your body and brain have fully matured. Crucial to understanding weight control, disease and aging, the last step of ATP production can only use regular H^+ hydrogen atoms. That's because the fifth structure of mitochondria's electron transport chain – called the "ATP synthase" – has a turbine motor head designed to fit H^+ as tightly as gears of a watch.

The electron transport chain ends with the 5th cytochrome, also called the ATP synthase, or ATPase. Powered by protons running through its turbine motor, the ATPase completes ATP production.

But, like fabric getting jammed in a zipper, the deuterium atom, with an extra neutron attached to its proton, is too big to fit through the spinning head of the ATP synthase. This gums up the tiny rotor so it doesn't spin properly, which breaks the ATP synthase and lowers ATP production. The electron transport chain then makes more oxygen radicals, which contribute further to the aging process.

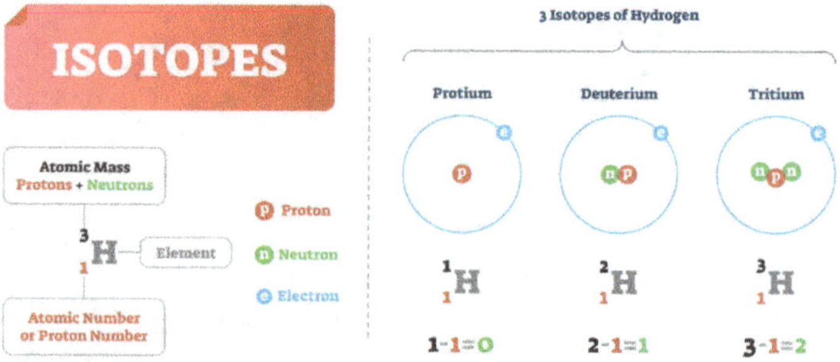

Isotope: Elements with a different number of neutrons than their basic variety – often making that isotope radioactive. For example, the radioactive tritium is a hydrogen atom with two extra neutrons.

ATP synthase blockages also decrease the magnetic field around mitochondria, because the fewer ATP synthases you have spinning vigorously in a mitochondrion, the weaker the magnetic field around it. Lower magnetic field weakens blood flow, hormone delivery and DHA to the cell, as well as oxygen supply to the mitochondria, because they're all drawn to magnetic fields. Deuterium also inhibits the formation of e-zone, because its extra neutron doesn't fit in the e-zone's crystal matrix. That's how deuterium in the diet is like pouring syrup into a car's engine instead of oil. The net result being, ATP production suffers, as does blood flow, oxygen utilization, hormone delivery and regeneration. What's more, when deuterium clogs up the ATP synthase, the body can't fix them. If certain species of bacteria aren't available to unjam the ATP synthases, they have to be replaced.

To recap: Deuterium is like a hormone we get from our environment that regulates: (1) current energy consumption *vs.* storing it for later use, (2) maturation *vs.* delayed development, and (3) how our bodies interact with the seasons. Discovered in 1931, deuterium controls biological programs mostly by regulating mitochondrial production efficiency and hormonal growth programs. The body uses their difference in mass to sort protium atoms from deuterium in processes driven by light.

When deuterium becomes a bad actor

Deuterium (aka "heavy hydrogen, symbol D, hydrogen-2, or ^{2}H") is a stable (non-radioactive), naturally-occurring isotope of hydrogen that derails ATP production, because the ATP synthase can only use regular H^+ hydrogen to make ATP. With this drop in ATP output, not only do you lose your ability to regenerate cells and fight disease, but you also make more fat from the food you eat. And that turns a lot of the fruits and vegetables we eat *thinking they're non-fattening and anti-aging* into fattening and age-accelerating foods due to their high deuterium content. Pretty crazy, huh?

Even though fruits and vegetables may provide antioxidants, vitamins, fiber and other good stuff, those benefits are diminished, or completely overshadowed, by the drop in energy production from the mitochondria, which is *the* single biggest factor in how healthy or sick you are. And that's what makes deuterium a major concern to mitochondriacs.

Found in a variety of foods, a high deuterium level:

- increases inflammation
- makes you more sensitive to nnEMFs
- screws up enzyme function
- is associated with cancer, MS, osteoporosis and autoimmunity

- reduces ATP output, which then increases your chances of having fibromyalgia, brain fog, Lyme disease and autism
- blocks hormone production by disabling cholesterol.

And that's why surplus deuterium is Nature's fat-maker, energy-decreaser and age-accelerator.

Deuterium changes gene expression through epigenetics

Some educators still believe disease is caused by gene defects when, in fact, probably more than 90% of diseases are caused by epigenetic alterations to the way genes build a human. This makes it look like a disease is being caused by the genes themselves misbehaving, when it's actually deuterium and other influences that are responsible for activating or suppressing genes, which then turn into symptoms we see.

To illustrate how DNA works, genes are preceded by segments of DNA that *activate* the gene – called a "promoter" region – as well as DNA that *suppresses* gene expression – called a "silencer." When deuterium collects in the promoter region, its proteins change shape, which alters their resonance and function. That gene then gets over-activated ("over-expressed," as it's called), which can result in elevated levels of a disease-causing protein in a cell. Similarly, when a suppressor gets loaded with deuterium, that gene may get more turned off. In this situation, fewer protective agents are made to combat disease. To make matters more interesting, varying degrees of both effects can happen at the same time.

In other words, a buildup of deuterium in DNA's transcription regulators (promoters and silencers) alters gene expression so those epigenetic alterations are often misconstrued to be genetic mutations that cause disease. This is one (unrecognized) way gene expression changes in response to your environment (aka epigenetics). Unfortunately for us today, that means deuterium predominately corrupts gene expression for the worse. And it's a major reason geneticists are tricked into believing a disorder is being caused by genes that look defective when, in fact, it's just epigenetic misfortune that all of those afflicted have in common.

Extremely common examples today are glyphosate (active ingredient in the weedkiller Roundup) and childhood vaccine exposures triggering a cascade of: leaky gut and autoimmune issues; poor digestion with pathogen overgrowth; mineral and vitamin deficiency; pitiful detoxification; and protein misfolding (particularly in the brain). This tidal wave of chronic issues leads to autism and other GAPS conditions. Deuterium plays a role in much of that because it's tied to energy in many ways – detoxification and metabolism, in particular. And it regulates our biology. The net result being, genes don't do much in comparison to epigenetic influences such as deuterium, glyphosate and toxins in vaccines.

Cholesterol can't be turned into hormones when its deuterated

Normally, cholesterol gets broken down into pregnenolone (a master hormone), which gets converted into sex steroid hormones such as estrogen, progesterone and testosterone, as well as vitamin D and bile salts (which break down fats). But when cholesterol has too much deuterium in it – when it's "deuterated" – it can't be used as a raw material in the production of those crucial hormones and biochemicals. That's because carbon-deuterium bonds in cholesterol are seven times harder to break than carbon-hydrogen bonds in hormone production. Whatever the reason, **when cholesterol is not being converted into vitamin D and hormones efficiently, your cholesterol level goes up. That's an important reason so many people have high cholesterol today.** Its normal pathways get clogged by dehydration, nnEMFs, lack of UV on the skin and hormonal disconnection. With outputs blocked, cholesterol then floods the system. And we blame it on animal fat.

Why "local" produce is good for you and "non-local"... not so much

Nature designed us to eat what's available in our local environment by yoking deuterium in our food to UV light exposure of the seasons. That means the more deuterium your food has, the more UV light exposure you need to offset the decrease in mitochondrial efficiency. When you *eat* fruits and vegetables in weak sun weather (i.e., late fall and winter) that were *grown* in strong sun weather (e.g., imported from sunny places), you create a seasonal mismatch with insidious health consequences. That's because fruit is loaded with deuterium. That's bad for your health... or it would be bad without ways to offset deuterium's adverse effects. Fortunately, Nature gave us the solution which required no thought or effort on our part: UV light. UV light makes your mitochondria more efficient at producing energy.

Concerned about *organic* vs. *conventional* vs. *GMO*? Instead, think about the water used to grow that fruit or vegetable

What's the most important factor in food quality to a mitochondriac? It's the quality of water that food was grown with – specifically its deuterium concentration (for most mature adults). That determines how efficient your mitochondria are at making ATP. Plants grown in coastal regions of California and Florida, for instance, tend to have the most deuterium because the heavy hydrogen of deuterium is released by rain clouds before regular H2O, as they travel inland. And deuterium naturally collects in waterways at lower elevations because it's twice as heavy as regular hydrogen and last to evaporate. Therefore, plants grown closer to water bodies tend to be higher in deuterium. Likewise, plants grown toward the equator are naturally high in it, while those grown further North, or

at elevation, tend to be lower. Consequently, an avocado grown in California is not as good for you as one grown in Mexico City (elev. 7,349 ft.), in terms of deuterium.

Those with excellent mitochondrial capacity, and those living in stronger sunlight, can get away with eating more high-deuterium foods, because their mitochondria typically have more efficiency to spare. Meaning, their metabolisms are better equipped to offset a drop in output. Higher redox potential also makes them better able to withstand the burdens that GMOs, glyphosate and poor-quality food place on their detox systems. Free radicals are usually less of an issue for them too.

Conversely, those with poor mitochondrial production are hurt more by high-deuterium foods and water, because they're already low on power. They're likely to have more inflammation. And so they don't repair, renew or maintain weight as well. For these reasons, a cucumber that's loaded with deuterium can be a highly fattening food for the wrong person – especially when it's eaten out of season for your locale. And that's why deuterium level in water, and in produce, is a bigger factor in people's health than those grocery store labels we've blindly been trusting to bring us wellness.

Big Ag loves deuterium

Farmers and commercial agriculture companies (Big Ag) love what deuterium does for them, because it helps plants grow faster. Whether they realize what they're doing or not, deuterium is a profit-maker for producers because Nature uses it to accelerate a plant's maturation.

A higher deuterium concentration goes hand-in-hand with

- mass-production practices such as irrigation and hydroponics (unnatural hydrological cycle),
- greenhouse growing and two-crop seasons (accelerated growth cycles),
- grow lights (altered light and photosynthesis),
- as well as GMOs and the crop amendments that go with them (more growth aids, less microbial life in soil to sequester deuterium).

Crop amendment: Material applied to a crop and/or soil to improve its physical or chemical properties.

Unfortunately, when you eat plants that are loaded with deuterium, if you're done growing *upwards* deuterium encourages you to *outwards* (as in getting fat) due to the higher deuterium-to-nutrient/calorie ratio. And it ages your cells and mitochondria faster. That's not winning.

Deuterium depletion

Deuterium depletion starts in photosynthesis. Bacteria then continue it in the gut

Plant chloroplasts start reducing deuterium levels in the food chain through photosynthesis. Recall that photosynthesis converts water and CO_2 into sugars using sunlight. Chloroplasts separate helpful hydrogen from harmful deuterium by taking advantage of the difference in bond strength between each of them and oxygen. In a water molecule, deuterium holds on to oxygen ten times stronger than common hydrogen. It has to do with deuterium's mass. So when light hits water in the photosynthetic process, hydrogens break off more easily to move the photosynthetic process forward, while deuterium tends to stay attached so it can be dealt with as a hazardous substance.

This enables chloroplasts to start the multi-step process of discriminating and fractionating deuterium from hydrogen. Later, in the gut microbiome and other tissues, bacteria filter deuterium from the body by running the ATP synthase in reverse. This is why recent research shows that breast cancer is linked to an *absence* of certain bacteria strains. It's because bacteria are actively depleting deuterium levels in breast tissue mitochondria, thereby supporting metabolism, detoxification and renewal programs locally, which protect the individual from cancer.

TCA cycle, aka Krebs cycle, citric acid cycle. See page 50 for diagram and description.

Glycolysis and the TCA cycle decontaminate water of its deuterium to protect the ETC

Deuterium is so destructive to mitochondrial function that our cycles of metabolism (e.g., glycolysis and the TCA cycle) employ what seem to be unnecessary steps just so they can remove and examine hydrogens from their intermediate substrates to make absolutely sure they're not deuterium before reaching the ETC and breaking the ATP synthase. Substrates are base molecules that are added to, or taken away from, to make other molecules in a chain of reactions (think "precursor"). Example: adenosine diphosphate (ADP) is a substrate for the production of adenosine triphosphate (ATP).

See diagram of glycolysis on page 49 to understand the transformation of one molecular substrate into another.

Reducing deuterium level is so important in keeping energy production up, and disease processes at bay, that the body performs action steps that appear to be overly complicated and wasteful when seen through a *'mitochondria only make ATP'* lens. When, in fact, glycolysis and the TCA Cycle have a total of 19 enzymatic steps not to befuddle pre-med students, but rather to purify the ETC's fuel multiple times to prevent deuterium from jamming the ATP synthase.

Pyruvate: An end product of glycolysis, pyruvate is a chemical compound that fuels the TCA cycle (aka the Krebs Cycle, citric acid cycle). See diagram of Cellular Respiration on page 48, and the citric acid cycle on page 50.

In other words, glycolysis and the TCA cycle are setup the way they are not to maximize energy production, but do the best job possible at depleting deuterium. That's the real purpose of glycolysis and the TCA cycle. Researchers just assumed that ATP yield is the one and only goal of metabolism. This lends credibility to Dr. Boros's assertion that mitochondria's most important function is making deuterium-depleted water, while ATP is more like a "handy helper" in bodily function.

Here's the short version of how these two cycles of metabolism separate H^+ hydrogen from deuterium to keep it out of the ETC: Enzymatic stages of glycolysis (the steps you see on page 43) pull hydrogens off sugar substrates and put them in a pool of water in the cell. A tiny percentage of these hydrogens are deuterium. Some steps in glycolysis have H^+ hydrogen added back to their substrates, while other substrates are repleted with H^+ later in the TCA cycle. Deuterium stays in the cytoplasm where it can't break the ETC in the mitochondria.

This process of glycolysis has been described as "dehydrating" glucose, meaning removing *hydrogens* themselves, not water. The end product of glycolysis, which is deuterium-depleted pyruvate, then goes on to feed the TCA cycle in the mitochondria. Next, three enzymatic stages of the TCA cycle take low-deuterium metabolic water from the mitochondrial matrix and attach it to the cycle's substrates. Another stage of TCA makes water and puts it back into the matrix.

Through these processes, any deuterium that started out on glucose molecules stays in the cell where it can't harm energy production, while substrates holding the desired H^+ hydrogens move on to drive the electron transport chain in mitochondria (which live inside cells). So you can think of the glycolysis-TCA pair as "washing" intermediate metabolites clean of deuterium like a washing machine. Glycolysis starts removing deuterium from substrates with a drying process in the cell's cytoplasm. And the TCA cycle completes the cleaning process by "rehydrating" with deuterium-depleted water from the mitochondria's matrix. That's how glycolysis and the TCA cycle repeatedly reuse metabolic/matrix water's purity to keep substrates clean of deuterium.

The takeaway: The body goes to extraordinary lengths to make sure matrix water is extremely low in deuterium because it absolutely devastates mitochondrial function and disease resistance. That's why high-carb/low-fat diets are bad for you, while nutritional ketosis (fats and proteins) can alleviate conditions such as diabetes and neurodegeneration. It all comes down to how much deuterium the cycles of metabolism are required to process. Carbs are naturally higher in deuterium, while animal fats are low in deuterium by nature.

Depleting deuterium can improve athletic performance

In the 2016 NFL season, the Los Angeles Rams finished with one of the worst records in the league, at 4–12. It's been reported the Rams owner, Stan Kroenke, met with Dr. László Boros (a leading researcher on deuterium and deuterium depletion) who coached the team on the benefits of drinking deuterium-depleted water. The next season, the Rams finished with an 11–5 record. So they went from near last place in the league, to near first place, in one season. They followed it up in the 2018 season by going to the Super Bowl.

Now, no one in their right mind would claim deuterium depletion alone was the sole reason for the stunning turnaround, but it does make you wonder how big a role deuterium in their water did play. I'm sure we'll learn much more in the years to come about drinking deuterium-depleted water to reduce recovery time and increase stamina. Breaking news: The Rams won the 2022 Superbowl just days before the First Edition went to print. They must be doing something right.

Deuterium ranges

Geologic records tell us deuterium in the environment was 10–15 parts-per-million (ppm) lower 15,000–20,000 years ago. That would have to mean human physiology is evolutionarily adapted to handle that concentration of deuterium with no problem. With that exposure level, you'd expect our detox systems to be able to remove it fast enough to keep us in a healthy range. This then is a sensible goal we can all aspire to as a general guideline: 10–15 ppm lower than our current level – both in the body, and from the food, water and air we're exposed to.

Unfortunately, society's ways, in addition to geologic trends, have raised non-threatening Ice Age levels up to the more problematic levels that we're exposed to today. To our biology, that means the higher up you go above that historic baseline (127–137 in the body), the harder it is for your depletion systems to get rid of deuterium fast enough to avoid energy deficits and organ dysfunction. Don't forget: dose makes the poison. And toxicity increases rapidly for every 5 ppm above the 130 range, because our deuterium depleting mechanisms get overwhelmed above a certain threshold.

Testing deuterium level

As part of the detoxification process, the body purposely moves deuterium from certain places where it's harmful to others where it's harmless or even beneficial. So you have to be consistent about which body compartment you're measuring in order to extract meaning from the numbers. One place you can measure your deuterium level is in your expired air. In one test, you breathe into a test tube, and the condensation in your exhaled air gives you a good idea what your tissue fractionation level is. Another one tests your saliva. The difference between breath and saliva tells you how effectively your body is excreting deuterium. A 7–10 point difference is good. A 3–5 point difference is on the low side. Any less than that and it looks like a weakness in your system is causing you to detox deuterium slowly.

Other tests measure deuterium in the urine or blood. Still more lab tests are being developed to measure it from other body compartments. Your outermost body compartments such as saliva should be highest, while your innermost compartments like the mitochondrial matrix should have almost no deuterium at all. Deuterium level should be somewhere in-between those two extremes for middle body compartments such as the blood, lymph and interstitial space.

Interstitial space: The fluid and structural environment between blood vessels and cells.

Center for Deuterium Depletion breath and saliva deuterium guideline levels:

- **Saturated.** Their "red" zone is over 150 ppm.
- **Elevated.** Their "yellow" zone 130–150 ppm.
- **Desired.** Their "green" level is 130 ppm and below.
- **Average deuterium level** in a human today is 137–147 ppm.

Generally speaking, it gets exponentially harder to achieve and maintain a level below 130 ppm in the body; 100–75 ppm is very low to extremely low. On the other hand, the more you exceed 130 ppm, the exponentially worse off your mitochondrial fitness and overall health are likely to be. A level of 150 ppm and above is cause for concern, because that's usually a sign you have an underlying medical condition that's preventing your system from getting rid of it. That means you're retaining it through some combination of poor metabolism, excessive intake and toxic environment.

Dr. Jack even goes so far as to predict that if you have a deuterium level in your tissue above 130 ppm, you're at risk for cancer and autoimmune conditions. If it's below 130 ppm, you will not have those diseases, period. For their part, Dr. Boros and Dr. Que Collins (Center for Deuterium Depletion) do agree with the "line in the sand" concept. But they'd choose a lower number. They say diseases like these would be extremely unlikely to occur at perhaps 120 ppm or below in saliva.

Some supplements are loaded with deuterium

Experts in the field now think high deuterium levels in supplements, drugs, vitamins and health products are secretly undermining the benefits you get from these ingestible products – potentially turning interventions meant to help you into a mixed bag of some good effects and some bad. In fact, the folks at The Center for Deuterium Depletion believe deuterium level will soon be recognized as one of the most important factors to monitor and manage in determining any therapy's risk-to-reward value. They believe supplement companies will eventually need to demonstrate low deuterium levels, perhaps be certified, for their products to be considered top-grade. For their part, the USFDA is now looking into testing deuterium level as a factor in determining a pharmaceutical's toxicity and efficacy.

Depleting your deuterium level

As a general rule, unless a person under twenty years of age has chronic health problems, there's no reason to think they should proactively take steps to deplete their deuterium level. You'd expect their mitochondria to be young and in good shape. Their percent heteroplasmy rate should be low. And deuterium does support human development earlier in life. So they probably don't need to go on a deuterium-depletion program. But when you get up into your late 30s and beyond, or you start to have health problems before that, you may want to consider taking active measures to reduce your deuterium load. Our mitochondria do decline naturally with age. Gumming them up with deuterium then accelerates that process. And, after your mid-thirties, your body and mind are well past any need for maturation at that point.

Plus, our ability to deplete deuterium also declines with age. Therefore, deuterium level almost always rises over time. So depleting the body of it through daily practices, or a dedicated program, can help you be the youngest and healthiest you can be, because it keeps your mitochondria as productive as they can be. Indeed, experts in the field believe most diseases afflicting modern man today are caused by deuterium upsetting local energy production in mitochondria, which triggers clusters of adverse effects we call disease. Specifically, when our deuterium-depletion processes fail, metabolic and hormonal diseases such as cancer, Alzheimer's, dementia, diabetes, heart disease and obesity can occur.

For these reasons, most treatments and modalities (including "standards of care" in mainstream medicine) work better when you lower your deuterium level first. That's because the body then has more endogenous energy to fix itself. Same thing applies to keto and paleo diets. They work precisely because they deplete deuterium when you eat fats and proteins, which are low in deuterium – instead of sugars, grains and starches, which are high in deuterium.

Standards of care: Normal, acceptable ways to treat disease in the medical system, according to medical boards and public health agencies. When conventional doctors step outside these boundaries, they risk being sanctioned by their boards.

Your total body burden of deuterium is a function of your intake *vs.* excretion rate

Incoming exposure comes from the food, water and air sources listed in the next section. While the body gets rid of deuterium through breathing, sweating, urinating, defecating, metabolism and in reproductive cells (sperm, egg and uterine lining). That means sleep, exercise and reproduction are the three activities that deplete deuterium the fastest, as far as excretion is concerned. You burn fat when you're sleeping and in ketosis. Fat is the best macronutrient with which you make metabolic water. Metabolism increases when you exercise, which also makes metabolic water. And reproduction requires lots of deuterium to accelerate growth.

More generally speaking, it takes energy to deplete deuterium. But the funny thing is, the cycles of metabolism make ATP, and they make metabolic water. Both ATP and metabolic water deplete deuterium. But as deuterium level increases, metabolism crashes because deuterium derails steps of metabolism in glycolysis and the Krebs cycle. The stronger your mitochondria, the more energy your Krebs cycle, urea cycle and ATP synthase have to process the deuterium they encounter and make metabolic water. Unfortunately, low energy availability makes you detox deuterium more slowly in a vicious cycle, which clogs several steps in metabolism.

That means you can either deplete deuterium through a protocol first, and mitochondrial metabolism improves as a result. Or you can fix your mitochondria first, and deuterium tends to come down in-kind. You're either pushed toward greater health or functional decay, depending on your deuterium level, because water from mitochondrial metabolism is what dilutes deuterium best.

Deuterium levels in the environment and food today

- Average deuterium level at lower latitudes and elevations – both oceans and rainwater – is 145–155 ppm.
- Glacial melt water averages about 125 ppm.
- Extremely low levels from snow or spring water at high altitude is 25.
- Unsaturated fat is 110 ppm.
- Animal fat 118 ppm.
- Butter 124 ppm.
- Olive oil 130 ppm.
- Table sugar 146 ppm.
- White wheat flour 150 ppm.
- Some seed oils are as high as 250.

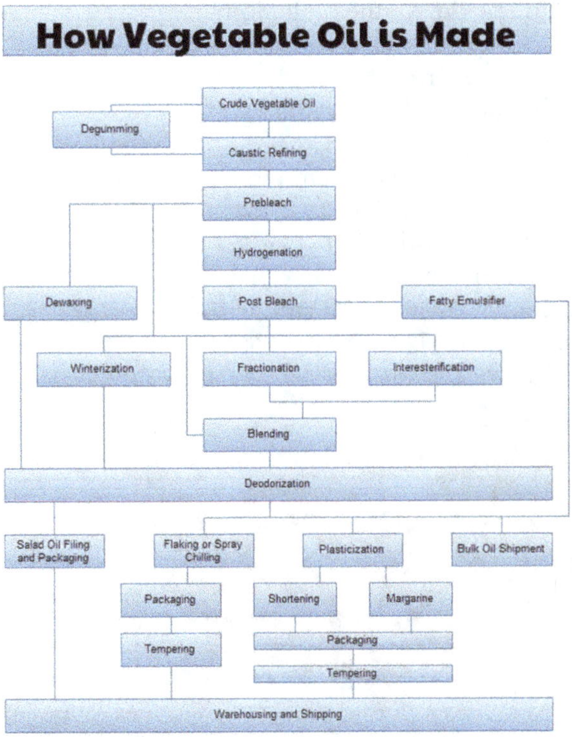

Deuterium depletion dos and don'ts

Consume natural oils and fats. Drinking a natural oil, such as olive oil (not that I'm recommending you do), is theoretically the perfect way to make "matrix" water that's low in deuterium. You can think of the technique as active dilution of deuterium. Eating animal fat such as bacon fat or lard does the same thing. **Avoid refined seed oils and carbs.** Both groups are loaded with deuterium because plants concentrate it as they're growing, and/or we process them commercially.

Eat naturally-raised animal products. They're lowest in deuterium because an animal's mitochondria make low deuterium water, which dilutes levels systemically. And their detox systems are designed to remove deuterium all the time. That's why naturally-raised animal products are an asset in lowering your deuterium level – particularly fat. In contrast, grain-fed animals are higher in deuterium because they consume more of it in their food and water. **Adopt a ketogenic diet.** Animals, like humans, have systems that deplete deuterium naturally. So a ketogenic diet high in fats and proteins helps lower your deuterium level, in addition to periods of ketotic fasting.

Reduce vegetables grown abroad. Vegetables are typically higher in deuterium than animal products because they deplete deuterium differently than animals do. Chloroplasts lower a plant's deuterium level by segregating then concentrating it in their seeds, fruits and vegetables, rather than diluting it through metabolism. Plants also need deuterium for rapid growth cycles, so they have less need to get rid of it, as an animal would when it is mature.

Avoid processed junk food. Junk food contains ingredients such as grains and sweeteners that are already high in deuterium to begin with, and we concentrate them more. You then get the worst of all worlds: (1) grains, oils and sugars that are (2) refined and concentrated; (3) grown in poor quality water; and (4) often GMO. This stacking of the biological burden is arguably the most harmful aspect of why junk food is bad for you: It gums up your mitochondria, lowers your energy level, and hastens the development of disease and dysfunction by loading you up with deuterium. **Avoid GMOs.** They're designed to grow as fast as possible, regardless of environmental conditions. That gives them more deuterium, whether on purpose or by accident.

Eat more plants grown inland. Plants grown near the coast tend to collect more deuterium, because deuterium-laden water is heavier than regular water. So it evaporates last to form clouds, then precipitates first. That's why organic produce grown in California and Florida may not be as healthy for you as conventional produce grown inland.

Avoid fruit that doesn't grow locally (at that time of year). Naturally highest of all whole foods, fruit is a virtual deuterium bomb, because Nature concentrates deuterium in a plant's fructose. And many fruits are darn sweet. That means eating imported fruits out of season can be the most fattening of all food types, because they've got

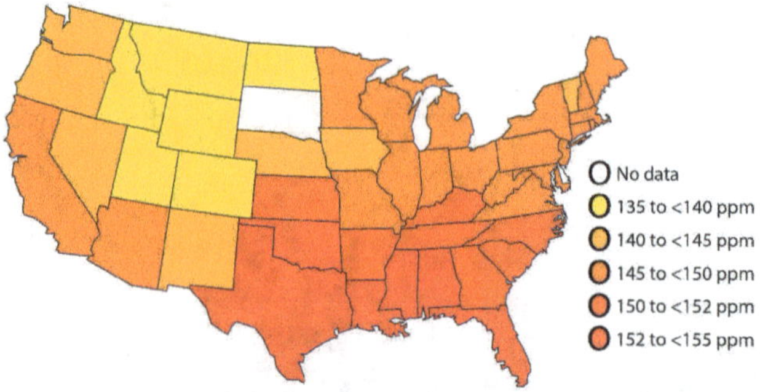

Deuterium levels across the United States

both the sugar and deuterium contributing to calorie retention. Shopping at local farmer's markets is the easiest way to make sure the produce you're eating was locally grown.

Get more sun. Sun exposure, even when it's cloudy, helps reduce your deuterium level by supporting metabolism and detoxification pathways. As examples, red light invigorates cytochromes III, IV and especially V so more ATP gets made. UV light raises melatonin levels to enhance autophagy and apoptosis. **Heat, sweating and exercise reduce deuterium through perspiration and urination.** These have the additional benefit of enabling you to harvest more light because the water that remains is easier to charge-separate. **Avoid grains when you're not getting UV light.** Carbohydrates are among the most deuterium-heavy food types. From late spring to early fall, you can get away with it, because long light cycles make your mitochondria more efficient at producing energy. In other words, strong UV light exposure helps you burn off deuterium without adverse effect.

Drink deuterium-depleted water. Deuterium-depleted water is now being used as a health supplement and therapeutic intervention. In fact, 25 ppm water is a registered drug that veterinarians in Europe use to treat cancer in dogs. So far, super-low levels (<50 ppm) are available on the Internet but expensive. **Drink water from higher latitudes and elevations.** Water from polar regions like Iceland is naturally lower in deuterium (~125–135 ppm) than water from tropical regions near the equator (~145–155 ppm). **Drink natural spring water.** Water that's low in deuterium (and lighter in weight) is the first to come out of natural springs. Check out the website "findaspring.com" to locate natural springs all over the world. **Soak in hot springs.** Water that's low in deuterium is the first to evaporate from hot springs. So the spring water itself is low in deuterium and the vapor is even lower.

Ovulation and menstruation (in women). Women have built-in deuterium-detoxing advantages due to their monthly cycles. Their bodies accumulate deuterium in the egg and endometrium (which are put there to facilitate growth). So, with every monthly cycle, they're releasing loads of deuterium. This is an unrecognized reason women tend to live longer than men. **Have more sex (men).** Deuterium collects in germ cells to support maturation of a fertilized egg. So the release of sperm is one way to deplete deuterium. **Finally,** deuterium collects in dead skin cells, so use a brush or loofah to remove deuterium-ladened cells.

Mind your melanin. Melanin is a universal (but little-known) detoxifier. It grabs hold of deuterium and escorts it out through dead skin cells or colon. Use of this route is indicative of overloaded detox pathways when employed so extensively practitioners can see melanin expressed the colon (e.g., on a colonoscopy). Use of this pathway can be problematic because it puts you at risk for running out of melanin in other places along the ectodermal tract. In other words, you may be running out of melanin elsewhere because the body only resorts to using the colon for this type of detoxification when it's desperate to get rid of deuterium, heavy metals, free radicals, and other toxins.

12

YOU ARE WHAT YOU EAT?

Food's few effects that make a big difference

Food takes its orders from physics

Natural healers have been telling us for decades that food is the root of exquisite health and a resilient body. They say what you eat determines how long you're going to live, and how healthy you're going to be in that time. But that claim is not as valid as you might think. Food is indeed important to health and healing. But it's further down the list than most realize. In fact, it's probably fifth or sixth for most modern humans – after real sun exposure, water quality, magnetism, man-made EMFs, stress level, earthing and cold exposure.

The reality is, food does not act alone in feeding the body as health gurus want us to think. Instead, physics govern how food affects the body. Light, water and magnetism control how well or poorly that food is received by your mitochondria and turned into resources either supporting your biology or sapping it of life force. Perfect example: UV light suppresses appetite and controls mitophagy, thereby enhancing energy balance and efficiency, no matter what you eat. On the other hand, food is not as good for you when consumed under conditions that conflict with Nature. To the misfortune of millions, eating deuterium-ladened fruits in winter detunes your mitochondria and contributes to weight gain.

So, basically, your environment makes food better or worse for you, depending on how true-to-Nature it is. That's why this chapter is shorter than health advocates might expect. Food (except for DHA) is not really that important, compared to the physics of life. Here are some pertinent aspects of that conversion of food into biologic function, as supported or inhibited by the physics of life:

We say young people have a "fast metabolism"

That's not entirely accurate. It's not simply that young people burn calories faster. It's more accurate to say young people can eat whatever they want and not gain weight because their mitochondria **operate more efficiently.** Their mitochondria produce more ATP from the food they eat, with less waste – meaning fat and free radicals. That means fewer calories in, and fewer calories absorbed, to get your supply of ATP.

On the other hand, as you get older, you lose mitochondrial efficiency, which means you need to eat more food, and metabolize more calories, to make the same amount of ATP. That releases more toxins through digestion. And it places greater burden on metabolic processes. For instance, high insulin takes a toll on your body. As you age, your mitochondria lose the ability to operate on poor-quality food too. Processed foods hit your detox organs harder, and they damage your mitochondria more, leading to more loss of performance. Like a car when it ages, your engines run dirtier, and you make less power. So you need more gas to run your vehicle. The difference is, cars release inefficient combustion as air pollution. Whereas, our extra calories become weight gain, toxin load, inflammation and free radicals which lead to aging, DNA damage and disease.

Carbohydrates in excess raise positive charge

As we've already discussed, carbs eaten out of season *appear* to be bad for you because, on the surface of it, carbs look as if they contribute to metabolic issues. However, we learned the real reasons carbs cause problems is not primarily the carbs themselves. It's UV deficiency creating infradian conflicts between the light signal contained in carbs and that hitting our surfaces. The dirtier combustion from carbs also produces fewer electrons compared to fat metabolism. That equates to relatively more protons, which raise positive charge through a hormone in your brain called neuropeptide Y. Unfortunately, if you don't have sufficient negative charge to offset a high-carbohydrate diet, those excess carbs increase acidity and inflammation, while decreasing redox potential.

In summary, carbs turn into bad guys when eaten without Nature's complement to carbs: UV light. Eating muffins, pizza and sugary drinks at the wrong time of year not only reduces your healing capacity by limiting your redox potential, but carbs in winter or late fall also cause your mitochondria to make less ATP, electrical charge and magnetism. This is why Nature made carbs ripen during the long-light cycles of summer to fall, when the strong sun makes up for the lower power density of carbs and all the potential problems that go with it.

But too much electron flow can uncouple your mitochondria

On the other hand, it's not a good idea to eat a carb-free diet 365 days a year because autophagy never gets activated. When you eat fats and proteins all the time – especially when you're not burning that energy as soon as it's made – your metabolic pathways naturally uncouple your mitochondria in an attempt to get you into autophagy. In other words, if you fat-burn too much, and for too long, your mitochondria will purposely turn down their efficiency in order to temporarily mimic a carb-burning state, with its benefits.

Autophagy: The controlled breakdown and replacement of damaged cellular components to keep the cell running well.

Lowering mitochondrial efficiency in this manner is an adaptive program designed to keep your power production as high as possible in the present AND your mitochondria strong long-term (when you may be forced to eat whatever you can find). So, no, it's not healthy to be ketotic all the time. You actually need cyclical periods of lower power production from carbs in order to induce autophagy to optimal effect.

In addition, some people have such low DHA, electric charge and magnetism that they can't make the free radical superoxide. This too blocks autophagy, causing your cells and systems to hobble through life on worn-out engines. This tends to show up as obesity, digestive and hormonal problems when you're younger, heart failure and neuro-degeneration when you're older.

To put this bio-program in perspective, this is one of the body's coupled systems that makes you pay a price when you eat a contrived diet that's different from what grows in your area. Yes, you're supposed to eat carbs when and where your environment grows them. But you miss out on the opportunity to freshen up your mitochondria when you eat a concocted diet year 'round that was invented by marketers to sell stuff – including keto, vegan/vegetarian, high carb/low carb, Atkins, raw, fruititarian and paleo diets. The flaw in these unnatural ways of eating is they don't take seasonal rhythms of light, electrons, protons, free radicals and autophagy into account when their authors made them up. Again, physics of Nature leads and biochemistry follows.

The happy middle ground

Carbs are Nature's way of slowing down the electron transport chain to renew mitochondria. However, they have to be eaten in a balanced fashion with fats, and they have to be eaten in season, to be of greatest benefit to your metabolism and mitochondria. Like spring cleaning for your metabolism management, carbs produce a far smaller stream of electrons through the ETC than fat does. In this colder, dirtier burn mode, mitochondria make more free radicals from burning those carbs. It's this free radical signal that triggers highly beneficial mitochondrial renewal – provided it happens in a controlled manner.

Meaning, you don't want to get stuck in this low power production mode all the time because it makes excess free radicals. Accordingly, it's good for you to eat some carbs mixed in with fats and proteins in the summertime because it: (1) reduces mitochondrial burn efficiency; (2) makes a manageable amount of free radicals from the ETC; and (3) activates autophagy. That's the ideal diet for preserving the vitality of your mitochondria: Eat the food that grows in your environment.

Why ketosis matters

Ketosis is basically the burning of fat and protein rather than carbs, because fat is more energy-dense (as electrons and protons) than carbs. Fat makes 135–147 units of ATP per unit mass, compared to the 37 of carbs. Those extra protons and electrons moving through mitochondria produce more electrical current, and a stronger magnetic field, to power cells and organ systems. All of those good things previously mentioned – blood flow, oxygen, DHA delivery and hormone balance – are raised when you're in a ketotic state.

On the other hand, carbs expand the respiratory proteins of mitochondria, making electron tunneling less efficient. That means carbs not only make you consume more calories to make ATP, but they also increase heteroplasmy rate long-term. That's one reason why excessive carbohydrate consumption erodes mitochondrial vitality and energy production over time. In summary, optimal health requires you to have high mitochondrial efficiency the majority of the time, interspersed with pulses of low mitochondrial efficiency and higher free radical formation. This combination supplies your body with the energy it needs now, and keeps your mitochondria strong long-term by inducing autophagy.

Iodine helps normalize thyroid function, fat-burning pathways, and weight

Most women today, and a lot of men, have depressed thyroid function. This "hypothyroidism," as it is called, makes it easy for you to gain weight and hard to lose it. In addition, you may experience other unwelcome effects of the thyroid's energy conservation efforts such as thin hair, brittle nails, cold extremities, dry skin and chronic fatigue.

Hypothyroidism can be caused by

- fluoride accumulation in the thyroid
- leptin resistance
- autoimmunity to your thyroid
- circadian disruption
- nnEMF exposure
- radioactive isotopes.

Regardless of cause, iodine may help you recover thyroid function. It can turn on your ability to burn fat and lose weight by **elevating thyroid hormone levels** to an optimal range. These stimulate the thyroid to optimize the body's "idle speed" and energy expenditure. Iodine also increases fat-burning by **turning on uncoupling proteins** that let mitochondria burn calories as

Image used under Creative Commons 2.5 license. Author: Dr. Dennis Cronk.

Brown fat: A specialized type of fat whose dense mitochondria populations burn it to make heat when you get cold.

free heat, instead of those electrons and protons going through default pathways such as the Krebs cycle, glycolysis and ETC. Brown fat stores are burned as free heat through this pathway.

Iodine makes you a better fat-burner by fixing your thyroid so its weight-control feature works properly. It's broken in a lot of people. At the same time, iodine can shift your metabolism *away* from the tightly-coupled energy production of an equatorial person, *toward* the heat generation programming of a Northerner, when needed. In doing so, iodine helps mitochondria turn calories directly into heat, while avoiding the usual downsides of metabolism, including weight gain and ROS. Iodine is extremely important to your endocrine hormonal system for these reasons and more. You can think of it as a secret weapon in your battle of the bulge. Get it from ocean plants such as seaweed or kelp, if you can, or supplements, as a second option.

DHA

DHA is a very special fat the body preserves and concentrates in almost all our cell membranes because of what it, and only it, can do. It's the only nutrient found in food that's truly irreplaceable. That's because DHA (docosahexaenoic acid) is the only substance that can convert light into DC electricity and back again. Far more than an ordinary omega-3 fatty acid, DHA is the only food constituent that can turn light directly into electricity in order to power our biochemical processes. No other fat can do that directly – not even the fatty acid DPA, which is almost identical in structure.

Nature has not found one suitable substitute for DHA since it first started using it 600 million years ago

DHA is essential to all members of the animal kingdom. In fact, almost every cell, in every eukaryote, uses DHA in its membranes, including humans. We actually hoard three pounds of it in our nervous system, yet it's never used as a fuel source like all other fats are. Instead, the body retains every last bit of it we can get and recycles it through the eye, which is called the "short loop" or the liver, which is the "long loop."

In contrast, a shortage of DHA starves the body of electrical charge, which uncouples light receptors such as the retinal pigment epithelium from supervisor organs like the SCN. This reduces the coherence of timing mechanisms and thus the competency of organs. For instance, circadian disruption, metabolic syndrome and infertility can be caused by a shortage of DHA. DHA deficiency causes supervisor glands like the hypothalamus to run low on electrical power, which makes its clocks run slower than subordinate glands – not good for quantum coherence.

The science behind the situation: Individual electrons don't actually travel the length of a (semi)conductor to conduct electricity. Instead, one bumps into another, which bumps into another, like a lineup of billiard balls. Individual electrons hardly move. Therefore, electrical wires – whether carbon-based and organic, or metallic and inorganic – can only carry a current when they're densely packed with electrons. And DHA, with its twenty-two carbons, has lots of electrons to carry a current.

To illustrate the value of DHA in action, consider conditions we think of as allergic reactions: eczema and allergies to seafood and eggs. They go away when you raise your DC electricity with DHA, get more sun and reduce nnEMF exposure. Which is all to say, most of the conditions we think of as diseases are completely reversible when you stop polluting your body with unnatural frequencies and live the way Nature intended. DHA plays a key role in bringing your DC current back up to full power and reestablishing hormonal communications. The biggest enemies in this fight being blue light and wireless microwaves.

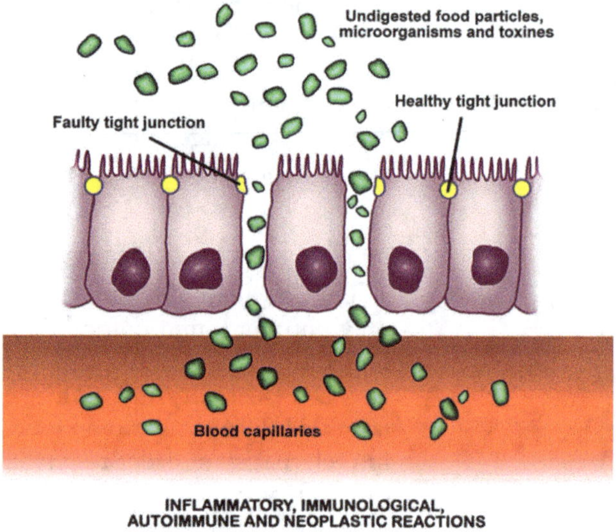

Image used under Creative Commons 4.0 license. Author: BallenaBlanca.

DHA is paramagnetic

DHA is attracted to the magnetic fields generated by electron flow in mitochondria. That means DHA is most attracted to the organs with the densest mitochondria populations and strongest magnetic fields, which are the heart, brain and hormone/endocrine organs. High energy-consuming organs like these collect more DHA from the bloodstream than relatively inactive organs, in order for their cells to convert light into electrical current and magnetic field. The more DHA a cell membrane has, the more electricity it makes, and the stronger the magnetic field around the cell and organ.

A primary cause of leaky gut is DHA deficiency

Cells of the gut lining (enterocytes) are loaded with DHA for three important purposes: (1) digestion releases photonic energy from food; (2) enterocytes only live 1–5 days (depending on who you ask); and (3) protein recycling is one of the most energy-consumptive processes the body must do. Those three factors contribute to a common need:

enterocytes require DHA to make lots of DC electricity, so they can do lots of work. DC electric current gives cells the power to break down decrepit enterocytes and tight junctions (which are like Velcro for gut cells), and build new ones in their place. And it gives peripheral clock genes in the gut the juice they need to keep accurate time to circadian rhythms. Otherwise, enterocytes don't know what time it is and when to replace themselves with new ones. Aged enterocytes then promote leaky gut, metabolic syndrome, IBS and Chron's disease.

Unfortunately, when you lack DHA, you can't make enough electricity to run the biochemistry that recycles enterocytes and tight junctions. Hence, they don't turn over as fast as our toxin load demands – with gluten, glyphosate and some prescription drugs being the biggest threats to integrity of the gut lining. That's a leading contributor to leaky gut: cells and proteins not being renewed as quickly as they're wearing from lack of electrical conductivity. DHA can help with that.

DHA is Nature's way of amplifying the solar yield up North

It does this through DHA's special ability to convert photonic energy into electrical power. Just think: fish and seafood, which are high in DHA, grow best in cold, arctic waters. At higher latitudes, sunshine is scarcer and weaker. So the most readily available food that Eskimos and Scandinavians have/had to eat in their local environment contains a substance that helps them extract more solar energy from sunlight: DHA. In effect, they were given a solar collection booster to offset the scarcity of natural sunlight where they live. Dang, Nature thought of everything.

There's good DHA, and useless DHA

In order for DHA to enter the central and peripheral nervous system, and do what it's supposed to do – which is a dense electron cloud facilitating DC electricity production – it has to have a special planar structure, which is the "SN-2" position. This flat shape makes DHA paramagnetic, giving it its quantum conversion ability. Otherwise, it can't get into nervous system cells like the brain. And it has no ability to convert sunlight into direct current.

You see, fats have three linked carbon atoms on their glycerol backbone. And the DHA found in natural fish oil can only convert sunlight into DC electricity when it's "ordered" in the middle SN-

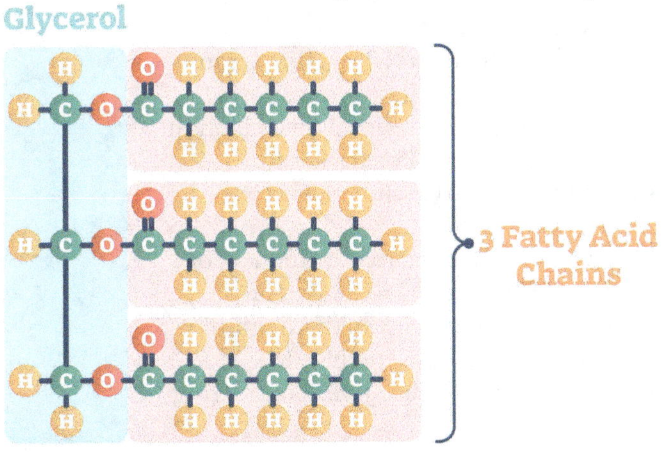

2 position of the three carbons. To a lesser degree, grass-fed beef also contains DHA in the SN-2 position. But to our disadvantage, processed fish oil supplements have their DHA mostly in the outer SN-1 or SN-3 positions, which makes most of it incapable of producing DC electricity.

The way DHA is made in nature is this: fish eat algae that contains DHA mostly in the SN-1 and SN-3 positions. Fish convert algae's DHA to the SN-2 position, where it can produce a DC electric current. Otherwise, it's useless. What's more, DHA is a polyunsaturated fat that can cause inflammation when the body is not able to use it correctly.

The takeaway: You can get a little SN-2-DHA in supplements. But you've got to make sure it's made, shipped and stored in a cold, dark environment because DHA is highly susceptible to temperature and photo-oxidation, especially in supplement form. So, if you're trying to fix some kind of health problem, the best practice is to get as much DHA as you can from oysters and cold-water fish, and not rely on supplements. Farmed fish is better than the best grass-fed beef liver. Raw and wild is best, but farmed fish is better than nothing. The body can convert 1–3% of alpha-lipoic acid (ALA) to DHA, which is not much.

Should you worry about mercury in seafood?

This question comes up a lot because seafood is the best source of DHA. No food group is higher in it. However, we've been told to limit our consumption of seafood because some species of fish are high in mercury. So, what should we be more concerned about: mercury or DHA? A: No, you don't have to worry about mercury in seafood… when your redox potential is high. On the other hand, when your redox potential is low, you need to be careful about what you eat, as you work to bring it up.

Here's the conflict: Certain species of seafood give you both mercury that toxifies, and DHA that helps you detoxify. So you can raise your redox potential with DHA. But if that effort fails to get rid of mercury faster than it's coming in, for whatever reason, neurotoxicity can be an issue. The solution is two-fold: First, stay away from things that assault mitochondria function in a big way, such as cholesterol-lowering statins, high stress or chain-smoking cigarettes. Risk factors like these can put your detox efforts in a pickle. Then, don't binge on seafood that's high in mercury but low in DHA, while your redox is low. For instance, swordfish and tuna are notoriously high in mercury, while raw oysters are a go-to source for DHA.

To illustrate these ideas in action, when Dr. Jack started eating lots of seafood, he started to notice symptoms of heavy metal toxicity. But they went away after six weeks because his redox potential went through the roof with all the things he was doing to improve his mitochondria. If anyone should have mercury poisoning it would be Dr. Jack, considering how much seafood he eats. But he doesn't – showing you mercury in seafood isn't a problem when your redox is high. On the other hand, low redox potential causes you to retain heavy metals because you don't have the positive charge in cells to push out positively-charged metals. Bottom line: You need DHA. Seafood is the best source. So seafood with moderate to low mercury levels is still better than the best grass-fed meat as a source of DHA.

PART 3

The Electromagnetic Spectrum

13

HOW FRIENDLY FREQUENCIES SUPPORT OUR BIOLOGY

**Electromagnetic frequencies native to this planet and solar system*

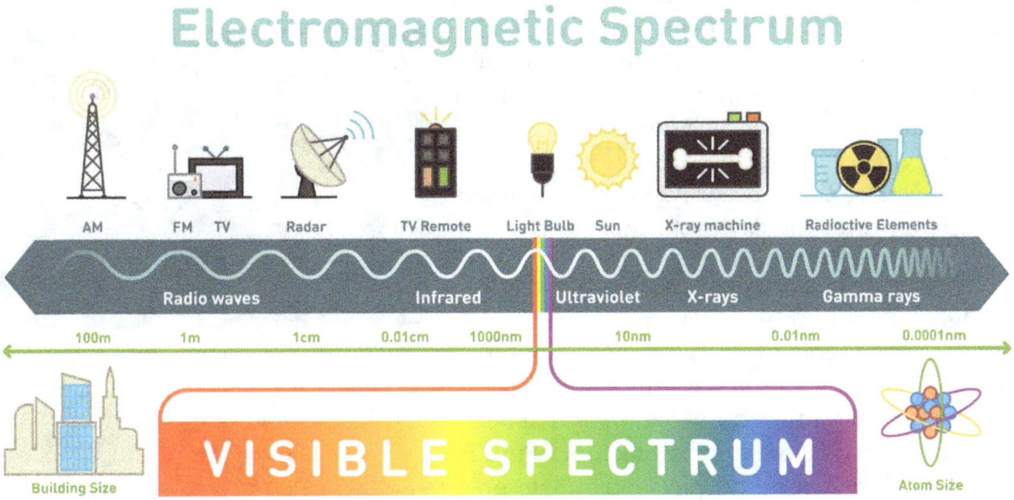

Many kinds of electromagnetism occur naturally on earth and throughout the universe. All living things use oscillating frequencies to form their structure and sustain their function. In fact, we can't live without electromagnetic fields – EMFs, as they're called – because everyone and everything in the universe is vibrational in nature. Meaning, vibration creates all matter in the universe, all forms of energy, and even carries consciousness itself.

One of the most rudimentary ways that vibrational energy supports our well-being is this: At rest, it's *frequency* that "teaches" brain and nervous system cells the vibrational rate at which they work best – similar to the way a metronome helps a musician stay in rhythm. The ultimate example is the "Schumann resonance." Oscillating at 7.83 cycles per second, it's said the Schumann resonance originates from energy waves such as lightning strikes and cosmic rays circumnavigating the globe. As the quintessential friendly frequency, Schumann's gentle electromagnetic waves are like the heartbeat of the planet. They effectively synch our vibrational frequency to the optimal rate for brain and body cells to thrive in, which happens to be the same as that of planet Earth.

Having lived in kinship with the earth for so long, we've synched our internal biorhythms to the slow, rhythmic pulses of the planet, because 7.83 Hz is our brain and body's neutral calibration point. It's what you might call the body's baseline frequency between the normal 9–12 Hz of daytime, and the 8-to-under-1 Hz of nighttime.

Two other native frequency ranges that support human biology are infrared (IR) and ultraviolet (UV). Below the visible spectrum, infrared powers cells and mitochondria. And it controls regeneration programming. Above the visible spectrum, ultraviolet light makes biochemicals, powers cellular activity, regulates biorhythms, and beneficially stresses our proteins when received in moderation. That is, UV gently oxidizes proteins such as collagen and elastin in the skin, which can be good or bad for you, depending on how well your body rebuilds the corrosion.

Let's explore the many ways light has uplifted our biology since the dawn of time.

Sunlight is far more than it appears

Sunlight allows us to see. It warms the planet. And it sets the food chain in motion via photosynthesis. But those aren't the only ways sunlight builds our reality and regulates our biology. For starters, sunlight isn't just one color, as it appears to us. It's a blend of many colors mixed together. Remember Pink Floyd's album cover for *Dark Side of the Moon*? It depicts white light going into a prism and splitting into six colors of the rainbow. Well, what you might not have noticed is that effect is actually bi-directional. White light can be split into six colors, or the six colors can be combined to make white light.

So sunlight is actually an amalgamation of colors that appear white. What's more, real sunlight's composition of colors changes over the course of a day, and over the seasons, as the sun's rays penetrate the atmosphere from different angles. Six of these colors are visible when isolated and seen independently, while two at the far ends of the spectrum are not. But, even more interesting to mitochondriacs, each color contained in sunlight controls different aspects of our biology via photoreceptors in the eyes and skin, with the help of regulatory organs such as the suprachiasmatic nucleus (SCN).

To illustrate, early morning sun in most parts of the world has all the colors of the visible rainbow – including blue light at 1800° Kelvin, 42% infrared light and no ultraviolet. This mixture of light, with its high IR content and no UV, preconditions the skin to accept the UV that comes later so you can tan without burning.

Over the next 3–4 hours, the color temperature (intensity) of the blue ramps up from 1800°K to 5700°K. This spectrum of light in particular tells your internal pharmacy to start releasing stimulatory hormones such as cortisol, and neurotransmitters like

Kelvin (light): A measurement of a light's color, ranging between 1000°K and 12,000°K or higher. Traditional incandescent light is considered "warm" at about 2700–3000°K (yellowish), color temperatures around 5500–6500°K approximate daylight (white), and modern, energy-efficient LED lights and screens are 6500–9000°K (blue).

adrenaline, to help you get through your day. When late morning rolls around, blue light begins to peak and UV-A arrives. UV-A tells your internal pharmacy to turn off hormone production.

Then, as evening turns into dusk, the termination of blue light turns off cortisol and turns on melatonin, which gets you ready for sleep and regeneration.

On the flip side, there's artificial light, which *looks* similar to real light, but is actually worlds apart in both composition and effect. In fact, the differences may seem trivial at first, but it would be fair to call manufactured light "synthetic," since the negative connotations are so fitting. To summarize the difference between real light and fake light...

Natural sunlight is like a whole food. Man-made light is like junk food

To the average person, sunlight and artificial light are the same because they look the same. But nothing could be further from the truth. In reality, natural sunlight contains a variety of ingredients (like co-factors) that need to be present to give you a healthy, balanced effect. The best example is blue light. Blue light from the sun can potentially over-stimulate the regulatory system if you get too much of it, and at the wrong times of day. As a natural antidote, the regenerative effects of red counteract the potentially harmful effects of blue. For these reasons, Nature always supplies red with blue, whether that's in sunshine or from fire.

In contrast, man-made light is like a refined food. Over the last few decades, our science and industry have learned how to make light that's composed almost entirely of single, pure frequencies because they're more energy-efficient. So now our lights and tech devices contain concentrated doses of blue, while they leave out purple and red frequencies that give you a balanced response. Basically, artificial light is like junk food that's high in sugar (the stimulus), but low in fiber (the moderator).

But the sun causes cancer, wrinkles and eye problems, right?

Well, not exactly. Dermatologists and oncologists are partly right, partly wrong, about UV because they don't understand the full story behind sun exposure. The truth is, sunlight energizes us. It regulates our biorhythms. And sunlight heals us. Our bodies are designed to be out in the sun. So anyone who thinks sun exposure is bad for you has frankly been brainwashed by false conclusions going way back, and a prevailing paradigm few think to question today.

In fact, at least eight large-scale meta-studies prove that **all-cause mortality goes down the more sun exposure you get**. Dermatologists and ophthalmologists can't explain these results. But the simple fact is, in every disease ever studied, sun exposure has been found to help people live longer, healthier lives. So not only is sunlight not bad for you, it's actually the single best promoter of energy, health, longevity and mood. We are dependent on the sun for our very existence.

However, **not everyone can capture sunlight and use it the way Nature intended** – hence all the confusion and controversy. True, you can damage your skin and eyes from too much sunlight too quickly. Indeed, the UV-B in sunlight does cause ionizing damage to skin cells (that we repair with good redox potential). And, yes, infradian mismatches can damage your mitochondria, which then cause disease. But sun exposure is just like exercise: You need to work your way up to higher intensities and longer exposure times to receive benefit without hurting yourself. You need a little knowledge and strategy to reap the rewards and avoid the risks. After all, sunlight is the strongest medicine Nature has given us. So you need to be smart about it and treat your light exposure with the same respect/reverence that ancient peoples did.

When you're sensitive to bright light

Bright sunlight hurts your eyes, and you tear up. You squint. You need to blink. You feel a strong urge to put on sunglasses. What's really happening here? A: You're not getting enough ultraviolet and infrared light. A quantum biologist will tell you that blinking covers the cornea with tears. That limits exposure and cools the surface of the eye. Cold temperature increases electrical conductivity, which helps you absorb more UV light. So, believe it or not, blinking in strong sunlight is a biophysical program hardwired into mammals to help us capture more sunlight when we're solar deficient.

Bottom line: **When your eyes are overly sensitive to sunlight that means you're solar deficient. You need more sun, not less.** Unfortunately, dark sunglasses do exactly the wrong thing. Shade is much better at helping you moderate and acclimate. However, it's not just the *quantity* of sunlight that gives you the benefits you want. It's how well your physiology and mitochondria are able to *incorporate* the sunlight you do get, which includes staying away from bad influences.

Each kind of light penetrates tissue a different depth

People are surprised to learn how deeply sunlight penetrates tissues.

Infrared: The light that penetrates deepest of all is infrared light (IR). IR penetrates soft tissues a whopping 10–30 centimeters, depending on intensity. That means it can go all the way through the skin and subcutaneous fat, reach internal organs, and charge up the water in your cells and mitochondria with electrical potential. That's a very good thing.

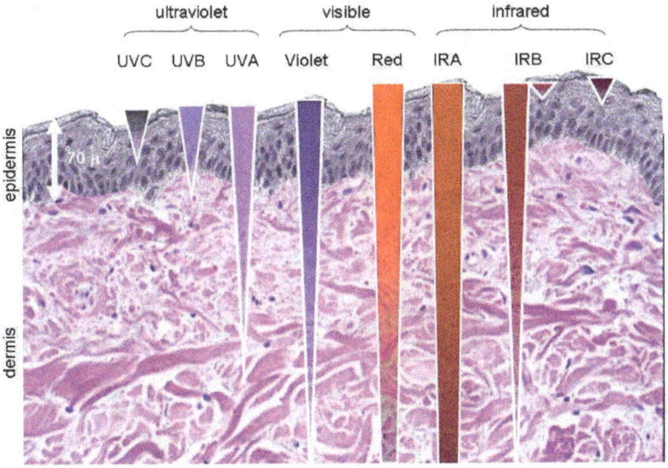

Source: Health Effects of Artificial Light. Author: Scientific Committee on Emerging and Newly Identified Health Risks.

Blue light penetrates the skin 3–6 centimeters. That's really bad news when it comes to absorbing artificial blue light from tech devices and light emitting diode lights (LEDs). For social media junkies, software developers and women in general that means the thyroid, which sits less than a centimeter below the skin surface, is bathed in blue light pretty much every waking hour from your phone and lighting. That stresses the thyroid into releasing hormones non-stop, which can turn into adrenal dysfunction, leptin resistance, weight gain, hormone imbalances, and all kinds of metabolic problems.

Ultraviolet: UV-A can only penetrate the skin 0.3 mm, UV-B only goes 0.1 mm deep, and the very limited amount of UV-C that makes it through the atmosphere penetrates the epidermis about 0.05 mm. (Although a lot more UV-C has been reaching ground level as the atmosphere is ravaged by geoengineering.)

Light produces non-linear effects in biology

Another reason we don't realize light can influence our biology so profoundly is because *linear effects* dominate the human experience. *Linear* means an effect is more or less proportional to the amount of input or stimulus you put into a system (e.g., small input, small effect; big input, big effect). Nowhere is this phenomenon better exemplified than in eating and weight loss. We're conditioned to believe we need to eat less and exercise more in order to lose weight: That is, each calorie we consume potentially makes us gain weight, while each calorie we burn potentially causes us to lose weight.

So we think radical changes are required to produce radical results. That's a linear effect. But that's not the way light affects our biology when it comes to metabolism, hormones, restoration, and daily rhythms of sleep and stress. Instead, even modest exposure to certain frequencies of light can produce sweeping effects, and cascades of effects. So even though the lens of the eye filters out 97% of UV-A and 99% of UV-B, those small doses still produce prodigious effects. For example:

- Your "internal pharmacy" of biochemicals closes for the day when UV-A shows up in late morning.
- Only 1–3 minutes of morning sun exposure is needed to set your internal clock for the day (when done consistently).
- Specific frequencies of light control some 100,000 biochemicals and their linked reactions.

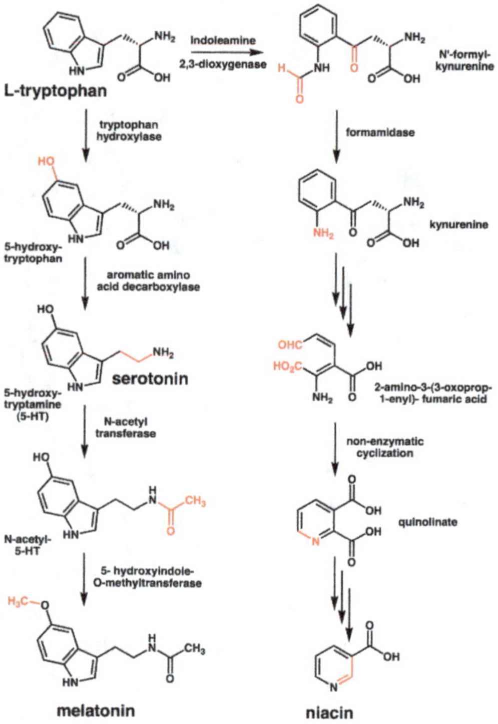

Public domain work. Author: Boghog.

- Many diseases such as type 2 diabetes get worse when eye surgeons remove cataracts and replace them with artificial lens implants that block all UV and half of the blue.

The concept to plant in your subconscious is that a minority of light's effects are indeed linear, like a tan being produced over 10 to 60 minutes of exposure. But the majority of light's effects are non-linear. Meaning, sensors in the eye, skin and gut read the light in your environment and trigger a cascade of effects as if you flipped a switch. And most of it is beyond our conscious awareness.

The body incorporates light in many ways

The human body uses some 100,000 linked biochemical reactions to run its operations. But, even more impressive than the quantity, specific frequencies of light (out of 81 x 10^{36} frequencies near the visible spectrum) control individual steps in the pathways that turn one biochemical into another (like the tryptophan pathway above). What other force in the universe has the power and range to accomplish so much with so little? Simply put, our biology uses boatloads of biochemicals to run countless processes. And it's light that controls most of the proceedings. Striking example: When sunlight hits your skin, it makes about a thousand metabolites as LDL cholesterol is turned into vitamin D.

Major pathways through which light becomes energy, matter and physiologic function

Chromophores. Originally, *chromophore* meant that part of a molecule which gives a material its color – based on the band of color frequencies absorbed and emitted. Today, mitochondriacs focus less on what a chromophore makes a material look like, and more on which color(s) enable the body to do certain things. So, to mitochondriacs, the term chromophore is not about what we see; it's about how certain color bands affect biorhythms and bio-energetics. The frequencies of light that a substance absorbs is called its "absorption spectrum."

For instance, water is the perfect red-light chromophore. Water absorbs all frequencies of red from 600 to 3000 nanometers, and circulates the conjugation around the body as the largest constituent of blood. Water also harnesses IR to build e-zone. Other examples of red light chromophores are the protein subunits of Cytochromes III, IV and V.

Made by the body, these chromophores use the power of red light to tune up the electron transport chain for maximum efficiency.

Photoreceptors. Old definition: Photoreceptors are sensors in the eye that convert light into an electrical signal for the brain to decode into pictures and information. Decades ago, biologists thought of the rods and cones of the retina as the body's only photoreceptors. But the definition has evolved since then for both researchers and mitochondriacs, as we learn more about how light runs the biology within us.

Modern, expanded definition: Photoreceptors are simply receptors for light stimuli. They are proteins that produce sensitive and specific responses to light. In other words, **photoreceptors are sensor/switches found in the eye, skin, fat and blood vessels, whose signal helps control daily and seasonal cycles.** As examples, leptin and vitamin A each absorb light and trigger chemical reactions that activate/deactivate physiologic responses. Thus, each is a mitochondriac's photoreceptor.

Opsins (a class of photoreceptor) are complex molecules composed of seven helical coils that straighten out when exposed to light. This produces a reverse type of signal by decreasing the flow of glutamate – a stimulatory neurotransmitter. Less glutamate = a stronger signal, basically. Research into opsins is exploding right now because they are frontline sensors that control our chronobiology. Examples:

- **Melanopsin.** Found in the eye, skin, fat and blood vessels, melanopsin is of great concern to mitochondriacs because of its light-sensing capabilities. However, vision is not involved. Instead, melanopsin absorbs blue light to control circadian rhythms. The problem is, when you get unfavorable blue light exposure – in source, time and duration – the weak bond between melanopsin and vitamin A breaks. Leptin signaling, mitochondria function and vitamin A function then suffer. Melanopsin is a blue light detector.
- **Neuropsin.** Discovered in 2009, neuropsin is a UV-A detector that's found in the eye and skin. It helps control biorhythms through its response to UV-A light.
- **Rhodopsin** is a pigment found in the rods of the retina that absorbs mostly green and blue light, and is responsible for black-and-white vision in low-light conditions.
- **Photopsins (aka cone opsins)** are found in the cones of the retina. They help us see colors. The L-cone opsin (aka Photopsin I) is the red-cone opsin that absorbs mostly yellowish-green frequencies. The M-cone opsin (aka Photopsin II) is the green-cone opsin whose absorption peaks in the green range. And the S-cone opsin (aka Photopsin III) is the blue-cone opsin that does most of its business with bluish-violet light.

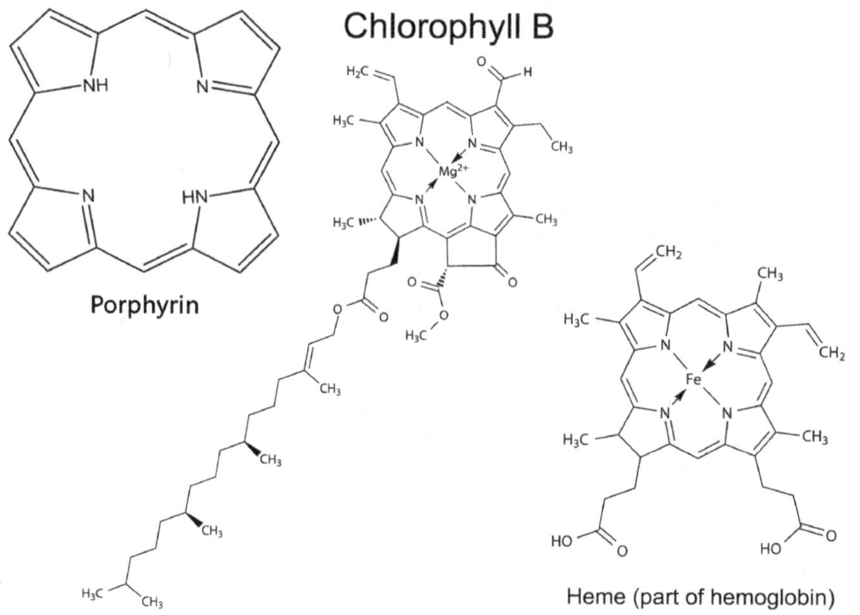

Chlorophyll B

Porphyrin

Heme (part of hemoglobin)

Porphyrins. Porphyrins are a class of molecule with a ringed structure and a spot in the middle around which more complex molecules are built in both plants and animals. In plants, the porphyrin molecule is called chlorophyll. And that space in the middle is occupied by magnesium (Mg), which captures sunlight on its electrons. Now since magnesium absorbs mostly red and blue light, while reflecting green, this gives chlorophyll and plants their green color.

The animal kingdom, including humans, employs the porphyrin structure in the heme subunits of RBCs to form the pigment hemoglobin. Hemoglobin's distinguishing feature is an iron atom (Fe) in each of its four heme cores. Fundamental to all complex life forms, it's (positively-charged) iron that enables each heme subunit to carry a (negatively-charged) O2 molecule around on red blood cells. There are 270 million hemoglobins in each RBC, and each hemoglobin can carry four O2 molecules. Interestingly enough, when hemoglobin is holding oxygen, it absorbs blue-green light, as well as yellow, while transmitting red. This is what gives oxygenated blood its bright red color in arteries.

Through the respiratory process, when the hemoglobin molecule is in an acidic environment – like when you're out of breath – the hemoglobin changes shape, which ejects the oxygen molecule. The positive charge of acidic tissue is the electromagnetic force that helps unload the negatively-charged oxygen. This is how oxygen is delivered to tissues that need it, and not to tissues that don't.

After releasing its oxygen, hemoglobin returns to its original shape, which changes its absorption spectrum to green, while transmitting red and blue light. This is what gives venous blood its burgundy color on the way back to the heart. But, even more brilliant, Nature gave hemoglobin the ability to absorb four non-visible frequencies of light we desperately need: three frequencies of IR and one frequency of UV. What are the chances that was just a random accident?

So that's how blood carries light around the body: both on the hemoglobin (porphyrin) in red blood cells, and in the chromophore that water is. Red blood cells are also shaped like two-sided satellite dishes to maximize the visible frequencies they can absorb. **These are reasons why Dr. Jack says red blood cells are ferryboats for light that wirelessly connect the sun to our mitochondria**. Last, but not least, cytochrome complexes in the ETC use porphyrin-based heme proteins to ferry electrons. Porphyrins are key purveyors of health because they absorb all frequencies of UV light.

Fluorophores. A fluorophore is an organic molecule that can absorb one frequency of light and emit it at a different wavelength – mostly in the UV range. The main molecule of respiratory complex I, for instance, is NADH. It's a fluorophore made from tryptophan. Dentin and tooth enamel are also fluorophore proteins. They fluoresce when cells around them emit certain frequencies of light.

Important biochemicals are first made in the eye. Dopamine, melatonin, serotonin and melanin are first made in the eye from aromatic amino acids and UV light. When UV light hits the retinal pigment epithelium (RPE) of the retina, it makes its dense core granules spin, which makes biochemicals such as melatonin to run sleep programming. That's why getting full-spectrum sunlight on the eyes and skin is so good at improving sleep quality. Finally, tyrosine, tryptophan and phenylalanine each have a **benzene ring** (a ring of carbons in the center of a benzene molecule) that traps photons. The ring captures light, particularly UV and IR, to program these important precursors with the light of the morning sun.

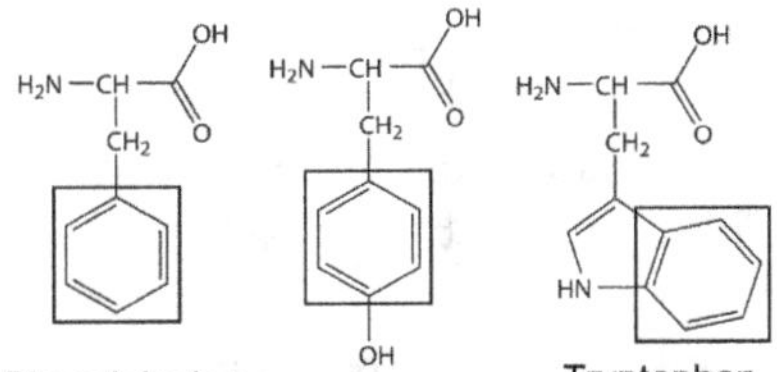

Phenylalanine Tyrosine Tryptophan

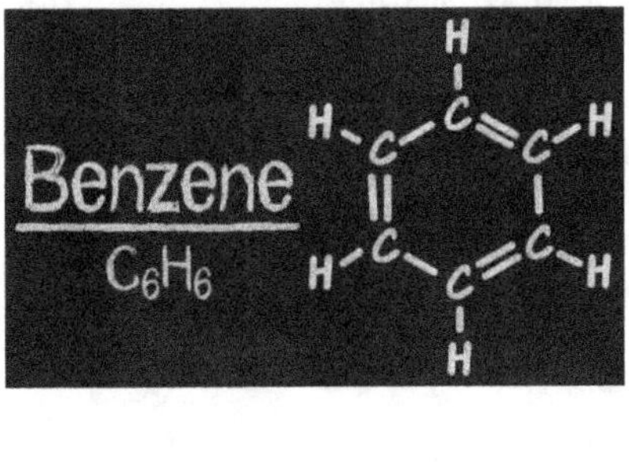

An interesting idea to reflect on: What happens when you eat plant-based supplements that are grown, not synthesized? Some say the benefits that vitamins, herbs and supplements offer are not all chemical in nature. Rather, they work partly by storing sunlight for later use. Aha! So that's why some plant-based supplements work well, while others work poorly, or not at all – even when the ingredients on the label are identical. Oftentimes, it comes down to where those ingredients were grown, and **the solar exposures contained within.**

Blue light

Blue light fundamentally has a stimulating effect on our biology. Blue frequencies in morning sun tell your pituitary it's time to wake up. It's time to get moving, use that brain of yours and get things done. It turns on the daytime portion of our circadian cycles, based on its color temperature (think brightness of color). The lighter the blue, the more stimulating it is. Serving as your internal pharmacy, the awakened pituitary coordinates the release of hormones and neurotransmitters that fuel mental and physical activity. Cortisol, dopamine, adrenaline and sex hormones such as testosterone are called into action, along with everything related to the stress response. A few hours later, UV light starts showing up to tell the pharmacy when it's time to stop dispensing for the day.

Unfortunately, modern humans are doing a great job of messing up their daily biorhythms by exposing themselves to too much blue in their technology, and for far too many hours per day. As a result, your system gets over-stimulated chronically, and you get depleted of biochemicals that run your endocrine organs and brain. Blue without red and purple basically "rev up" your adrenal engine constantly, without ever allowing it to rest and recharge.

Infrared (IR) light

As an integral part of the stimulation/activation process, Nature designed sunlight to give you equal parts blue and red, because infrared is Nature's antidote to the stimulatory effects of blue. Infrared protects your circuits from blue light overload by activating regeneration programming. After blue light reduces mitochondrial function by stretching out your respiratory proteins, the heat of infrared raises mitochondrial efficiency back up by absorbing into water and shrinking the respiratory proteins. Infrared forms e-zone, which holds electrical charge and healing capacity. IR also spins the ATP synthase faster all by itself, thereby making more ATP. And red light controls mitophagy. ...Yet more examples of how Nature's brilliance is found in its balance: blue light activates, while red light restores.

In case you were wondering, heat is the vibrational movement of atoms and molecules (kinetic). This motion emits IR light at earthly temperatures, which you see as colors on a thermal image. Very close but not the same, infrared light is an electromagnetic oscillation only. It is pure photonic energy that is not heat in itself, but rather it *can* increase the temperature of materials by making their atoms vibrate (same as microwaves). In other words, heat at earth-type temperatures makes matter emit IR frequencies. Whereas IR becomes heat only after it hits atoms and makes them jiggle.

Red light makes the ATP synthase spin faster

Red light has a superpower that makes it especially valuable in elevating your energy level, healing capacity and resistance to aging: It actually makes the fifth cytochrome in the electron transport chain spin faster, thereby **making more ATP without any electrons from food, brown fat or grounding.** In fact, the red part of the spectrum makes the ATP synthase spin twice as fast: 3,000 protons/second revving up its quantum nano rotor motor *vs.* 1,500/second normally. This means red light is a substitute for food, to some extent. It's one reason we feel charged up when we get out in the sun: Red light, whether absorbed directly through the skin, or captured and carried around in the blood, is literally increasing energy production, along with all the benefits that more ATP offers.

Red light biohacks

SaunaSpace makes shielded and grounded near-infrared home saunas (faraday saunas) and "Glow Infrared Therapy Lights" that recharge and regenerate using near-infared light, aka "photobiomodulation" (PBM). See Recommended Resources for contact info.

- **Infrared lamps.** Most modern humans have no choice but to work under LED or fluorescent lighting for many hours a day. We then go home and watch TV at night. You can neutralize much of the damage that bad light does to your mitochondria and circadian rhythms by manually adding back infrared. Simply shine an infrared (heat) lamp into your room or work-space when you're getting zapped by the blue. In addition to offsetting adverse effects of blue light we just talked about, red also fills in the gaps when your AC-powered light strobes at 50 or 60 times-a-second.
- **Red light therapy.** Many devices on the market use visible and invisible red light to give you therapeutic benefits. Some target specific areas to increase circulation (e.g., diabetic neuropathy). Others use red to improve mood. While mitochondriacs use red light devices for their mitochondrial, healing and circadian benefits. Joovv is one such company making red light devices for clinical-strength therapy.
- **Gold foil.** If you have a home sauna, you can amp up your infrared exposure by lining its walls with real gold foil. Real gold is an ideal infrared reflector, not foil that's just gold-colored.

Ultraviolet (UV) light

Ultraviolet is the most villified type of light in the modern world. It's certainly the most underappreciated, considering all it does for us. You see, most consumers believe incorrectly that UV light can only harm our skin, eyes and man-made materials (as oxidation). When, in fact, human biology has evolved to use UV light in the weirdest, most wonderful ways.

UV frequencies give us more biophysical benefit than any other by virtue of their unsurpassed energy within the friendly parts of the electromagnetic spectrum. Integral to its benefit, UV contains the most powerful wavelengths our biology can possibly tolerate before hormesis (beneficial stress) becomes irreparable injury. These frequencies help build and run the body because the more energetic the light frequencies you put into your semiconducting proteins, the more DC electricity can be produced as an output to accomplish work around the body. In particular, the biochemicals and redox potential our bodies make with UV control our thoughts and mood, power our cellular machinery, and regulate our biorhythms. Here are four ways you can feel the effects of UV, without knowing a lick of biochemistry behind how they work:

- UV light exposure gives you a natural high by making a biochemical called POMC. Beta-endorphin separates from the POMC molecule to make people feel good in the sun.
- UV naturally lowers your blood pressure by dilating blood vessels with nitric oxide.
- UV calms down the sympathetic stress response from the paraventricular nucleus (which is chronically overstimulated in most people). At the same time, it raises parasympathetic tone from the vagal motor nucleus, which controls your rest-and-digest response.
- As part of the preceding process, UV is a natural calcium-channel blocker. It reduces the stress response.

A sympathetic state is alert, active, and focused.

Parasympathetic is the state of rest, digest, and calmness.

Now you know why UV light relaxes you in friendlier ways than pharmaceuticals. But, too bad for us, over the last century, medical science has observed some of the destructive qualities of UV and proceeded to condemn it with its teachings and technology, while also ignoring its gifts. Thus, our science and industry now tell us UV light is toxic and dangerous, to be avoided at all costs.

Well, let me take this opportunity to set the record straight: ***They're wrong:*** **UV light is not bad for you.** It's simply misunderstood. Doctors, public health agencies and corporate interests don't have a clue how UV works in the body. Hence, modern society denounces and mismanages it. Clearly, we're ignorant about UV, and that's why our relationship with it is broken. Failing to realize what UV does for us, we've all but deleted UV from our lives. But that doesn't mean *you* have to going forward. Instead, start doing your homework on UV to figure out how to reap its rewards and avoid its hazards. Dispense with the false assumptions of science-past, and trust that Nature did not bless us with UV light by accident. Learn what it's designed to do in you, and for you, and you will live as biologically-empowered a life as you were meant to.

Sunglasses, contact lenses and glass block UV light

In 1969, Philip Salvatori (trustee of the Environmental Health and Light Research Institute) did an experiment that compared different types of contact lenses under different conditions. In the eye of one subject, his team put a contact lens that blocks UV light. In the other eye, they put one that lets the UV light shine through. The results shocked and amazed them.

Indoors, under artificial light with no UV, pupil dilation was the same. But, in outdoor sunlight, the pupil that was blocked from getting UV was more enlarged than the one that let the UV through. Therefore, even though we can't see UV frequencies with the naked eye, photoreceptors in the eye do, in fact, pick them up. And the pupil responds to these unseen wavelengths by constricting. So, **when we wear sunglasses or contacts that block UV, our pupils expand more than they should and let in more visible light than our eyes are meant to tolerate.**

Of course, a modern human's natural response is to reduce all light entering the eye with sunglasses. That ought to protect your eyes from overexposure, and make you look cool in the process, right? Unfortunately, this is one more instance of man underestimating Nature… and paying the price for it. What we failed to realize is that sunglasses don't fix the real problem, which is inadequate pupil constriction when we circumvent the UV signal.

No wonder so many people are light-sensitive and prone to sun damage: We've been *over*exposing our eyes to full-spectrum sun, while at the same time *under*exposing them to UV. Secretly sabotaging our health all this time, we've uncoupled *cause* from *effect* in our biology. You see, *X* amount of sunlight (UV) is supposed to constrict the pupils *Y* amount. But we've messed up our photo-regulatory systems by changing Nature's formula before knowing the rules. And now we're suffering the consequences without realizing why.

Similarly, normal window glass blocks about 75% of UV-A light and almost all UV-B. Car windshields, being multi-layered, are specially-treated to block UV-A. But side and rear windows are not. They allow most UV-A to pass through. Plastic, such as polycarbonate, is the exception: it does not block UV light. Furthermore, most sunglasses block all the UV they can, while standard contact lenses in the US block all UV-A and UV-B, as well as half the blue.

The point to remember here: When we filter the UV out of sunlight, it's actually over-exposing our eyes to all the other wavelengths – as we do indoors, for instance. Bizarre to think: **just being indoors automatically makes you blue-light toxic** the more time you spend behind glass, because window glass is blocking the UV, and some IR, while letting all the blue through. However, as we discuss elsewhere, mindlessly reducing sun exposure across the board with sunglasses, sunscreen or sun avoidance is not a sensible strategy either. Indeed, it's a health hazard with serious consequences. It's quantum malnutrition, plain and simple.

These are important reasons why our relationship with UV has become so dysfunctional: (1) We purposely omit UV in all our *artificial* sources; (2) we block it in our *natural* sources as if were poisonous; and (3) we're over-exposing our eyes to other frequencies when we do block UV. As a result, our biology is *uncoupled* and *out of balance* with our light environment. These are unrecognized reasons why we're so sick and getting sicker.

Sun *exposure* is different from sun *damage*

Sun exposure can make your skin tone look more uneven short-term – some would say "older" looking – because it exaggerates pigment level variation, freckle darkening and dead skin layer formation. To the uneducated, irregular skin tone after being out in the sun might appear to be premature aging. But this is not necessarily the case. Rather, it is use and conditioning. These are just Nature's ways of helping you collect more sunlight from your solar panels – UV in particular. It is part of a comprehensive natural system that helps humans incorporate more light into our physiology.

The net result on a population basis is, mortality from all diseases declines the more sun exposure you get. That includes skin cancer: It goes down when you get more sun. Think of it like this: If you cover a car and let it sit for thirty years, it might still look fabulous on the outside when you take it out and give it a wash. But if you value life over looks, it's the condition on the inside that counts. For in that time, its internal parts will have rusted, rotted and seized from neglect. The engine and transmission will actually be worse off mechanically than if you drove it every once in a while.

Well, you and your skin are no different. Use your solar panels regularly and you may see that they've been used. However, your organs, tissues and mitochondria will be in better working order than if you had avoided the sun at every opportunity. Fix your relationship with light, and your skin will naturally look better long term. Consider getting early-morning sun on the eyes and skin, unblocking UV and IR in your life, drinking good-quality water, avoiding blue light at night, protecting yourself from nnEMFs, eating deuterium-depleting foods, getting more sulfur in your diet, and doing all the things you can to tune up your mitochondria.

All that being said, wrinkles, dryness, fragility and uneven texture are indeed signs that skin has lost health. See the section on "Bad Skin," Chapter 18 (page 333), to learn more about skin problems.

Primer for UV assimilation

At the top of the bio-friendly spectrum, UV light contains the most energy. It's the most potent of the bunch, in good effects and bad. So our biology developed techniques to modulate our UV absorption. When you need less UV, Nature first created dark skin. Dead skin cells on the top layer are another UV-protection factor. They are a natural sunblock.

When you need more UV, and photoreceptors sense it on the skin, the circadian system releases nitric oxide. Nitric oxide dilates blood vessels to bring more light-capturing blood to the surface. Since UV only penetrates a fraction of a millimeter, the body brings RBCs containing hemoglobin to the surface to deliver one wavelength of UV and three of infrared to cells around the body. Similarly, after eating, extra blood is directed to the gut to extract the light that microbes liberate from food.

Along the same lines, you know how very light-skinned people get red after being out in the sun a short while? Everyone thinks that's a sunburn, but it's not. It's actually the body bringing red blood cells up to the surface to help harvest UV light. The process just persists longest in fair-skinned people as an extra effort to capture more sun. This is why anemics have trouble in this area: A deficiency of red blood cells (hemoglobin) slows down the transit of UV away from the area. Likewise, the reason Irish people have more freckles than any other is that melanin in their freckles help them harvest what little UV light they get in their cloudy climate. Melanin is a storage protein for UV light. It absorbs UV light during the day. Then, at night, the body offloads the stored light to porphyrins in red blood cells.

How much UV is available where you live?

People living at higher latitudes are seasonally disadvantaged when it comes to collecting UV because, in some places, there might not be any of it reaching ground level in the shortened light cycles of fall and winter. In fact, the angle that sunlight travels through the atmosphere can block all UV in the darkest months, at the highest latitudes.

But you may be able to hack your way around that seasonal limitation.

- Higher elevations let more UV light through. So skiing on a mountain lets vacationers offset their UV deprivation in winter.
- Aluminum foil is a perfect reflector of UV light (picture old face-tan reflectors). Yet another practice that might look silly to the uninitiated but will, in fact, give you the last laugh when you escape the diseases of civilization that those around you come down with.
- Cloudy days are no excuse to skip your daily dose of sun. Clouds do substantially reduce solar yield, but a lot of sun still makes it through. And a little UV can go a long way, since light acts in a non-linear fashion with our biology.

8 tips to better harvest sunlight

- **Sun Secret #1 – Build your solar callus.** Infrared-A in the morning prepares your skin to receive UV later in the day. Early-morning sun contains 41% IR light that increases tolerance to UV-B rays that can cause sunburn. There's lots of IR in late-evening sun too. Even extremely pale people of Irish descent can tolerate many hours of intense sun exposure without burning... after they've worked their way up to high solar yield through "hybrid tanning," as Dr. Alexander Wunsch and Dr. Jack calls this IR pre-conditioning strategy.

But what if you can't get early-morning sun? Get an infrared-A therapy light. It's not as good as full-spectrum sun, but it will pre-condition your skin indoors to receive more UV light outdoors. Carbs are meant to do something similar. As one of Nature's energy-balancing mechanisms, carbs naturally thicken the skin to help you capture more summer sun. However, eating more carbs than you should just to get this effect (whether in season or not) is not recommended for reasons mentioned throughout.

- **Sun Secret #2 – Adequate hydration.** When your skin is dehydrated you can't make vitamin D. That prevents you from harvesting sunlight appropriately. The four biggest causes of dehydration are microwaves and blue light, low fluid intake, diuretic beverages such as coffee, tea, soft drinks and alcohol, and weak mitochondria.
- **Sun Secret #3 – Sulfur.** Sulfur helps us capture good frequencies of sunlight, while reflecting the bad. For instance, sulfur helps us collect IR and UV light close to the visible range, while limiting our exposure to IR and UV extremes like UV-C. Unfortunately, cholesterol, vitamin D3 and DHEA don't have as much sulfur as they used to.
- **Sun Secret #4 – Avoid polyunsaturated omega-6 fatty acid (linoleic acid).** Most health educators know that too much omega-6 vegetable oil is bad for you – including seed oils of corn, safflower, sunflower, cottonseed, soybean and canola. But most don't know just how bad they are at increasing inflammation and oxidation in skin cells, predisposing you to sunburn. Dr. Joe Mercola believes linoleic acid is far worse for you than sugar. He says, it's much harder to get sunburned when you don't eat these vegetable oils. Dr. Kruse says linoleic acid does contribute to inflammation and sunburn, but it doesn't play as big a role as Dr. Mercola posits.
- **Sun Secret #5 – Natural sunblock.** Dead cells on the top layer of skin unwind their DNA, which acts as a natural protectant against overexposure.
- **Sun Secret #6 – Avoid chemical sun blocks.** Avoid enemies of your biophysical state such as sunscreen. It blocks the UV you desperately need. Same thing with most types of makeup: They're designed to block UV in the hopes it will reduce the signs of aging. Unfortunately, by avoiding all UV-induced wear-and-tear on your

skin, you also depress cell repair and replacement. Just as bad, most sunscreen contains chemicals that are toxic or even cancer-causing.

- **Sun Secret #7 – Avoid UV-blocking materials.** Just because you can see through it does not mean you're receiving full-spectrum sunlight through it. So look for ways to remove the presence of any material that alters the light spectrum your eyes receive. Here's one for the road: When you're driving, at least *crack* a window, and/or open your sunroof. That's enough to let some sun in.
- **Sun Secret #8 – More electrons.** The more electrons you have coming in, and the more you retain, the better you're able to capture and use sunlight. Remember, light can only interact with electrons.
- **Mitochondriac's Sun Secret #9 – Dermal pooling.** When skin is exposed to direct sunlight, vessels dilate to increase the surface area of your solar panels. Indeed, 60% of total blood volume is deployed to the front lines to capture as much sunlight as possible through the body's technique of "dermal pooling," as Wunsch calls it.

Set a goal for yourself: raise your melatonin level with more UV-A and infrared-A light – particularly from morning sun. Melatonin repairs the damage that every photoreceptor in the eye incurs through daily use. And melatonin recycles mitochondria, which protects circadian clock functions of the eye and mitochondria.

Sun-permeable bathing suits and clothing

For those who have areas of the body "where the sun never shines," remember, those areas suffer energy deficits from a lack of sun exposure just like other body parts. UV- and IR-deficiency basically forces those organs and tissues to use energy collected through other means, instead of those cells and mitochondria harvesting energy directly themselves. Second-hand supply means fewer frequencies, and less of them. To put it simply, solar malnutrition of your privates does contribute to reproductive dysfunction (e.g., erectile dysfunction, prostate problems, lack of libido, and infertility in both sexes). And remediating that deficiency can help them get their mojo back.

Help from the mitohacker's toolbox: A number of swimwear manufacturers such as Kiniki® and CoolTan® now make UV-permeable bathing suits and clothing that allow a portion of UV and IR rays to go right through them. This "tan-through" clothing is a great solution for the more bashful among us that badly need the UV and IR healing frequencies, but don't want to be cruising around with our privates out in public.

Fall and winter sunbathing, made more pleasant

Another challenge that Northerners have in getting adequate sun exposure is the cold. Many Northern cities do get direct sunlight in the fall and winter that could be used for our biophysical benefit. But many people find it too cold to go shirtless and pantless below 60°F.

Consider a Biomat™ far-infrared heating pad to help you stay comfortable when it's cold out. The Biomat™ is filled with electrically-heated amethyst crystals and activated carbon that give you relaxing and therapeutic far-infrared rays which feel truly heavenly. Like any electrical device, the Biomat™ does create an electric field when powered on. But the manufacturer has designed it with a built-in grounding mechanism to minimize harmful effects. To reduce that to zero, you can always heat it up beforehand, and turn it off just prior to use.

When you lay on the mat, the heat from below will help your backside stay warm, while your front side is warmed from the sun. When your top side gets cold, flip yourself over. Like other minerals, the crystals are dense enough to retain heat for fifteen minutes or so. Infrared heat lamps and propane-powered patio heaters are other ways to eliminate excuses from your regimen. Whichever way you can manage to get full-spectrum sun on your skin and eyes, do it – be that electrically-powered infrared light bulbs, or propane-powered heat/light.

See the Recommended Resources section at the end for more information about the Biomat.

To summarize our connection to light

We need sunlight as much as plants do. We just have more ways to get it, and better backup systems to cope with its absence for longer. However, a deficiency of real, natural sunlight hitting our eyes and skin is still increasingly detrimental to our energy level, health and healing the deeper the deficit goes. We've just never realized its significance before, because we've been misguided into believing so many half-true, partially-effective paradigms for so long, instead of where the real fault and solution lies: in the quality of our light, water, magnetism and mitochondria.

John Ott said it best, based on his research from the 1930s onward: A shortage of certain frequencies causes a deficiency of biochemicals and hormones in both plants and animals that he described as a state of "mal-illumination" comparable to malnutrition. This deficiency of natural light is about half the story of why we're so sick and tired, and mentally unwell today. It's a huge, previously untold story that Dr. Jack Kruse and fellow mitochondriacs are piecing together as we speak – and telling everyone who will listen. The other half of the story is how man-made, foreign frequencies foul up our energy production, our circadian and infradian rhythms, and our biochemistry. That's coming up in the next episode.

John Ott, researcher and author. As detailed in his book, Health and Light, *Ott developed time-lapse photography techniques to study how light affects plant growth, in addition to how their chloroplasts utilize light in photosynthesis, and how children became hyperactive when exposed to artificial light in classrooms.*

14

HOW MAN-MADE ELECTROMAGNETIC FREQUENCIES HURT YOU

Earth used to be a nice, quiet place to live, energetically speaking

Whatever solar radiation the earth received was moderated or virtually eliminated by the atmosphere before it got to us. That means terrestrial biology, having evolved alongside these frequencies for eons, has the ability to use these waves as energy and information. So if an electromagnetic frequency was found on earth a thousand years ago, we can use it for our benefit, or at least tolerate it. That made the earth a sanctuary that let in good solar frequencies that helped us, while keeping out bad energies that hurt us. This, more than anything in the terrestrial experience, made life on earth possible. Sunshine through the earth's atmosphere pampered us with friendly frequencies, while protecting us from alien electromagnetic fields intrusive to our experience. That was life on earth, in a nutshell.

However, the message within the message here is that our biology is not built to tolerate frequencies and intensities that were not present on ancient earth. If we didn't grow up around it, our biology is not designed to be around it. As a result, foreign frequencies are stressful at best to torturous over time. And that's exactly the situation we find ourselves in today. We've contaminated our once-pristine electromagnetic environment to the point where more microwave EMFs are trying to get out of the earth's atmosphere, and can't, than those trying to get in.

Man-made frequencies have turned a supremely nurturing environment into hostile territory, antagonistic to all biology. We're now swimming in a sea of invisible, unhealthy, non-native electromagnetic radiation that clashes with the Schumann resonance and overpowers it. This deluge of faster, more intense man-made EMFs interferes with mitochondrial function in mischievous ways. And it corrupts our brain and body's electrical system through effects we can't imagined.

Many think of radio and microwaves as "electromagnetic pollution"

Tech devices bombard our cells with harmful frequencies that conflict with those native to our planet and beneficial to our biology. Using frequencies in the millions or billions of cycles per second (megahertz {MHz} or gigahertz {GHz}, respectively), our cell phones, Wi-Fi networks and smart devices use frequency ranges radically faster and more powerful than those our brain and body are meant to receive.

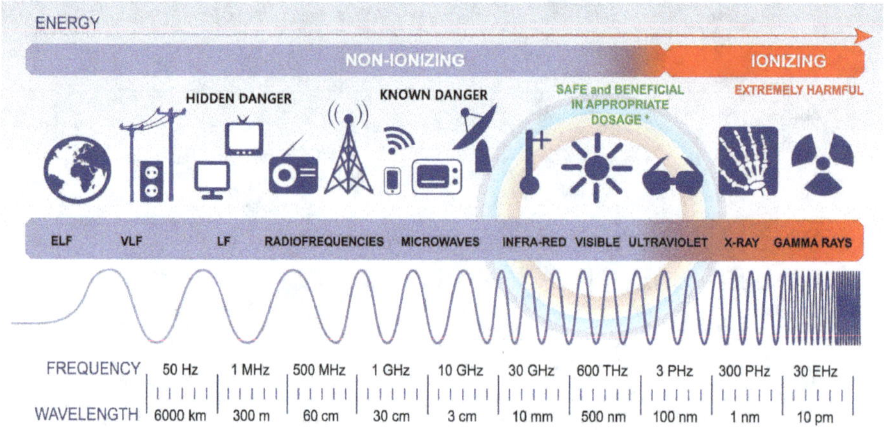

Take "smart meters," for example. They're an almost ubiquitous health hazards now toxifying our airwaves. Digital power meters wirelessly report electricity usage from your home back to the power company. And they do it in the microwave/radio-frequency range – supposedly to help you use electricity more wisely. But what they really do is collect data about you and your family so they can sell that information to marketers and government agencies – including which devices you use, when you use them, and even what TV shows you're watching (based on the waveform of electricity used).

Privacy issues aside, probably the most harmful aspect of having a smart meter on every home and business is that each one transmits microwaves to and from a relay station many times per hour. That means in populated areas you could have hundreds of these things transmitting microwaves through you every few minutes, if not more. This chronic bombardment of high-energy EMFs undermines human health by drowning out the body's own electrical signals which are much weaker. Non-native EMFs basically spoil your internal environment by making your cells vibrate to an alien tune, instead of Nature's own supportive frequencies.

But even worse are cell phones due to their proximity. Cell phone signals agitate cells all over the body with their microwave radiation – especially the brain when you hold the phone against your head, or use a Bluetooth headset. This can disturb your sleep and concentration. It's known to cause headaches, fatigue and hypersensitivity to all nnEMFs. And cell phone microwaves can depress your immune system, lower fertility, and even cause brain tumors. And we aren't alone. Lower life forms are even more profoundly affected by nnEMFs because their biology is simpler, and more closely attuned to nature. The metabolism of bees, for example, is spoiled by nnEMFs so they have great difficulty living near transmitters. Non-native EMFs stunt plant growth, particularly trees. And nnEMFs aggravate our gut flora too.

But, despite decades of data to the contrary, tech companies still claim that modern communication devices are not dangerous because the EMFs they emit aren't strong enough to heat tissue, or shake electrons loose the way radioactivity does. At the same time, corporate executives and engineers acknowledge behind closed doors the subtle health dangers that *non-ionizing* EMFs present to all life forms. They're just keeping it quiet as long as possible because wireless technology is driving most of the "progress" and profit in commerce today… so much so that experts on both sides of the argument now consider the crowding of our airwaves unstoppable, no matter how bad it is for people and planet.

So, as a smokescreen, industry spokespeople argue that microwaves are a natural phenomenon… that we shouldn't promote fear-mongering because the environment has always borne microwaves throughout earth's existence. And they would be right. But that's not the whole truth. You see, the part they neglect to mention is that total microwave radiation our biology must withstand today, compared to 1950, is… wait for it… **a quintillion times what it was back then!** And it's expanding rapidly with nothing but awareness efforts like this to slow the spread.

Bottom line: nnEMFs are far worse for you than you might imagine. Yet it can be difficult for the average person to trace his symptoms back to their source. That's because the mitochondriac community is pretty much the only group of people actively studying and using biophysics as a frontline modality to combat modern disease. When you learn for yourself the full extent of the damage that electrosmog inflicts upon people, you'd be shocked how little is being done to protect human health and safety, despite decades' worth of evidence of harm. That means you're going to have to take steps to protect yourself and your loved ones, because no government organization or mainstream doctor is going to do it for you.

Alternating current (AC) electricity

Your average American doesn't believe electric fields harm their health as much as they do because they never heard it from "the experts." But awareness is spreading as more people come down with chronic diseases and realize nnEMFs could be to blame. Without a doubt, stray voltage around wiring, as well as fields around the devices, are more harmful to human health than we ever thought. It just didn't pose great threat before Wi-Fi and 4G because most people had healing capacity to spare. Thus, we could tolerate environmental stress without showing overt symptoms.

Unfortunately for us, alternating current in our walls and devices disturbs the physics of the body. In particular, America's 60 Hz electricity messes with mitochondrial function below your conscious awareness. That makes obesity, neuro-degeneration, and cancer more likely to occur. While on the other side of the pond, Europe's 50 Hz 230 V power causes more electro-hypersensitivity to nnEMFs. Even low-voltage direct current from battery-powered devices is damaging when used close to the body. Continuous current is not nearly as bad as oscillating current. But it does indeed disturb biology – especially since portable devices are often used inches from sensitive organs.

Nikola Tesla carries a lamp a few meters from the generator, but it continues to shine! Author: Napoleon Sarony, circa 1898.

EMF injury started with Tesla's electric power

The story of unfriendly frequencies stressing our biology started in the 1890s when Nikola Tesla invented alternating current (AC) electricity. Now, no one would argue with the fact that Tesla's AC current is technically superior to Edison's DC current at powering our electric grid. However, if we take a moment to consider the health effects of alternating current, we see downsides we never noticed.

As a "foundational disturbance" of sorts, alternating current upsets biologic systems more than DC, because AC oscillates faster than the frequencies at which our bodies recharge and regenerate. In contrast, direct current operates on a continuous flow of electrons that's far less disturbing to the body's electrical system. Of course, few paid any attention to the health effects of these frequencies through the early 20th Century, because the cumulative dose was small enough that most could cope with insults and not feel compromised. AC power just wasn't a huge issue across the population.

But that started to change when radio, television and radar signals began filling our airwaves. In the ensuing decades, we found every imaginable use for electricity in our lives. So now, our obsession with technology has led gadget-makers to turn to wireless as society's mega trend to sell stimulation-seekers lots of new stuff. Certainly, battery power helped get the wireless party started. And artificial intelligence is about to raise the roof on the saturation of our airwaves – in self-driving cars and smart cities, to name two applications. All of that is rapidly depleting whatever regenerative resources we have left.

"Dirty electricity" causes most of the harm

Even worse than "clean" electricity, morbidity risk rises when irregular high-frequency oscillations piggyback on the primary 60 Hz waveform – as it usually does in our power grid. Looking somewhat like fur on a camel's back, most 60 Hz power supplied to our homes can aptly be called "dirty," because variations in the waveform are *the* most harmful aspect of electricity to our cells and systems. Dirty electricity, which used to be called "electromagnetic interference" (EMI) or "high-frequency voltage transients" (HFVT), is caused by:

- irregularities in how electricity is generated at the power plant
- appliances drawing power from shared lines (mostly power supplies)
- airborne nnEMFs brought into the home through power lines.

Image used courtesy of Dave Stetzer of stetzerelectric.com.

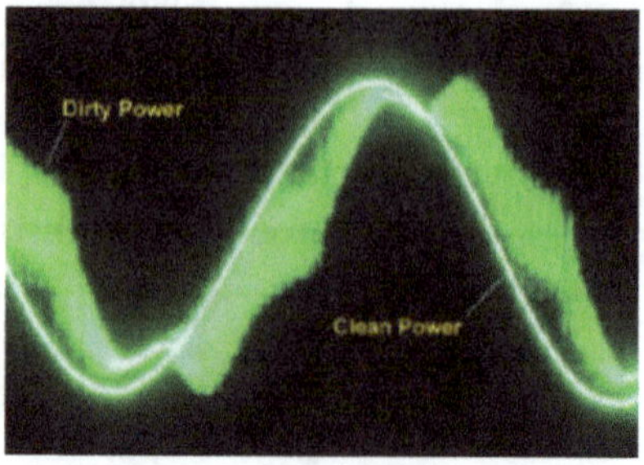

Disturbances like these form tiny anomalies on the waveform which sound like crackling noise when heard over a speaker. The result being exaggerated wear-and-tear on your appliances, electronics, cells and mitochondria. Like countless tiny speed bumps, dirty electricity violently shakes things up, making them and you break down faster. Basically, any sort of intermediate frequency noise between about 60–2000+ Hz can infiltrate power lines and travel downstream to your outlets or upstream to "infect" your home's circuits. It's this dirty electricity that contributes to in-home EMF injury – once fairly uncommon years ago, but all-too common today. Without a doubt, dirty electricity has become a much bigger threat to the population now that we all must endure wireless stressors irritating our anatomy virtually non-stop.

Common sources of dirty electricity

- compact fluorescent light bulbs
- smart meters
- dimmer switches on LED lights (they delete portions of the waveform without replacing them, resulting in abrupt on/off transitions)
- household appliances such as microwave ovens, refrigerators, plasma TVs and computer power supplies
- variable speed HVAC, well and swimming pool pumps
- solar power, wind turbines, cell phone towers
- electric vehicle battery chargers
- home wiring errors
- even trees brushing up against power lines.

Dirty electricity remediation

For under $200, you can buy a meter and see for yourself how bad your home's dirty electricity is. You can hire a specialist to give you a more thorough analysis. Or you can start off by seeing how dirty electricity actually affects you. Simply turn off circuit breakers at night and see if your symptoms go away (assuming you have rooms NOT affected by neighbors' electric fields). Once you know your threat level, most homes can be cleaned up fairly easily by installing filters in your outlets or at the breaker panel. You just plug handheld boxes into strategically-chosen outlets around your home.

These filter out extraneous high-frequency noise emanating from lines, the devices plugged into those lines, and ambient frequencies streaming through your home's circuitry. This effectively eliminates the majority of stress-causing noise that makes electricity harmful to you, your pets, and your devices. For best results, hire a nnEMF remediation specialist, because there can be intricacies involved both in cleaning up the electropollution and in how the signal impacts your health, healing ability, and cognitive function. Specialists often find threats you missed.

How AC power corrupts healing and regeneration

The fields around AC power shake our cells at rates and intensities that conflict with the brain's own regenerative frequencies of between 12 Hz to under 1 Hz. In doing so, electricity can overpower the body's own message frequencies because cells then can't hear the more subtle signals the brain sends to organs and tissue. As we examined, when the AC signal is stronger than the brain's outgoing message frequencies, stem cells don't know where to go, and what to turn into, when they mature. That interferes with cell signaling for repair and replacement. At the same time, the brain doesn't receive accurate information about the needs of tissues around the body that normally report their status back to the brain to call for service. Astrocytes are then left in the dark about what cells need.

Plus, in half the countries around the world, 50 Hz electricity oscillates at a frequency that closely competes with the mitochondria's optimal fat- and protein-burning frequencies of 100 Hz. That's troublesome for energy production in mitochondria, resulting in more free radicals, less ATP, and greater vulnerability to nnEMFs due to lower healing capacity. In this state, your defenses such as redox, magnetism and glutathione are weakened, so foreign frequencies can cause you problems as electro-hypersensitivity. In short, AC power fields surrounding your wires, walls and devices shift your metabolism in the wrong direction and diminish your redox reserves. Daily exposures then become more serious threats to your regenerative capacity. And, let's not forget, electricity does not act alone any longer. The stress that AC power places on the body is cumulative with all the other non-native frequencies we use. Future generations of wireless technology then take that damage to another level.

The ETC burns fat best when cytochromes oscillate at 100 Hz

Electrons from carbs tend to enter the electron transport chain at Cytochrome I, while most fat and protein electrons enter the ETC at Cytochrome II or III. The thing is, Cytochromes II and III need to oscillate at 100 Hz to work best. That means 100 Hz, which is the second harmonic of 50 Hz (power), messes with mitochondria's ability to burn fats and proteins by screwing up the middle portions of the electron transport chain with static interference. Hence, throughput of the ETC gets restricted, the body resorts to less-efficient processes such as glycolysis for energy, and symptoms of low energy output develop.

On the other hand, you could say that 60 Hz power helps Americans over-indulge in carbohydrate consumption, because 60 Hz electricity interferes less with Cytochrome I and the ETC as a limited harmonic of 100 Hz (60 x 5 = 300 = 100 x 3). However, we disturb our mitochondrial efficiency through other means – such as too much deuterium from processed foods, more GMOs, eating foods out of season, dirty electricity, and blue light stretching out the respiratory proteins.

Most of Europe uses the GSM standard of cell phone service, while the majority of US cell service is CDMA, with the minority being GSM.

The point to make here is that the two paths of mitochondrial corruption make two different sets of free radicals, leading to different symptom clusters. Europeans, with their 50 Hz 230 volt power, and GSM cellular standard, experience more electro-hypersensitivity. On the other hand, Americans with 60 Hz power, and predominantly CDMA cell service *combined with* GSM, tend to be heavier. We also have more cancer and cognitive decline as Alzheimer's, Parkinson's and dementia.

Home wiring errors can immerse you in bipolar magnetic fields

In America's 60 Hz supply wiring (called "mains" electricity in the UK), the "hot" and neutral wires each create a magnetic field that cancel out each other when side by side. But when the wires get separated, the magnetic field between them can increase 17-fold. This can create a hazardous situation whereby the field from a wall in front of you interacts with the field from a wall behind you, creating a bipolar magnetic field that varies throughout the room.

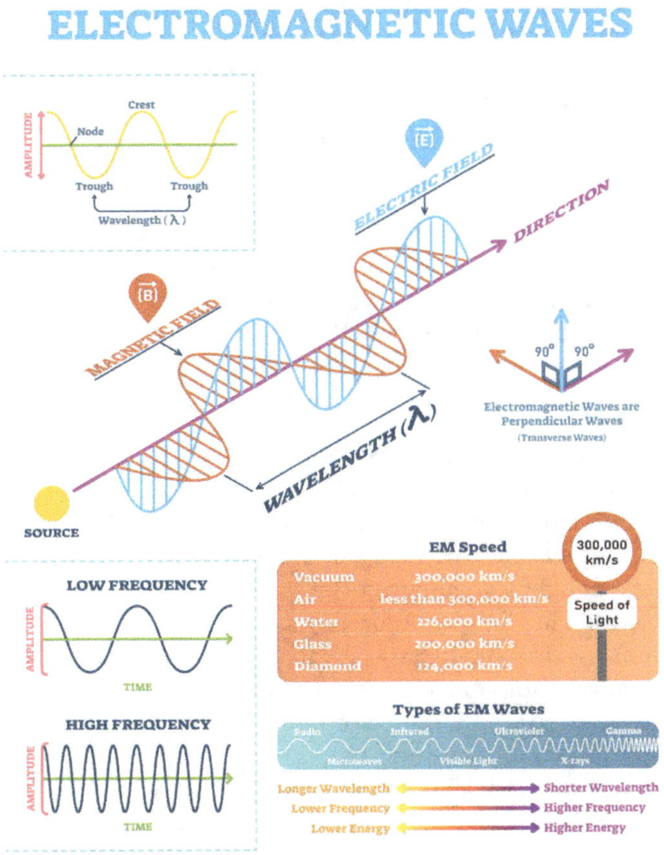

That's terrible for you because *stable* bipolar magnetism is bad enough. It can set off an emergency response in the body that can help temporarily, until the brain runs out of chi/vitality. And it can also decimate your mitochondrial energy production, hormone levels, enzyme activity, and overall biochemistry, as well as DNA replication. But in a *fluctuating or pulsed* magnetic field, the mix of polarities and intensities creates even more chaos in the body.

That means something as simple as moving from one side of your couch to the other can change the polarity and intensity of magnetism running through your body from what it was just moments before. It's these unstable, stratified zones of magnetism that do serious damage to your organs and tissues by disconnecting them from one another. Organs and tissues then operate out of synch with each another as they struggle with varying degrees of function to dysfunction. That can easily lead to fibromyalgia, cancer, autoimmunity or neuro-degeneration after your astrocytes run out of electricity to recharge cells.

For your information, "live," unshielded electrical wires produce an *electric* field around them, whether current is being used or not. That's not good for you. But AC lines produce a *magnetic* field around them only when current is running through the wire. And that's horrible for you. You can think of the danger posed by *electric* fields as a function of proximity times voltage in the line – whether the line is being used or not. High voltage is why kids have developed leukemia living under high-tension power lines. On the other hand, the health risks associated with non-uniform *magnetic* fields rise in relation to current running through the line and proximity. It's the flow of electrons that produces a magnetic field at a 90° angle.

Ground current

Stray electricity in the earth beneath your feet is called "ground current." It's caused by improper wiring in buildings, poorly-designed recovery of electrons from the power grid, and a buildup of static electricity from 5G. First researched as a result of unhappy cows, ground current was shown to lower milk production, cause foot sores, and trigger miscarriages when cows basically got mildly electrocuted non-stop just standing over stray current. Ground current is a major reason it's not safe for people to earth themselves in some populated areas anymore. There are just too many places that have excessive stray voltage to ground yourself safely. You can test for current in the ground by measuring magnetic field with a very sensitive gauss meter (sensitive to at least 0.01 milligauss {mG}). A safe level is a magnetic field below 0.02 mG.

Blue light

Artificial blue light is ruining our circadian rhythms (daily cycles of sleep and repair) chronically more than any toxin known to man. That's because *natural* blue light is supposed to control the body's daily-nightly stress levels and mitochondrial fitness. But *artificial* blue light doesn't have the red and purple light present in real sunlight to counteract the potentially harmful effects of blue. *Natural* blue light exposure, in isolation, is bad enough for you. But *man-made* blue light is profoundly more damaging to our biology because: (1) computer screens and modern light bulbs radiate blue light that's three or four times the color intensity; (2) we get it way too often; (3) we're exposed to it far too long; (4) we get it at the wrong times of the day; and (5) it's usually present alone, without its natural counter-balancing frequency of red.

To give you some perspective, not too long ago, all we had were incandescent light bulbs. They make light that's kind of like the sun's. It contains almost the full spectrum of colors, including some IR, with a tad more blue than real sunlight. But, unfortunately for our health, the industry moved away from incandescent light bulbs, to fluorescents, and now LEDs. They sold us on their energy efficiency, while ignoring the health decay. You see, to be energy-efficient, LEDs don't waste energy making light you don't see, and wouldn't think to miss. So modern LEDs produce light that's almost entirely blue, and not much else.

Let's take a moment to define blue light. When mitochondriacs talk about artificial blue light, most of the time we mean white-looking light with a very high color temperature (>4500°K) so it appears bluish, instead of having a yellowish hue like traditional incandescent light (<3000°K). To explain how the idea of color temperature came about, the nomenclature is modeled after the color at which tungsten glows when heated to the corresponding temperature. The lower the temperature that tungsten is, the more yellow/orange/red it appears, while the hotter the temperature, the more bluish-white it appears. Color temperature is very important because our circadian system is controlled by these frequencies of light hitting our photoreceptors.

Another reason we call artificial light *blue* is because it fundamentally begins that way. There is no such thing as an inherently white LED. LEDs can only produce one color, whereas real white is a mixture of colors. So most of our tech devices create "white" light by coating blue LEDs with yellow phosphors to make them glow white in color. Unfortunately, blue light from our gadgetry is many times more intense than the blue that's mixed in with natural sunlight. It's often 6500° Kelvin or more in LEDs *vs.* 1800°K in early-morning sun. Artificial blue light is also used everywhere, in pretty much every light-emitting device made – from smartphones and LED lights, to street lamps and modern TVs.

But to compound the catastrophe, we're psycho-socially addicted to blue light because it stimulates our hormones and nervous systems, and because of the digital content that's delivered with it. That means a typical millennial–GenZ'er looks at their cell phone more than 150 times a day, which depletes these social media addicts of dopamine, cortisol, and sex hormones through constant activation of their adrenal, pituitary, and psychological reward systems. That's the single biggest reason why children and young adults today have adrenal dysfunction, brain and behavior problems like attention deficit disorder, metabolic dysfunction like diabetes and obesity, and fertility problems… as well as being addicted to social media, chemicals, and ***insert your favorite addiction here***.

Blue light basically tells our internal pharmacy to release chemicals so we can concentrate and deal with daily stresses. But we're doing this all day and night, non-stop, so our pharmacy runs out of chemicals to give us to meet normal daily demands. To show you that Industry knows this, my old computer warned: "Screens emit blue light, which can keep you up at night. Night Light displays warmer colors to help you sleep." Most people never notice how much that blue light is stressing their system and inhibiting sleep because they're running on emergency reserves every day. But if you ever try falling asleep well before your body is exhausted, you'd be surprised how jacked up the blue is making you feel.

You wouldn't believe how toxic blue light is to our biology

- **Increases heteroplasmy rate.** Blue light expands the respiratory proteins in mitochondria, thereby lowering mitochondrial efficiency.
- **Dehydrates you.** Sluggish mitochondria make less water. That results in weaker e-zone, slower detoxification, and impediments to dozens of bodily functions due to protein malformation.
- **Activates the stress response.** Blue light in the eye stimulates the adrenal system into releasing cortisol and dopamine to prepare us for daily activities. The stress response also raises blood-sugar level, and dampens digestion and metabolism. In doing so, blue light chronically stresses the body and mind, upsets our hormonal systems (especially the adrenals), and contributes to type 2 diabetes.
- **Wrecks DHA function** in the eye and brain. Weak mitochondria (from heteroplasmy) make a smaller magnetic field. That reduces DHA delivery because magnetism moves DHA around in blood and offloads it into cells.
- **Gives you brain fog and dumbs you down by "over-spending" your dopamine supply.** When your dopamine level drops, you can't concentrate as well, you don't make mental connections as well (creativity), you lose motivation, you lose empathy, and you lose patience. You become a slave to stimulation in your environment. In short, that's ADD.

- **Disturbs melatonin and sleep.** Blue light interferes with sleep cycles by damaging melatonin via melanopsin-vitamin A degradation.
- **Impairs thyroid function.** Blue light penetrates the skin 3–6 cm. So, when you spend many hours a day with your throat area exposed

to intense blue light – as gamers and social media addicts do – that blue light can wreck mitochondria function in your thyroid, trigger inflammation, and give you hypothyroidism... or even Hashimoto's, in severe cases.

- **Upsets testosterone production and ovulatory cycles.** This causes hormone imbalances, sexual dysfunction and sometimes infertility.
- **Increases reactive oxygen species.** Stretched-out mitochondria make more free radicals, which perpetuates inflammation and exposes you to biologic aging.
- **Exaggerates metabolic problems.** Blue light raises glucose level and contributes to type 2 diabetes.
- **Promotes vitamin A deficiency and obesity.** Blue light breaks the bond between melanopsin and vitamin A (retinol), turning retinol into a toxic version of itself: retinal. Especially significant to those with weight problems, vitamin A-deficiency is linked to obesity. That's a little-known reason blue light contributes to obesity.

Blue light stretches out your mitochondria (heteroplasmy)

The most harmful thing that artificial blue light does to your long-term health is it stretches out the respiratory proteins in your mitochondria. In the same way that increased distance makes it harder for static electricity to jump from finger to doorknob, swelling of the respiratory proteins makes it harder for electrons to jump from one spot to the next along the ETC. That lowers ATP. Cells then don't work as well. DC electricity drops, as does the magnetic field that moves blood, oxygen, DHA, and hormones around the body. A less-efficient ETC, combined with reduced oxygen delivery, then creates more free radicals. All of that describes how blue light directly depresses mitochondrial function.

How blue light destroys photoreceptors

Photoreceptors are proteins that absorb light to help regulate circadian programming. In humans, photoreceptors are loosely bound to vitamin A (retinol). Vitamin A/retinol must stay bound to a photoreceptor to maintain its biologic benefits. Unfortunately, when blue light or other non-native frequencies hit photoreceptors, the loose covalent bond that holds vitamin A to its photoreceptor disintegrates, and vitamin A becomes toxic. Retinol turns into a rogue version of itself – an aldehyde called "retinal" (aka retinaldehyde or retinene, which are poorly named since they're easily confused with each other, and things related to the retina).

With nothing to disable its destructive qualities, retinal blows other molecules to bits, with insidious consequences (imagine formaldehyde circulating freely). Uncoupled from a photoreceptor, **retinal runs amok in the body, trashing any photoreceptors it encounters.** For example, researchers have reported seeing retinal eat up the rods and cones of the eye in its unchaperoned form. For this reason, rogue vitamin A (combined with poor regeneration of photoreceptors and mitochondria, due to low melatonin and dopamine) is a primary cause of macular degeneration and cataracts that are increasing by leaps and bounds in recent years.

This scenario also explains some confusing literature that said vitamin A is harmful to human health. It's because the design of these studies was fundamentally flawed in more ways than one. First, researchers didn't distinguish between vitamin A in supplement form (which could be retinol or retinal) and the real thing operating under natural conditions – which is coupled to vitamin D usage, circadian rhythms and opsins. Second, mainstream studies don't control for light, water or magnetism. In other words, the effects that you see in a double-blind, placebo-controlled study are not the same as what happens in real life.

Dissociated vitamin A (retinal) attacks all photoreceptors

Melanopsin and neuropsin. Crucial to circadian biology, disassociated vitamin A destroys one of the main photoreceptors the CRP uses to control daily and seasonal cycles: melanopsin. When melanopsin isn't working correctly, light frequencies such as UV lose control of daily and seasonal rhythms of stress, sleep and metabolism.

Melatonin and dopamine. Through retinal toxicity, blue light also ruins *melatonin* function via collateral damage of *melanopsin*. That's really bad for human health because melatonin dysfunction disturbs sleep, mitochondrial recycling, and regeneration of all photoreceptors.

Melanin. Melanin absorbs almost all EM frequencies – both internal and external – to regulate all five senses, the gut, and hypothalamus via the retinal pigment epithelium. Blue light-induced retinal breaks down melanin to ruin signaling of these pathways.

Nitric oxide. Retinal trashes NO, and vasodilation is affected. Our response to UV-A is then impaired.

Respiratory proteins. Dissociated vitamin A also attacks mitochondrial cytochromes. They too have photoreceptors. That's bad news for energy production because broken-down respiratory complexes ruin the ETC's handling of electrons, protons and photons. So, to those who think a special diet such as keto, paleo or vegan will fix a calorie, insulin or hormone problem: It won't, because most weight and metabolism challenges aren't primarily a macronutrient problem (carbs *vs.* fats and protein). Special diets may help in alleviating symptoms, but blue light-induced heteroplasmy is a primary source of metabolic dysfunction. That's what's causing diabetes and obesity, more than anything. So it doesn't matter what you eat; the efficiency with which your mitochondria process food is the real problem in both weight gain and insulin resistance.

Blue light inhibits regeneration of melanopsin

Photoreceptors that capture daylight in the eye are optimized to work around the yellow frequency of light, which is near the center of the visible spectrum. So the cones of the eye use mostly blue, yellow, green and red to see with during the day. This is the daytime color vision we know and love. But what most people don't realize is that our night vision is run by the photoreceptor melanopsin. The rods of the eye use melanopsin to convert the light and shadows we see in dim settings into an electrical signal our brain uses to see with in black and white. Most important, melanopsin is optimized to work around the blue frequency.

Here's where the plot thickens: You regenerate melanopsin during the day, and the cones at night. So when modern humans inundate their eyes with hot blue light during the day, melanopsin receptors get beat up from overuse at a time when they should be resting and regenerating. To make matters worse, melanopsin receptors aren't limited to just the eye; they're also found in the skin, blood vessels and subcutaneous fat. They too get fatigued when blue light hits that anatomy.

All of that's extremely bad news for mitochondrial function because the melanopsin system is what controls the central retinal pathway (CRP). It's the CRP that controls mitochondrial density almost everywhere in the body. Crucial to your constitution, melanopsin damage from blue light messes up your mitochondria. That's a major reason why everyone's health today is so fragile and their thinking so foggy: Communication devices and artificial lighting expose you to blue light that ravages your mitochondrial function – not to mention causing hormone imbalances, a non-stop stress response, and all that that entails. That's why restoration and metabolism are crashing in people like never before.

That's the secret, subversive reason why people are so weak and susceptible to sickness today. It's a chain reaction of: (1) too much blue light crashing the melanopsin system; (2) impaired melanopsin crippling CRP function; and (3) a disabled CRP wiping out mitochondrial populations and productivity. As a result, people are living on mitochondria that should have been retired long ago.

Blue light causes near-sightedness (myopia)

What causes myopia? Of course, there is a genetic component to near-sightedness. But the overlooked factors underpinning most myopia today are artificial light and nnEMFs. Poor eyesight would not be nearly as common or acute without the presence of blue light from smartphones, computers, TVs, fluorescent lights and LED lights. To understand how myopia happens in a person, look at a rainbow of colors coming out of a prism. You'll notice that blue light bends more than most colors (see page 178). Now, by nature, our eyes are built to focus using full-spectrum

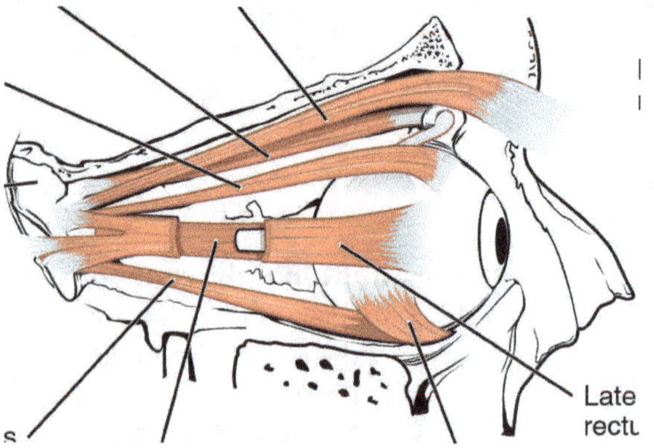

Source: Textbook OpenStax Anatomy and Physiology. Used under Creative Commons license 4.0.

Keratoconus: Condition in which the cornea thins and bulges outward into a conical shape.

light. This puts the eye's point of focus more forward, toward the center of the eyeball, which means a rounder eyeball and better vision. Conversely, artificial light contains mostly blue. Under fake blue light, our eyes get used to forcing their point of focus further back to compensate for the fact that blue light bends more than other colors. The longer this goes on – blue, instead of full-spectrum – the more pressure there is on the eyeball to elongate and stay that way.

Dopamine depletion compounds this effect. Blue light exposure diminishes dopamine, which impairs full constriction of the muscles that keep the eye in place and control its aim (see image). Reason is, dopamine controls skeletal muscles, and these muscles are skeletal. So when ocular muscles lose the tension that keeps the eyeball from elongating, the eye expands forward. That's myopia or/and keratoconus.

Too much blue, too little dopamine cause short-sightedness – at least until you shift your "light style" toward natural light, and away from the blue, which is what I've done. Since I began learning from Dr. Jack how light works, I've made a concerted effort to open my sunroof while driving, avoid glasses and contacts when I exercise outside, open windows, and get sun in my eyes. And the results are striking: My eyesight improved so much, so quickly I suddenly needed a new prescription. I can see further than ever with my old prescription. And I can see better without glasses – especially after getting real sunlight the day before, and avoiding artificial light at night. But I can't see things close up through my glasses anymore; my prescription is too strong. So now my vision is caught in a tug-of-war. It gets better when I get UV exposure, and it gets worse with glasses and artificial light.

But the hard lesson some people learn from myopia is this: Having bad eyesight is much more than buying glasses or contacts and you're good to go. **Myopia is actually an early-warning sign of mitochondrial distress and diseases of low-energy supply. It's "canary in a coal mine" predicting the occurrence of other diseases,** as this chapter should make abundantly clear.

How blue light creates insulin resistance

Of great concern to diabetics, the blue light and retinal-toxicity story is a major factor in the development of insulin resistance – not just carbs overloading the system with glucose (although that isn't helping). Instead, the bigger bottleneck is happening downstream. Blue light creates insulin resistance because it breaks your mitochondrial engines by stretching out the cytochrome proteins. Mitochondria then can't metabolize glucose as efficiently because retinal damages respiratory proteins.

So insulin injections force more glucose into cells faster. But that only improves throughput marginally because the bottleneck is glucose metabolism at the cell and mitochondria level, not so much getting glucose into cells. Hence, those with type 2 diabetes tend to get worse over time the longer they take insulin... unless they fix their mitochondrial dysfunction first. Once again, mainstream medicine is extremely concerned with pacifying symptoms without actually fixing a damn thing. However, I do believe this is a rare case of legitimate ignorance because they don't know biophysics like a mitochondriac.

Jack Kruse tells Aaron Alexander how blue light and nnEMFs are used to dumb people down, get them addicted, and harvest their money

From Aaron Alexander's "Align" podcast, episode 144, titled: "Dr. Jack Kruse: Longevity Secrets, Cell Phone Effects, Obesity Paradox."

AA: "Looking at the world from a dystopian perspective... it's almost like a mental illness towards staring into things that end up zapping you of energy. Or eating sugary things that zap you of energy... It's like being in an abusive relationship."

JK: "If you could see my face, I've got the biggest shit grin smile on it. ***You're using this as a metaphor. It's not a metaphor. It's actually really what it [technology] does.*** *That's what nnEMF does. It lowers your dopamine and melatonin to make you addicted. Guess what, Google owns the patents on this. Facebook has patents on this. This is how Alexa works.*

What you don't realize, when your dopamine level drops, you effectively are becoming less able to [make mental connections] in your environment, and you become more controllable. The more you use technology, the more you're going to fall in line with everybody else. And that's exactly what they want you to do, because that's how they harvest money. A lot of people think I'm a giant asshole for pointing these things out. I'm not doing this to be an asshole. It's because people are asleep. They do not see this going on. They don't even realize these patents exist. The guy that originally warned [us] about all this was Becker.

Not even Becker probably could fathom just how bad this is. The Navy in '73 asked him to review some papers that were never published about pilots [to study the safety of nnEMFs for a prospective radar installation]. You know what all the biochemical changes were? They had cognitive problems, they had symptoms of leaky gut, they also gained weight, and they looked like they were developing metabolic syndrome. This was in '73, before anybody even knew about this stuff. This was from pilots that sat in cockpits of planes.

Becker pointed out [to the Navy about the antenna they were studying] this had effects between 1–10 Hz. He goes, 'When you guys realize this antenna is over 60 Hz, that's exactly where the power grid oscillates to. So not only is the antenna a problem, but the entire electric power grid is a problem.' You [Aaron] are worried about the MacBook and the [iPhone]. These work on frequencies that would blow your mind. When you understand Becker's original work... then you begin to understand that this is not a metaphor. This is by design [the way technology devices dumb you down and get you addicted].”

AA: “So you think there’s actually like a physiological dependency here?”

JK: “Dude, I don’t think… It’s published. There’s so much data on this, it’s not even arguable anymore. Anyone that poses that question to me announces themselves to the world that they are fundamentally clueless about what’s in the literature. The guys at Facebook, the guys at Google, the guys at LinkedIn … There is not a social media empire that doesn’t know about this [phenomenon]. There is not a cell phone company that doesn’t know this. That’s the reason why every screen is blue lit.

How easy would it be Aaron to turn the screen to red? …Just make sure it’s blue lit [screens] and you thin the retina, affect their growth and metabolism pathways, lower their dopamine and melatonin, and make them more controllable, so that you can give ‘em suggestions, so that they follow along, and you can harvest all the money out of their wallet. This is not hyperbole. This is not Jack Kruse’s opinion. I could give you tons of patents that are tied to this.”

Video from Paul Joseph Watson sums up the societal impact of cell phone addiction: brighteon.com/5837731061001

Dr. Kruse goes on to say the #1 biggest risk to mankind today is a 1996 FCC ruling that sells large portions of the electromagnetic spectrum – particularly in the microwave range – to the telecom companies to do with as they please. He says (and I agree) it’s akin to letting the Big Pharma companies put any experimental drug they want in our water supply without testing it, and without telling us!

Blue light in a nutshell

As of 2020–2021, artificial blue light (absent UV and IR) is the most destructive toxin to modern man because it interferes with photoreceptor function, hormones, mitochondrial efficiency and energy collection. It ruins circadian programming that controls sleep, stress level/alertness, metabolism and renewal. Blue light is sneaky-disruptive, more than you can imagine. Yet modern medicine can’t see the connection because it’s totally focused on biochemical imbalances and resulting symptoms, not the true source of what’s upsetting our biology.

To put it simply, when all you have is a hammer (pharmaceuticals and surgery), everything starts to look like a nail. And nail us in the body, mind and pocketbook they will with all their off-target solutions. So that’s where we’re at with healthcare and intentional wellness today. However, blue light is soon to be dethroned as the most harmful toxin of them all as 5G ramps up in cities all over the world.

Non-native EMFs

Since the 1970s, supporters of industry have said the non-native EMFs used for communications can't hurt us because the radiation they emit is non-ionizing. They're not powerful enough to knock electrons out of atoms the way radioactivity does. While technically that may be true, it would be a mistake to think non-ionizing radiation is harmless. It just does its damage in other ways. For example, a single cell phone, Wi-Fi device, or smart meter can't shake large volumes of electrons free quickly and create free radicals. But what microwave frequencies can do is activate voltage-gated calcium channels, which raises intracellular calcium. That makes reactive oxygen species through a secondary, *oxidative* process, which is a chemical exchange of electrons, rather than *ionization*, which displaces electrons through bombardment of high energy particles.

In other words, ionization directly damages DNA by dislodging electrons from atoms, while too much calcium in the cell causes electron loss through oxidative chemical exchange. Two different mechanisms, similar amount of damage done. The net result being, the industry's flawed reasoning fails the real-world test. But that's all rhetoric. Actual EMF safety guidelines are based on thermal effects. Standard industry tests basically say if your cell phone signal can't heat water significantly with one call, you have nothing to worry about; it's completely safe. But real-world exposure is not limited to one call on one cell phone, or one smart meter, or one Wi-Fi signal. Nor is risk of injury limited only to thermal effects.

Rather, we're now swimming in a sea of always-on electromagnetic pollution everywhere people live and work. Even worse, people are using their devices with the frequency of a recovering surgery patient "jonesing" for pain medication – delivering tiny hits of stimulation every few minutes to keep their dopamine level up. Therefore, thermal tests are a completely inadequate reflection of real-world exposure. They're essentially meaningless... a sleight-of-hand trick, or industry con, if you will. Back to reality: Dr. Martin Pall's analyses of published literature reveal a multitude of effects that have nothing to do with heat, and at orders of magnitude lower than industry-designed and -sponsored safety guidelines:

Jonesing: Strong need, desire, or craving for something – especially by an addict.

- lowered fertility (18 reviews)
- neurological/psychiatric effects (25 reviews)
- 3 types of DNA damage (21 reviews)
- compromised apoptosis (13 reviews)
- oxidative stress/free radical damage (19 reviews)
- endocrine/hormonal effects (12 reviews)
- excessive intracellular calcium (15 reviews)
- cancer (35 reviews).

In a nutshell, the quantity, intensity, variety and duration of two-way signals you're being exposed to would simply blow your mind if you could see or hear them. At the same time, the wide range of effects being ignored or denied by the entire industry is equally hard to wrap your head around. So, for now, all you need to know is that the degree of damage caused by these artificial frequencies is determined by how antagonistic an electromagnetic wave is to *your* biology, times *your* total exposure.

Non-ionizing **EMFs may do more damage than** ***ionizing*** **radiation**

To most scientists, this concept may seem preposterous. But here's how Dr. Pall believes it could be true: Moderate doses of ionizing radiation, like that you'd get from an x-ray machine, give you one level of amplification that goes like this: High-energy photons knock electrons out of place. The photons and displaced electrons dislodge more electrons in a chain reaction, until the energy is used up and the chain reaction stops. Atomic bombs and nuclear power plants do something similar – only these reactions apply more energy than an x-ray machine, and their fuel is far more reactive than human tissue. That's *ionization* in essence. On the other hand, non-ionizing radiation, like that used in cell phones and other tech devices, damage our cells at three levels of amplification:

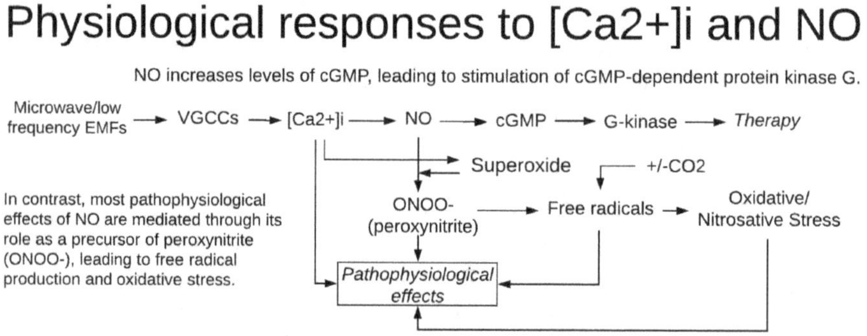

The diagram (left) is a reproduction of Dr. Martin Pall's work.

1. Microwave EMFs over-activate voltage-gated calcium channels (VGCCs), **allowing over a million calcium ions ($Ca2^{+}$) to flow into the cell per second.** VGCCs normally bring calcium (+) into the cell to draw in more oxygen (-) and help you cope with stress.
2. Calcium makes more nitric oxide (NO) and superoxide radicals.
3. The super-destructive free radical called peroxynitrite ($ONOO^{-}$) is formed as the product of nitric oxide concentration times superoxide concentration (NO x superoxide = peroxynitrite). Peroxynitrite is a **highly reactive oxidant** that breaks down cell walls and proteins.

So the immediate and direct damage to DNA may be greater when you get an x-ray (from the burst of medium-strength ionization). But the total damage you suffer from microwave transmissions may actually be worse, due to relentless oxidative stress.

World governments know how toxic nnEMFs are

- **China.** Pregnant women are required by law to wear EMF-shielding belly bands.
- **Israel.** New cell phones must carry this warning: 'Heavy use of this device, or carrying it close to your body, can increase the risk of cancer – especially in children.'
- **France.** Wi-Fi was banned in nursery schools in 2014.
- **Russia.** Premier Vladimir Putin has gone on record saying, 'We do not need to go to war with America. America is committing collective suicide by the way they are using electricity (nnEMFs). We just have to wait until they're all in the psychiatric hospital.'
- **World Health Organization.** In 2011, the WHO classified microwave radiation as a Class 2B possible human carcinogen, based on increased risk of brain cancer due to cell phone use. And that was at 2011 exposure levels.

In addition, the following countries have set their radio frequency safety limits 100 to 10,000 times lower than the United States (not that these limits are enforced): Italy, Switzerland, France, Austria, Luxembourg, Bulgaria, Poland, Hungary, Israel, Russia and China.

Some symptoms of nnEMF exposure

- headaches
- tinnitus (ringing in the ears)
- brain fog
- insomnia
- general fatigue
- irritability/stress

Some health outcomes

- sleep problems
- infertility, sexual dysfunction
- leaky gut, food sensitivities
- diabetes, obesity
- Alzheimer's, Parkinson's, dementia and other neurological problems
- DNA damage, cancer.

To put nnEMF damage in perspective, we already know microwave radiation is currently being used around the world in military weaponry and "active denial" crowd-control systems. And we know the US government has publicly accused Russia, Cuba and China of aiming microwave weapons at US embassies with the intent of causing serious disease in diplomats. This has resulted in at least three ambassadors dying of aggressive and rare cancers in recent decades. In other words, we know for certain that microwave radiation makes a phenomenal weapon. And you can't see it, hear it, or feel it. So why would anyone doubt that it's harmful to humans, animals, and plant life? It's willful ignorance, once you know the facts. Or you can say…

We're now meat in a microwave oven

Microwave ovens are notorious for dehydrating food as they heat it. They bombard food with microwave frequencies to get their particles moving faster. This motion is nearly synonymous with heat. However, there are unwanted side effects. So much energy is added so quickly, many water molecules get ejected from their affiliations. And water is where an organism stores energy and information from its environment.

Just as bad, a small percentage of electrons are lost from that food, which alters them minimally in form, but substantially in function. For instance, many vitamins, enzymes, antioxidants and fats are fragile. Some of their chemical bonds self-destruct when oxidized by oxygen or EMFs. So eating or drinking microwaved foods is not going to kill anyone right away. But, from a texture and nutrition standpoint, the food will never be quite the same again after it's exposed to concentrated microwaves. You can see this when you cook meat in a microwave oven without adding water. The longer you cook it, the more leathery its texture.

Well, guess what: Most of our communication devices transmit and receive using microwaves that can cook food in higher doses. Check this out: When the technology was first introduced (I think in the 1940s), enlisted men would play around with microwaves from their ship's radar installation. After discovering that they felt warmer when standing in front of their ship's radar dish, they'd put hot dogs on a stick and literally cook them in front of the microwave transmissions. That's where the idea for microwave ovens came from. For us, that means you can multiply the number of devices around you, times the intensity of nnEMFs they emit, times how long you use them, times how often you use them… add all of that up, and we've become the steak in a life-sized microwave oven.

To illustrate how the telecom industry agrees with the microwave heating phenomenon, yet still fails to create testing standards that accurately reflect real-life exposure, get this: The industry-standard test to check cell phone emissions involves operating a cell phone a few inches away from a container full of water shaped like a human head. If it doesn't raise the water's temperature appreciably, the phone's emissions are deemed to be safe for use without restriction or warning – as if one call, on one phone, could heat a balloon full of water... as if heat were the only risk factor microwaves pose to human health.

Bottom line: Living organisms – living precisely because natural frequencies control their biology – are much more disturbed by foreign frequencies than inanimate foodstuffs. And we have been witnessing the myriad consequences to human health ever since the FCC began licensing the electromagnetic spectrum out to telecom companies more than two decades ago. We just didn't know which changes in our environment were responsible for causing the explosion of illness we see today. Hence, we continue to expand our use of the spectrum AND believe industry propaganda beyond any semblance of reason, science, or minimal common sense.

Sources of electromagnetic pollution

1. Cell phones.
2. Smart meters (for electricity and gas). Smart meters relay signals through each other to communicate "back to base," resulting in between 17,000 and 190,000 transmissions per day hitting your average city dweller.
3. Wi-Fi (wireless computer network devices).
4. "Smart" appliances, aka "Internet of Things" devices.
5. Alarm systems.
6. Baby monitors.
7. Police and fire systems
8. Compact fluorescent and LED light bulbs.
9. Air traffic radar.
10. Cars (most new cars now have wireless systems installed).

And don't forget to multiply each category by how many devices of each are broadcasting around you.

How nnEMFs and dehydration conspire to corrupt our biology

The more dehydrated you are, the worse your EMF symptoms get, because when you're dehydrated: (1) you can't assimilate as much energy from light (as exclusion zone water); (2) magnesium doesn't work as well in enzymatic reactions; and (3) your skin can't turn cholesterol into vitamin D. Simultaneously, nnEMFs alone weaken mitochondrial function, (4) which is where the best water for the body comes from: deuterium-depleted "metabolic water."

In case you're just learning how to make sickness go away and good health stay as long as possible, those four things are intimately tied to making optimal health happen intentionally: (1) harvesting energy from light; (2) maximizing mitochondrial function; (3) improving enzyme function, which aids our biochemistry; and (4) having all the vitamin D your body needs to run the immune, endocrine and cardiovascular systems.

Crucially important, but not yet well-understood, water is supposed to envelope the highly convoluted shapes of proteins in soft tissues and biochemicals. The shape and structure of proteins, particularly their surface architecture, controls the vibrations they emit in response to frequencies around them. So a lack of hydration in and around a protein can substantially change its function.

But before proteins can even collect water around them, they need to unfold. That's the job of ATP. An important role of ATP involves loosening and expanding the folds of protein so more water can bind to them. Once proteins have a hydration shell around them, light can charge-separate that water – basically turning it into a battery – to power thousands of cellular processes. In other words, light from the sun, or IR/heat released from mitochondria, charges up the water around proteins like a battery, fully activating the protein. Water also stores electrons and protons made from the ETC. **So when you don't have water, you lose battery capacity.**

What stress does to you

nnEMFs cause a stress response – especially blue light

The second-worst effect nnEMFs impose on you is that they activate a system-wide chronic stress response. By over-stimulating the brain center responsible for managing stress (the "paraventricular nucleus" or PVN), non-native EMFs put your nervous system into a prolonged sympathetic state (as opposed to *rest and digest*, which is *parasympathetic*).

The paraventricular nucleus of the hypothalamus (PVN), in collaboration with the suprachiasmatic nucleus (SCN), controls homeostasis (normal operational state) by regulating a broad range of autonomic functions – including cardiovascular, thermoregulatory, metabolic, circadian and stress responses.

Now, limited, temporary stress during the day is perfectly harmless, because being awake and alert uses stress chemicals the body likes in moderation – cortisol, adrenaline and dopamine, to name three. But being in a high-stress state all the time activates the body's fight-or-flight mechanisms without ever allowing the body time to relax, replenish and recover. Stress is supposed to be cyclical, with both uptime and downtime. But modern humans are habitually running their adrenals near redline… which depletes biochemicals the brain needs to run high-end operations such as concentration and creativity.

The way blue light in particular activates the stress response is by telling the eye and the SCN it's midday all the time. With things to do during the day, the circadian system infuses stress chemicals into your system constantly. This raises your stress level and heightens your alertness. But "going to that well" too much wears on your adrenals, brain and heart like racing a car designed for the street. That's effectively what you're doing when you seek gratification from your smartphone, LED TV and artificial lights: You're poking and prodding your adrenals into releasing whatever cortisol, adrenaline and dopamine your endocrine system has available. Stress chemicals are supposed to start out high in the morning to wake you up, then increase on-demand when more focus is needed. But **we're basically using blue light all day long like a drug. We've become addicts. And our digital devices are how we're getting our fix.**

Working in tandem with the adrenals, the PVN does something similar when it's stressed by nnEMFs: The PVN reads nnEMF radiation in your environment and reflexively raises intracellular calcium as a false trigger for your stress response. So, like a chicken or egg scenario, it doesn't matter which comes first: the stress on cells and mitochondria by itself, or simply more calcium in the cell. Either way, nnEMFs trigger a cascade of stress effects. And, let's not forget, the other influences that activate a sympathetic state are far more than just psychological stress. Rather, stress takes many forms.

As this excerpt from *Gut-Brain Secrets* explains:

Mental/emotional stress: Much harder on the body are our modern forms of acute and prolonged stress that tend to be more mental/emotional in nature, happen more often, last longer, and lack a physical outlet for expression. Examples would be repeatedly facing gunfire on the battlefield, working in a high-pressure sales job, moving to a new town, finding a new job, going through a breakup, or the death of a loved one.

Important in understanding the body's stress response, most situations from man's ancient past that required adrenal intervention were employed by the body to produce both a physical and a mental response. That is, situations of-old were often both physical and mental challenges. For that reason, the adrenal system does not differentiate between physical, mental, or any other type of challenge today. It treats all threats the same by releasing the same biochemicals – including stressors that fit into two newer categories: chemical and environmental stresses.

(Modern) Chemical and environmental stressors: Exposure to stressors such as bisphenol-A, artificial lights, irregular sleep patterns, temperature extremes, and even high levels of fructose, create a persistent stress response from the adrenals, just like traditional kinds of stress do.

Past (and future) stressors: What's more, traumatic events such as early childhood traumas, rape, or war-related stress can permanently hyper-sensitize your stress response, unless/until you successfully process them psychologically. Obsessively worrying about future events, by itself, can also chronically raise your baseline stress level.

So whether threats are physical, mental, emotional, chemical or environmental... whether they're in the past, present or future... whether they're real or imagined... your adrenals are constantly being tasked with the job of keeping you on top of all the challenges your body and mind face on a daily and minute-by-minute basis. And the adrenals respond to these extra, sometimes extraordinary, demands on the body and mind by releasing more of the same stress-management chemicals that routine mental and physical challenges employ in order to activate certain mechanisms in the body, while deactivating others.

Key point: All these stressors have a cumulative effect on the adrenals and body. All these individual stressors "stack" up, one on top of another, to form your total stress load. The sum total of all stresses in your life(style) turn into chemical wear and tear on your body. When any stressor occurs too frequently or too intensely, the persistent presence of stress chemicals in the system disturbs the body's biochemical balance, which can lead to physical damage to cells, organs, and systems from the increased energy consumption (e.g., ATP and electricity from astrocytes) and decreased recovery (resonance).

Emergency energy consumption also tends to borrow energy and alertness from the future, which encourages you to keep borrowing from tomorrow in the form of more stimulants to get you through each day and each situation... or else suffer an energy shortage we call a "crash." Simply put, the use of your adrenals to survive emergency situations like these is increasingly stressful to sustain the longer it continues. And there's a variety of prices to pay for unrelenting stress in its many forms.

Biochemical stress responses

Calcium efflux. Stress takes calcium from where it's harmless, or even needed, and puts it/keeps it where it's destructive long-term. Depending on cell type, calcium is used to either transmit nerve signals, contract muscles, secrete hormones or express genes. So any stress response, whatever the source, takes calcium out of storage sites like the bones and inappropriately transports it to areas of activity through voltage-gated calcium channels (VGCCs). The process is called "calcium efflux."

VGCCs are located on the plasma membrane of cells, and are actuated by exquisitely sensitive voltage sensors that do exactly what you'd expect: they're electrically-activated gatekeepers. Unfortunately for us, nnEMFs activate the voltage sensors by accident. Calcium then wrongly floods into the cell, which begins a cascade of effects. VGCCs basically keep the stress response turned on, including the effects discussed in this section.

Another sticky situation occurs when mitochondria need to make ATP, but the positive charge at the cell wall isn't strong enough to pull in enough negatively-charged oxygen to accept electrons from the ETC. When that happens, the VGCCs draw in more calcium and other positively-charged minerals to raise charge on the cell wall biochemically, instead of raising positive charge the biophysics way, which is through electricity from the astrocytes. In other words, calcium efflux is a biochemical backup strategy to raise the positive charge in cell walls so more negatively-charged oxygen can be pulled inside.

Later, when stress subsides through the parasympathetic response, the VGCCs use ATP and adequate (negative) voltage to reverse the process and expel the calcium out of the cell, where it's not detrimental. Unfortunately, when your cells retain calcium for any reason: classic stress, stress from nnEMFs, insufficient ATP and/or low voltage – then extra calcium in the cells makes more nitric oxide, superoxide and peroxynitrite free radicals – as described by Dr. Martin Pall (page 212). That's the second major mechanism whereby nnEMFs do their damage (after dehydration): **Calcium in cells makes more free radicals** (particularly peroxynitrite) that increase oxidative stress and chronic inflammation. And it happens from extremely low frequencies, like 60 Hz electricity activating the VGCCs to beyond the microwave range.

One way we know for sure that excessive calcium in the cell is responsible for the majority of electrosmog effects is by seeing what happens when a person takes a drug that blocks calcium channels. A number of papers have shown that when you block activation of the voltage-gated calcium channels, *violà*, you block the adverse health effects too. In other words, when you prevent calcium from getting into cells, you make fewer free radicals and stop the most damaging effects of nnEMFs.

Dr. Pall also believes electromagnetic fields harm plants through similar mechanisms of calcium-channel activation, the same as animals. That's because plants need calcium to germinate. So Pall believes that nnEMFs disturb plant biology by stimulating voltage sensors by accident and depleting calcium needed for germination. This would explain at least one reason plants are struggling to grow in places teeming with nnEMFs. Just look at the plants around you for signs of distress; they're everywhere.

Moreover, when plants are stressed, not only is their growth impaired, but some of them dramatically increase their production of organic compounds which they use to defend themselves from insects and herbivores, called terpenes and terpenoids. The huge problem for air breathers is terpenes are highly flammable hydrocarbons that were used to make turpentine years ago. So terpenes, nnEMF stress, and the combustibility of geoengineering particles such as aluminum, barium, cadmium and strontium combine to produce record-breaking forest fires that can send smoke halfway around the world. Yes, I'm saying global-scale forest fires are being caused primarily by nnEMFs, geoengineering and glyphosate stressing trees to their limit. And it's being done on purpose by those pushing climate change agendas.

Heavy metals. Stress and low ATP also contribute to heavy-metal buildup in cells. Cells aren't programmed to distinguish between calcium and heavy metals, which are both positively-charged. So when cells pull in calcium to attract oxygen, heavy metals get pulled in by mistake. That becomes a problem when chronic stress keeps calcium and heavy metals in the cell on purpose, or when your ATP production isn't high enough to deport the heavy metals.

Digestion. High cortisol shuts down stomach acid production, because digesting food is less important than running or fighting for your life.

Glucose and insulin resistance. Stress and cortisol also flood the body with glucose. But, perhaps counter-intuitively, cortisol inhibits insulin production in order to prevent glucose from being stored away in skeletal muscles. Instead, the body needs glucose to be circulating in the bloodstream – ready to go for anything (like the heart and brain).

Immune system. Cortisol shuts down the immune system, because fighting infection is a longer-term priority than fighting an adversary that's right in front of you.

Blood flow. The stress response also directs blood flow away from visceral organs such as the stomach and intestines, toward peripheral muscles so you can fight or flee.

Sleep and repair. Cortisol and adrenaline inhibit sleep and healing. You short-change restoration when you stay in a sympathetic state.

To sum up biochemical stress responses, a chronic state of stress keeps a plethora of bodily functions turned on or off, inappropriately. That leads to continual over-activation of neuro-chemicals and calcium channels. It causes heavy metal accumulation. And it impairs digestion, metabolism and the immune system. Stress is draining on the body long-term. It is corrosive to cells. That's how nnEMFs do damage on a biochemical level.

Biophysical stress responses

Biophoton loss. Under normal conditions, cells release individual photons of light – predominately in the UV range. They do this all the time. But the more stress your cells are under, the faster these biophotons escape from cells – even though you can't see them without a photo-multiplier. This leakage of light, as you may think of it, is lost energy – like having a hole in your gas tank. It's energy inefficiency. And the loss of light contributes to dysfunction and decrepitude around the body. It is one more way nnEMFs and blue light discharge your mammalian battery. Same thing happens in mitochondria. High heteroplasmy rate lets more biophotons escape the ETC simply because the respiratory proteins are more stretched out and "porous." However, biophoton loss is not normally seen or studied, so mainstream medicine ignores its role in biology.

Electron deficiency. nnEMFs such as artificial blue light also cause electron loss. When blue light and microwaves stretch out your respiratory proteins, they make the mitochondria's ETC less efficient. That translates into poor conversion of energy from one form to another – such as metabolism turning food into electrons and protons; red light making the ATP synthase spin faster; or the mitochondria making metabolic water to store more electrons and solar energy.

Of paramount importance, inefficient power production means fewer electrons are available as a resource for healing, regeneration and basic cell function. nnEMFs basically gum up systems and processes all over the body for capturing, converting, storing and using energy. They interfere with the regulation of energy on every level from the biochemical, to the biophysical and psychological. And that hastens development of dysfunction, illness and aging.

How nnEMFs ruin sleep quality and deplete us of magnesium

"In a typical scenario, a person… in their bedroom, their millivoltage could be anywhere from 500 millivolts at the lowest… up to 10,000–15,000 millivolts on their body… all night long, while you're sleeping you've got this contraction of your muscles happening.

You just think about the way that electricity impacts the body. How do we restart a heart? We pump voltage into it. It makes your heart muscle contract… Calcium causes the muscle to contract. Magnesium helps it to relax. …There are some studies that actually show this too… one of the reasons we're so magnesium deficient in our modern culture is because we've got this constant exposure to this alternating current that's got these micro-contractions happening all night long, all day long in some instances, with our muscles. And it's depleting us of our magnesium, which is trying to relax our muscles.

So we've got this constant tension that's going on all the time. And if you have that at night when you're sleeping, your body can never really get into that full parasympathetic state so that you can actually do the healing and restoration that you need, because your body is so tensed up. Your cortisol's going up. Your melatonin's going down. And you're not getting that restorative detoxification of the brain, and all the lymphatics detoxing optimally.

You'll get a little bit… and a lot of people will be like, 'Oh, I sleep great'. Sleeping great is not a sign necessarily that your body is actually restoring itself… People that have sleep apnea say they sleep great… but they have this oxygen issue, and airway issue."

—Brian Hoyer of Shielded Healing.com.

Simple test reveals if your adrenals are over-stressed

Try this: In a dark room, shine a penlight in your eye. If your pupil can't hold the constriction, or it starts to pulsate, that's a sign of adrenal stress. This Pupillary Reflex Test, as it's called, is a simple, in-home test to tell if your adrenals are exhausted. It's caused by blue light and nnEMF toxicity (on top of stimulants such as coffee, lack of restorative sleep, and chronic daily stress) depleting the pupillary muscles of dopamine. The solution? Cut out the blue light at night, shield your sleep and work spaces from nnEMFs, and enjoy full spectrum sun on your eyes and skin.

The frequency of EMF determines the type of free radicals made

Each type of electromagnetism has its own range of frequencies, called its "spectrum," that energizes our lifestyle. Near the lower end of the electromagnetic spectrum, electricity operates at 50 or 60 hertz (varies by country). Above that, FM radio operates at 88–108 megahertz. The visible spectrum goes from 400–790 terahertz (or 380–760 nanometers). Above the visible spectrum, x-rays are 30 peta-hertz to 30 exahertz. And, within the microwave spectrum, the range of frequencies the FCC allocated to 5G is 3 gigahertz all the way to 90+ gigahertz – by far the broadest spectrum of any man-made EMF.

Wavelength is very important because the exact frequency at which an EM wave oscillates determines the type of free radicals the electron transport chain makes. And the type of free radicals the ETC makes controls: (1) how much the respiratory proteins expand or contract (their productivity); (2) mitochondrial density and recycling; and (3) turning on/off of your own genes. It's this specific heteroplasmy rate, and resulting DNA activation/deactivation, that controls which organs and tissues are affected by nnEMFs, what symptoms you get, and thus what type of EMF sensitivity you develop.

To put it more plainly, each specific wavelength of microwave radiation causes a different kind of mitochondrial poison to be made and, with it, a different illness. You can think of these electromagnetic hypersensitivities as allergies to certain frequencies. However, when 5G is fully implemented, it won't be the same from one place to the next, or one person to the next, because the lay of the land will vary, 5G frequencies will vary, as will free radicals made, and which area of a person's body is hardest hit by weak mitochondria. In other words, cause-and-effect will be hard to pin down with any consistency.

Heat shock proteins and DNA damage

A leading researcher in the field of nnEMFs, Dr. Martin Blank, found that even low levels of nnEMFs increase production of a group of proteins that cells release when they get too hot – called, not surprisingly, "heat shock proteins." Cells make heat shock proteins when they're

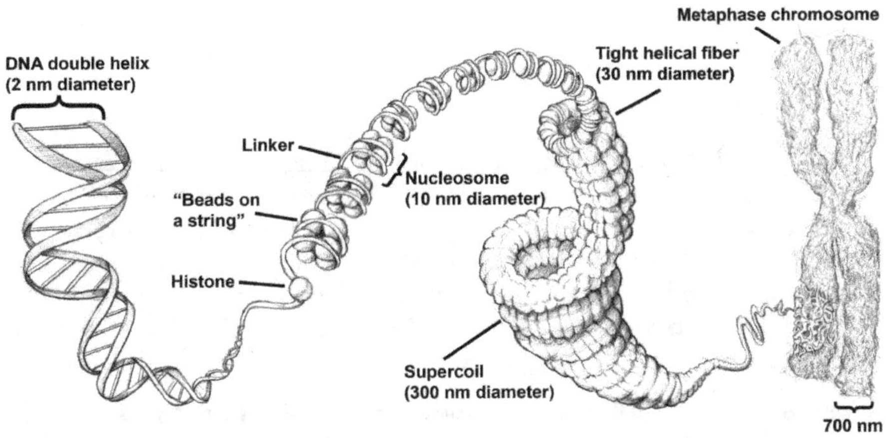

exposed to high temperature, acidity, heavy metals, or any potentially harmful stimuli – including nnEMFs. They're a big part of what happens when nnEMFs activate the stress response. In fact, Dr. Blank and colleagues even published a paper proposing that heat shock protein levels be used as a marker in EMF safety tests, because they are a clear indication cells are under stress.

Dr. Blank also found that DNA acts as an antenna for nnEMFs. Of course, we all know DNA is shaped like a twisted ladder, called a double helix. But hardly anyone knows that spiral shape actually twists upon itself to form a coil. And then that coil coils again, which is called a "supercoil." That's how chromosomes can stuff two meters' worth of genetic code into just nanometers. In doing so, DNA acts like a fractal antenna, capturing signals at three different frequency ranges. Meaning, its architecture replicates itself on three different scales, which makes it a good antenna for the top, middle and bottom portions of the EMF spectrum. The bad news is: some strands of DNA break apart when high-powered frequencies hit them. When they go unrepaired and replicated, these breaks become mutations that can get passed on to progeny.

So to sum up four sure signs of nnEMF exposure: We know that low-frequency nnEMFs are stressing our cells when **heat shock proteins** start showing up; **calcium efflux,** free radicals and the release of **biophotons** happen across the full range and intensity of nnEMFs; and higher power levels such as x-rays can damage the physical structure of DNA directly, causing **DNA mutations.** So clearly, nnEMFs are far more than just thermal effects and ionizing damage above the UV-A range.

How EMF sensitivities might manifest

A 5G cell tower from one company might be extremely efficacious at giving people cancer or autoimmunity, while another operating at a different frequency might cause Alzheimer's and neuro-degeneration, while yet another, in a different city, gives children close to it leaky gut, food sensitivities, obesity or autism. So one frequency may affect the brain; one with a slightly different waveform is more likely to cause problems with metabolism; and still other frequencies can mutate your DNA.

That's going to cause a lot of confusion and finger-pointing because scientists, government agencies, and watchdog groups won't be able to connect symptoms back to their real source. Remember: The current paradigm of randomized, placebo-controlled studies only tests one variable at a time in their safety and efficacy trials. That's the way they think, and how they operate. They don't examine multiple inputs and variable outputs... the way light and biophysics work in real life.

Women are more vulnerable to EMFs than men, children even more so, and babies in the womb most of all

Women are more sensitive to nnEMFs than men because the myelin sheath covering their nerves is thinner, as is the female skull. Thinner insulation means less attenuation/dissipation of harmful frequencies hitting nerves and tissue. Thinner shielding means nnEMFs do more damage. This applies even more so to children. Not only are their skulls thinner still, but their neurons are growing rapidly. Those neurons are less myelinated, and children are still developing cognitively. That significantly increases the net dose of EMFs children receive, as well as the damage done, compared to adults. One more reason to keep your kids away from EMF-emitting devices as if they were a health vampire... because that's what they are.

But, worst of all, nnEMFs are many times more damaging to babies in the womb than adults because the womb focuses foreign frequencies inward like a magnifying glass, and babies' heads are much softer than an adult's. In fact, nnEMF levels are shown to be twenty times higher inside the womb than out. So it's no surprise that the most reliable predictor of autism is nnEMF exposure in the womb – particularly the location where the mother slept when she was pregnant.

Nature made women more sensitive to nnEMFs for a reason. Women's mitochondria need to be acutely aware of environmental conditions that could give them a survival advantage – such as temperature, season, how they metabolize food, and even the emotions of others. The better adapted they are to their environment, the more likely they and their family are to survive and thrive, then pass their genes on to future generations. That's why women are more sensitive to vibrations of all kinds, both good and bad. And it's why women are more susceptible to dirty electricity, mismatched foods and weight gain, and especially alien electromagnetic fields and resulting EHS.

Why honey bees are disappearing

Prof. Neelima Kumar, PhD at Panjab University in India, explained a major (if not the #1) reason honey bee colonies are collapsing around the world. Her research shows that metabolism in bees crashes (and doesn't burn) when the bees are exposed to cell phone signals for ten minutes.

Carbohydrate concentration in their blood, called "hemolymph," went from 1.29 milligrams per milliliter to 1.5 after 10 minutes, and 1.73 after 20 minutes. Glucose rose from 0.218 to 0.277. Lipids rose from 2.06 to 4.50. Proteins rose from 0.475 to 0.825 mg/ml. And cholesterol skyrocketed from 0.230 to 2.565 mg/ml over that same time frame.

That means after just ten minutes of microwave exposure, a bee's mitochondria have less ability to turn sugars, proteins and fats into life-sustaining ATP. Their mitochondria are clearly affected by microwave radiation, which shuts down metabolism. However, because a bee's metabolism runs so much faster than ours, you can measure their energy reserves in minutes, whereas we can survive many years on fixed rations.

Many people can hear wireless nnEMFs with the naked ear

I know I can. And if you know what these frequencies sound like, you can probably hear them too. Older signals sound like an extremely high-pitched ringing in the ears – like tinnitus – and newer frequencies make me lose hearing for a second or two. Older frequencies are most noticeable at night, when all's quiet throughout the house and a smart meter or cell phone suddenly starts transmitting. It's as if a wall of ultrasonic sound suddenly sweeps through your body, and you go hard-of-hearing, or your ears start ringing like you'd been at a heavy metal concert the night before. Pay attention and it's obvious.

More nnEMF concerns

1. **nnEMFs open the blood-brain barrier and gut barrier.** Research shows another startling side-effect of nnEMFs: They break down barriers of the body, including the blood-brain barrier, gut barrier, and placental barrier. Perforating the blood-brain barrier is particularly bad news for those with autism, ADD, and children in general when trying to keep toxins out of the brain such as mercury and aldehydes.
2. **Microbes excrete biotoxins when they feel threatened.** For example, molds excrete 600 times more toxins when they're exposed to nnEMFs. They react as if they're being attacked, responding with the only weaponry they have: toxins.
3. **nnEMFs turn probiotic microorganisms into aggressive pathogens.** Just like molds, when symbiotic microbes feel they're under attack from nnEMFs, they too turn up their production of biotoxins, thereby turning friendly microbes into harmful pathogens.
4. **Nighttime exposure is much worse than daytime.** You're most vulnerable to the damaging effects of nnEMFs at night, when your parasympathetic nervous system is supposed to be renewing cells and replenishing biochemicals.

Every*body* is affected by nnEMFs

It's comforting to assume nnEMFs can't hurt you if it seems like you're not electro-hypersensitive. It's tempting to think if you don't exhibit symptoms, you don't have to worry about electrosmog. You would be incorrect. Everyone is affected by nnEMF exposure, but you may not realize which exposures cause which effects. More accurately, it's hard to tell when your healing capacity is being drained, before you come down with symptoms severe enough to diagnose as a disease. Perfect example: nnEMFs provably increase chronic inflammation on lab tests… even when you don't feel a thing. For instance, after installing a smart meter on a home or apartment complex, occupants' inflammatory markers such as TGF-Beta 1, MMP-9, and serum copper levels surge. Hormones and neurotransmitters are also thrown out of whack.

Changes like these lead to imbalance, then dysfunction, then disease. That's how disease develops. First your physiology is stressed and it employs corrective measures from its bag of tricks. If the exposure continues, your body's healing capacity is whittled down, until it runs out. Then you exhibit preliminary symptoms. And then, when your body runs out of regenerative capacity, your organ or system fails entirely. However, you might not notice anything except mild discomfort or "good days and bad days" along the way, before complete shutdown occurs. Hence, scenarios similar to the following have become all-too common: *One day my eight-year-old was perfectly fine, and the next day he came down with leukemia or kidney failure.*

Point being, nnEMFs disturb systems at their source, which is upstream from end-stage failure. So disease processes usually develop imperceptibly through poor hydration, mitochondrial dysfunction, circadian disruption, impaired vitamin D and A cycling, and reduced redox potential. That means everyone's physiology suffers adversity from nnEMFs, not just those that seem to be hypersensitive. However, mitochondriacs are the only ones seeing symptoms of crummy light, water and magnetism for what they really are, then addressing deficiencies and dysfunctions before they become serious complications. The early warning signs telling you disease is on the way are less obvious, and more easily dismissed, because many of the clues are different from what mainstream medicine trains you to notice. The symptoms are biophysical, rather than biochemical, so most people don't know what they're experiencing but not understanding.

Heavy metals scatter nnEMF transmissions

Heavy metals such as mercury and aluminum increase the injury that nnEMFs do because they break up the uniformity of electromagnetic signals and make them bounce around inside the body chaotically, instead of passing through more quickly and quietly. So instead of a unidirectional wave traveling through you, the wave becomes omnidirectional – ricocheting off in random directions after the wave hits metal. Shaking cells and mitochondria in all directions disturbs them even more than moving back and forth. It does not make them happy, which compounds stress and injury to cells.

If you've ever put metal in a microwave oven, you've seen what microwave radiation does when it strikes metal. You get sparks, violent popping sounds, heat buildup, and possibly a fire. On a microscopic scale, you don't get the same fireworks. But what you do get is microwaves and metal not getting along, in addition to (1) a buildup and sudden release of static electricity; (2) foreign frequencies made more destructive by multiplying their angle of attack; and (3) delicate biology in the middle of the mêlée – all reasons why heavy metals in the body amplify the damage done by nnEMFs. This is one more reason why vaccines are bad for you: aluminum-based adjuvants deflect biophotons, thus interfering with the body's own internal communications and magnifying nnEMF damage.

Metals can also act as antennae (transmitting EMFs to you)

Some common sources may shock you:

- **Tattoos.** Colored inks used in tattoos contain metals that don't play well with nnEMFs. Red ink, for example, is made with a lot of iron that interferes with the skin's ability to assimilate sunlight. In fact, tattoo inks diminish vitamin D production, redox potential and leptin reception. When that happens, mitochondria cannot maintain or repair cells as well. As a result, doctors examining x-rays often see metal artifacts in the tissues underneath tattoos, resulting in organ dysfunction or degenerated discs, as examples.
- **Mercury amalgam fillings, dental fixtures.**
- **Orthopedic hardware.**
- **Metal-frame glasses.**
- **Underwire bras.**
- **Metal belt buckles and metal jewelry.**
- **Heavy metals.** Toxic metals in the body respond to microwave and radio-frequency radiation. They receive them and react to them. In doing so, they interfere with the body's own regenerative frequencies.

The *real* reason the world seems to have lost its mind (e.g. 5G/IoT, prescription drug addiction, glyphosate, mandatory vaccination, desecration of the planet, water and electricity rationing, wild fires, homelessness, school shootings, mental illness, wokeism and cancel culture)

If wireless EMFs are so bad for us, why don't our elected officials and corporate heads do something about it? They wouldn't knowingly hurt millions of innocent people, would they? They'd purposely be destroying themselves, their children, and the planet by taking part in the plan. It would be suicidal. It would be like mass murdering billions – one bodily insult at a time. How could they?

You know what I say to people who live in such a profound state of innocence (to put it kindly)? *'Don't project your values onto other people – especially the ruling elite – because trusting the wrong people now can have deadly consequences.'* You see, these people were raised under a completely different moral code then you and me. They were taught a sense of entitlement since they were toddlers. So they grew up thinking they're better than everyone else. And now they believe they have a divine right to rule the planet, and do with it whatever they please – including exterminating whatever amount of humanity they need to accomplish their plans. Therefore, controlling and manipulating people doesn't pose a moral dilemma for them because, in their mind, they own us. We are their property. The higher up the pyramid you go, the more out in the open they practice this belief system.

The elite have grown up with all the wealth and privilege the world has to offer. And that takes a lot of the fun out of spending money on "ordinary people" pleasures like fast cars, nice clothes, exotic vacations, and the best of everything. So what happens? Having mastered money, they turn to controlling and manipulating people, systems and populations to get their kicks. Getting away with unspeakable evil and corruption on a grand scale is now their idea of fun, when just being rich doesn't do it for you anymore.

That means those under the top of the hierarchy are given an agenda to fulfill. There's a master plan created behind closed doors. And they each do their part in the plan, or else their income and privilege get taken away, and they go to work in retail or food service for worker's wages. That brings them in line pretty fast. In the wireless era, that means everyone from heads of industry, to "elected" officials, to end consumers do their part to move the wireless agenda forward. A few people know what their part in the plan is. Most don't. Another way to put it is consumers are played emotionally like a piano, while the elites of the world are incentivized with money, power and privilege as perks of being in the club. It's very simple.

Bottom line: 5G systems, artificial intelligence (AI), and the surveillance state are nothing less than tools of tyranny sold to us for their convenience and rewards. Along with injections and monitoring apps, these are the technologies that globalists are weaving into society to exercise power and control over We The People. In this socialist/communist/totalitarian state (aka "The Great Reset") the working class won't have real freedom and self-determination. What little you may have left is only an illusion. Can it be stopped? We'll find out in the next few years.

5th Generation wireless (5G)

As I write this in 2019–2022, 5G is set to become the most destructive force man has ever unleashed onto terrestrial biology. As it's rolled out in cities across America over the next 24–36 months, you're going to see unprecedented numbers of people suddenly coming down with serious unexplained illnesses, without any prior history or traditional risk factors. Conditions such as cognitive decline, autoimmune disorders, metabolic syndrome, cancer, suicide, psychosis, heart attack, stroke, and allergies will become frighteningly common. And they'll be hitting previously well people without warning. Pay close attention so you'll be able to recognize symptoms occurring in the people around you. Ignore it, or underestimate it, at your peril.

Dr. Jack said 5G may actually be a good thing. He's optimistic about 5G
Why in the world would anyone say that? A: Because 5G is the technology that will finally get skeptics and slow learners to realize that the explosion of chronic, degenerative disease is caused primarily by non-native frequencies – not just bad food, pesticides, heavy metals, bad bacteria, bad genes, or lack of exercise. It is the tipping point – the metaphorical slap in the face – that will finally wake people up to the devastating effects non-natural frequencies have on humanity… not to mention what it does to plants, animals and microorganisms.

But why now? Why 5G and not previous technologies? The main reason for the devastation coming our way is the nature of the 5G signal. Unlike 3G, 4G and Wi-Fi, 5G runs on a lower transmission power, higher density of signal – effectively creating a blanket of millimeter waves enveloping your physiology every minute of every day. Even worse, the waveform of the 5G signal is "micro-shaped," meaning smaller waveforms are embedded onto the primary sine wave so it carries at least one more level of data – very similar to the way dirty electricity rides on 60 Hz power… only with 5G, it's deliberate. This adds another level of damage potential to 5G exposure – possibly multiplying the harm it does. No one really knows for sure. But we can say dirty electricity is far more harmful than a clean 60 Hz sine wave.

Of course, the 5G signal itself is not immediately life-threatening the way nuclear radiation is. Instead, the damage is proportional to: (1) the number of nodes you're exposed to; (2) their proximity; (3) the number of directions from which they hit you; (4) the frequencies involved; (5) the damage potential of micro-structuring the waveform; and (6) the length of exposure. Most concerning of all, to give users "blazing-fast" download speeds, 5G has to surround you with transmitters every 500 feet or so. It's like cooking your brain, body and microbiome in a low-power, always-on microwave oven you can't feel, and may think harmless, because you trust those whose job it is to protect you.

Attack vectors of 5G

Wireless technologies prior to 5G were designed to connect you to one antenna tower at a time. But now, when you communicate over a 5G network, the data packets are broken up and routed through multiple cell towers, if possible. That's one of the reasons 5G can achieve such mind-boggling connection speeds: it's "gang-relayed" through multiple cell towers. More signal sources means that one call or data connection is now far more disturbing to the surfaces of your cells and mitochondria. So instead of ½–52 billion oscillations per second from a single source, 5G subjects your internal biophysics to hundreds of billions of oscillations per second. It's like you are being attacked on all sides with corruptive frequencies, instead of just one tower. The industry has not studied what multi-tower communications will do to a person's biology. But you can be sure it will cause more injury than any single-tower technology thus far.

But if there were any way the potential for injury could be worse, this is it: 4G transmitters can now be installed in 5G nodes (every 500 feet). Unfortunately for life on earth, the 4G signal travels farther than 5G, and penetrates much deeper into materials such as human flesh. That means we'll soon be getting the worst of both worlds: the virulence of the 4G signal, combined with the proximity of 5G. For many unfortunate families, that will put a high-powered transceiver within 25–100 feet of their sleeping space. And it will be camouflaged so people might experience adverse effects, but won't know what's causing it. How do you like your steak cooked? Because no matter whether the signal is designed to be a weapon, or just a way to stream movies, **microwaves maim regardless of purpose.**

Crazier still, 5G was designed specifically to carry the signals of Internet of Things (IoT) devices

So now, it's not just cell phone signals you need to worry about. When 5G really gets going, its signal will bombard your body with two-way microwaves from cell phones, laptops, tablets, TVs, refrigerators, washing machines, dish washers, microwave ovens, toasters, light bulbs, doorbells, baby monitors, surveillance cameras, speakers and everything else corporate interests and government can conceivably put on a network. And that's just one network among more than half a dozen equally corruptive waveforms that aren't doing your health and mental state any favors. Are you starting to see how living in a microwave oven is beyond a bad idea? But even more disturbing...

5G is designed for surveillance and centralized control

5G is not about connection speed, safety or security, because if those were its main purposes optical fiber is much faster, safer in terms of health effects, and more secure against hacking. Nor is 5G about making your life easier, being better connected, or other excuses like these. Rather, 5G was conceived and built from the ground up to spy on people, and ultimately control every aspect of your life and mine. Skeptics don't want to believe it, but that's exactly why puppet masters behind the telecom companies are pushing 5G as hard as they are.

It's military technology for the purposes of knowing everything about you – including what you're buying, who you're talking to, what you're saying, where you're travelling, what you're using your energy on and when... and, with AI technology, what you're likely to think and do next. It's wall-to-wall surveillance and control, with some benefits thrown in to get the mindless masses to accept their own enslavement by agreement, indeed pay for their own servitude.

But you've done nothing wrong. You've got nothing to hide, right? That's fine if surveillance is, and will only ever be, used to catch terrorists. But how would you like your bank accounts being docked for using electricity or water in unapproved ways? You can't undo the charges, or even question them, because penalties are automatic and irreversible. How would you like to be hit with civil disobedience penalties, possibly thrown in jail if you object to fluoridation, mandatory vaccinations, GMOs, taxes, government policies, critical race theory, or even the forced drugging of your children with stimulants such as Ritalin to attend public schools? How would you like to have your children taken away from you because you were teaching them traditional values, or the religion of your choice? How would you like to have your bank accounts turned off, **or your retirement accounts deleted**, when you speak out against injustice?... *Because 5G and IoT make all of that very easy.*

Don't care about privacy? Okay, why don't you post your passwords on your Facebook page? Why don't you record all your conversations and post them online so anyone can listen to them whenever they want? And while you're at it, why don't you tell the world when you leave the house, when you go on vacation, and where your children are at all times – on your Twitter feed?... *Because 5G is very hackable.* I'll say it again: Don't project your values onto other people and other groups. Just because you wouldn't use powerful surveillance tools like 5G for nefarious purposes doesn't mean others won't. That's like saying fairness and the rule of law always prevail over greed for money and power.

Here's the proof: China is already using a social scoring system, and Australia is rolling one out. For years now, the Chinese government has implemented state-sponsored surveillance and enforcement with their social credit scoring system. In this system, jaywalk and you could be fined electronically. Write articles that criticize the Chinese Communist Party and you could be docked social credit points. Protest, or break the law in any way, and you could be denied a bank loan, job, travel permit, credit card, social services, or even health care if your social score drops too low. Oh, but that's a communist country. It couldn't possibly happen here, in a democracy, right? That's naïve, frankly. All of that *is* happening as fast as we allow it to happen – that is, as fast as they can get away with. Some people just can't see it.

Bottom line: 5G is not inherently good or evil. It's just a tool. The intentions and moral compass of the people wielding that tool are what you need to worry about. So ask yourself this: Do your would-be rule-makers respect boundaries, or do they continually cross them, move the line, and then lie about it? Are they fair and honest people, or are they money-motivated, corrupt(able), and self-interested? Finally, do you trust them with unlimited access to your personal information, finances, life, and family's health? ...Because with every minute detail about you monitored and recorded for easy access, you won't have any privacy, and you must fit in at all times, or suffer the consequences. Or, an even simpler way to say that is *data is the new oil – specifically, your personal information. And it's being monetized now.*

5G is the demarcation line between an oil-based economy and one in which the information they mine about you is the world's most valuable asset. That means telcos are going to squeeze every last drop of profit they can out of selling your personal information to marketers and agencies that want to know what you're doing at all times. That's what 5G is really about. Invite it into your life through purchases or acquiescence, and that's what you're signing up for. But don't expect living in that world to be fair, to make sense like you think it should, and that it will never be used against you. That's fantasy. Indeed, one truth-teller called it brilliantly when he said his definition of *smart* is "Surveillance Marketed As Revolutionary Technology."

Watch out for freakish effects of 5G

Jump conduction. The power density of 5G is so much greater than previous generation it can actually accumulate on conductive materials and jump to other conductors. This can release shocking levels of static electricity, bordering on electrocution. For example, 5G frequencies can accumulate on street lights and jump to hand rails and manhole covers. Even worse, a home's electrical wiring can accumulate 5G frequencies and jump onto water pipes, gas lines, metal studs, and electrical wiring, thereby creating a seriously nasty field of dirty electricity right in your bedroom or living room.

Coronal mass ejections (CME) release large amounts of plasma, magnetic flux, and EMFs from the sun. Large CMEs can shut down power grids, cause fires, and electrocute people touching conductors.

Unexplained fires. It's been known for decades that coronal mass ejections from the sun threaten our power grid. Fortunately, in modern times, CMEs haven't caused any major outages in the US. However, 5G dramatically reduces the amount of solar energy needed to disturb the power grid so that minor solar activity could now cause massive damage. This is exactly the phenomenon that's believed by some mitochondriacs to have caused the Santa Rosa fires of 2017. Just days after 5G was activated in the area, huge changes to the electrical power grid were reported by PG&E. Shortly after that, fire ravaged Northern California's wine country, with Santa Rosa getting the worst of it.

But this was no ordinary fire. Some buildings were completely leveled as if an atomic bomb had gone off – even metal in the buildings. Aluminum engine blocks and metal appliances melted, while nearby trees were singed but would easily recover. Pictures revealed a nightmarish scene like a zombie apocalypse. The reason? The 5G signal, combined with solar activity, jumped onto power lines and was captured in devices and electrical conductors, essentially fueling the fire. On the other hand, trees and plants, being well-grounded, were able to dissipate excess EMFs back to ground and were spared terminal damage.

Freakish effects of 5G to look for: Water main breaks. Changes in earthquake activity. Underground fires. Fires at gas stations, triggered by 5G phones. (If you have one, put it on airplane mode when stopping at a gas station, because the energy these phones emit may now be strong enough to ignite flammable materials.)

5G will come from satellites

At the start of 2019, telecom companies announced plans to launch tens of thousands of satellites to carry 5G. By 2021, hundreds had already gone up. This will create a sort of simmering death ray, which will result in major health problems that doctors and government agencies can't explain. Retrofitting your home with metal roofing, or building from the ground-up with EMF-sheilding materials, may be the only way to protect yourself from this alien energy assault.

Telecom companies know the damage their products do

Telecom companies are fully aware that blue light and nnEMFs are destructive to human health, but they've worked tirelessly to hide the data and confuse people about what the science really says. All they've had to do is ignore the literature that's already been published, deny its validity when it's brought up, and confuse the issue as long as possible by funding questionable science that appear to support their safety claims. For example, designing cell phone studies to last four years when they know cancer starts showing up at the five-year mark.

Unfortunately, no single company or government agency is going to willingly stop their rollout of new wireless technologies, no matter how harmful, because sales and company stocks would crater. It would reveal a web of lies they've spun for decades. And it would crash economies around the world that depend on chronic consumption to stay afloat.

Furthermore, there are frequencies that are not harmful, or can even be used to heal. But the wireless industry has deliberately chosen frequencies that harm the human body. Sterilization is a prime example: A study done in orphanages, then psychiatric hospitals, found the best frequency to cause infertility is 2.4 gigahertz. And wouldn't you know it: That's the frequency the FCC chose to let companies use without license or active oversight – a kind of "free-for-all" frequency. Hence, 2.4 gH used in cordless telephones, baby monitors, Wi-Fi, Bluetooth, microwave ovens, car alarms and wireless microphones, among others. What a coincidence, that lines up perfectly with our plummeting fertility rates.

The tech industry installed liability shields

In 1996, after spending $50 million in lobbying, the telecom industry got the Telecommunications Act (TCA) passed. It prescribed into law that no one can prevent telecom companies from putting up cell towers. And no one can make them remove wireless service based on health effects, as long as they remain compliant with the prevailing FCC guidelines. Furthermore, an FCC law passed in 2006 exempts technology companies from all liability when their products cause death and disease, provided their products were legal to sell at the time of manufacture.

This proves they knew about the danger long ago and lobbied Congress to reduce their liability exposure as they ramp up the power and pervasiveness of electromagnetic pollution. However, those laws only protect telecom companies in the US market from liability. They have not been able to exert their undue influence as broadly in countries around the world. As a result, at least one retired executive from a giant telecom company says his company was putting away $250 million a quarter as a reserve for the coming cell phone lawsuits, despite the liability waiver they have stateside.

Industry leaders "in the know" keep their children away from nnEMFs

People idolize Steve Jobs for leading us into the wireless era. However, the story seldom told is how he and Apple have known for decades their devices are not safe from side effects, yet they've continued to claim their products are completely harmless. But how do we know for sure that *they know* nnEMFs hurt people? Because they've installed features in their products, and liability shields in the regulations, specifically to limit injury. You can deny the facts all you want, but the truth is they are purposely hurting children, women and men in order to sustain their market share and profits.

Case in point: For years, Apple has been putting infrared sensors in iPads to recognize when the device is close to a person's body while Wi-Fi is turned on but inactive. That way, Wi-Fi can stop transmitting when idling to reduce injury to a person's internal and reproductive organs. Shocking but true. They've gone to the trouble and expense of engineering a feature into their devices that helps protect people from harm. Yet they've never promoted that in their advertising. They must know about the health effects, but saying so would be admitting that they know. Despite FCC laws that shield the company from liability when legally compliant at the time of manufacture, they're still taking steps to reduce injuries. They are preparing for the inevitable disclosure that microwave EMFs and blue light are far worse than the industry lets on.

And what's the dead giveaway that Jobs knew? In the last few years of his life, he never let his own children use the wireless devices his company was making. What does that tell you? So, was his death from pancreatic cancer coincidence or was it karma? You be the judge. What's more, Steve Jobs is not alone. **Bill Gates did not allow his children to use wireless devices, either.** It seems he knows, as does the entire industry.

The pancreas sits in the abdomen – inches away from where a laptop signal on the lap, and a cellphone signal in the back pocket, would intersect.

As Dr. Jack Kruse suggests, every company of significant size and sophistication is fully aware of the damage that artificial blue light and microwave radiation does to our cognitive function, hormones, gut permeability, metabolism and mitochondria. Yet they do it willfully because repetitively stimulating your rewards system keeps you addicted. Depressing your dopamine and critical thinking ability helps them sell more stuff. It's a brilliant, albeit manipulative and self-serving business model. Finally, there are multiple reports of high-level telecom executives admitting privately that their companies know full well their products profoundly disturb our biology. But those insiders say they're terrified because whole economies are being supported by these new technologies, such as IoT devices. They're afraid to say anything.

In conclusion, anyone who believes the largest corporations in the world are innocent actors in the relentless march of progress simply doesn't know how the world really works. These are the type of sheeple who would believe the tobacco companies didn't know smoking was bad for you fifty years before they were forced to admit it in the 1990s. Honest mistakes of this magnitude are extremely rare when big money is being made. They know a day of reckoning is coming when the world realizes the full extent to which nnEMFs devastate human health, and they're preparing for it.

Lloyd's of London refuses to insure tech companies against 5G claims

Lloyd's of London – the international insurance company that's notorious for insuring things other companies won't touch – refuses to insure tech companies against health claims related to 5G injury. That tells you something. When a company world-famous for insuring everything under the sun refuses to take an entire industry's billions in premiums, it's because they believe the risks are too high. They anticipate more money being paid out to claimants than they could possibly recoup.

And it's not as if they simply don't know and are playing it safe when it comes to multi-billion-dollar risks. They already know the damage that 4G and other nnEMFs are doing, and can do with 5G. They absolutely, positively know. They've done the math and figured out they can't make money in the 5G business because they anticipate too many people getting sick and dying from 5G. Furthermore, their decision has nothing to do with fear in the way that the industry accuses consumer advocates of fear-mongering. Their well-reasoned decision is strictly business. It's their business to assess the risk *vs.* reward. And, based on what they know, they've decided the risks are too real and too high; they're staying out of the 5G business. Lloyd's is turning down billions of dollars in profits today, for a liability that may be years away. That's extremely well-informed analysis and rational decision-making we ignore at our peril.

We're moving into uncharted territory with 5G

As 5G is being rolled out across the globe in 2019–2023, no one really knows how bad the consequences will be to all life on earth because there is no research to say it's safe. We can only estimate, based on the science of how nnEMFs harm mitochondrial health and circadian rhythms, correlated with the damage that preceding technologies are proven to do. Early evidence is showing that 5G may be so blatantly bad for you that telecom companies will be forced to admit to it, and change it to make it less toxic to all life on earth.

15

NON-NATIVE EMF REMEDIATION

Strategies, products and practices

Where technology is headed and why

Since the 1980s, technology companies have claimed their devices are completely safe because their EMF emissions are not strong enough to heat tissue or ionize atoms like radioactivity does. They've avoided blame and financial responsibility for decades because, frankly, it's been hard to connect their products and practices back to individual consumer's health challenges. The connections have just been too hard to prove in a court of law or public opinion since most disease has multiple causes.

However, behind the scenes, most executives, engineers and elected officials are fully aware of the hidden dangers that *non-ionizing* frequencies present to all life on earth. They absolutely know. They're just keeping it quiet as long as possible because both industry and its supporters in government see wireless technology as the only future for sales and society moving forward. They consider 5G and the saturation of our airwaves to be inevitable, no matter what happens to life on earth.

So their thinking at this point goes something like the following: "We know the adverse effects are going to be unfortunate for some, and a living hell for others. But we'll just have to wait and see exactly how bad people's injuries are, and how much accountability we'll be forced to take. We'll deal with the backlash when it gets here. Until then, all opposition to 5G and the IoT must be stomped out by any means necessary, no matter what the consequences to them or us." No joke.

Becoming your own health boss

That means YOU are going to have to take steps to protect yourself and your family, because no doctor, three letter agency, or advocacy group is going to do it for you. YOU are going to have to: (1) know enough that you're not so easy to fool; (2) spend some money on your health and wellness from time to time; (3) try out some biohacks to find out what works for you and what doesn't; and (4) adopt lifestyle changes that attract good outcomes into your life, while repelling the bad. YOU are going to have to learn a little about light, water, magnetism, and your mitochondria to make more intelligent choices.

It's not like the good old days when you could leave your family's health up to a guy in a lab coat, an insurance provider, and supplement companies that claim to be looking out for you. Those days have been over since 4G and Wi-Fi rolled out… and since glyphosate started to be sprayed on crops by the billions of pounds. We crossed that line when communication went wireless (it's called disease statistics). We've just been in collective denial for more than a decade.

No one can honestly deny that human health has sunk to new lows over the last 5–10 years. Most now accept this without question. As a result, millions have become proactive with their wellness. People are making their health a priority with better commitment, more action. But is it enough? The sad truth: Millions still undermine good intentions with flawed information, deranged priorities, commercialism and complacency. To our misfortune, we've been fooled into believing the right diet, supplements, fitness program, and drugs will make us well and keep us well. When, in fact, they're more like a headwind or tailwind on our efforts, rather than the motor that drives us where we want to go.

For the record, exquisite health is built on a foundation of two things: getting your mitophysical exposures right – including light, water and magnetism – and ridding your living spaces of nnEMFs. As two sides to your wellness/illness coin, they are the most powerful approach to reversing illness, and the most direct route to attaining wellness beyond appearances. That means the benefits you get from doing what everyone else is doing pale in comparison to the gains to be made by upping the *native* EMF exposures in your life, and purging *unnatural* frequencies from your routine at every opportunity… starting now.

Get as much real sunlight as you can on your eyes and skin

- See the sun rise as often as possible. And see it set, if you can. It's Dr. Jack's #1 recommendation for improving whole health.
- Get as naked as you can when you're outside during the day, and let the sun hit your skin. Several companies make UV-permeable bathing suits and clothing so you can be more presentable in public places.
- While getting your daily dose of sun, forgo sunglasses, eyeglasses, contacts and window glass, because they all block UV and part of the IR. Getting UV and real sunlight also improves your eyesight.
- For every hour you spend inside exposed to blue light during the day, try taking a "sun break" outside of at least five minutes or more.
- Don't use sunscreen. Instead, acclimate to more sun exposure without burning by hybrid tanning, as described in chapter 13 (page 191).

Reduce your bad-light exposure

When you can't turn off the artificial light, swap it out, or avoid it, then block it through the eyes, and on the skin.

- Blue light is far more harmful after the sun's gone down than it is during the daytime. After dark, you need to block 100% of the blue to synchronize your circadian rhythms. You can buy pre-made blue-blocking glasses. Or you can add an amber tint to any prescription lenses through companies like BPI. But avoid wearing blue-blocking glasses during the day – when driving, for example – because they can make you sleepy by raising your melatonin.
- Next, most tech screens have settings or downloadable apps that shift their color balance away from blue, toward red. This can help more than you might imagine. But (good) blue-blocking glasses are considerably more effective at reducing your blue light exposure.
- The more indoor blue light you're in (intensity and duration), the more important it is to cover your skin with clothing. In particular, you want to cover your throat area when using your smartphone, watching an LED TV, or working on a computer, because blue light penetrates into the thyroid (which sits just below the skin surface). Remember, heteroplasmy and circadian mismatches from blue light are a major contributor to hypothyroidism. So button your collar all the way to the top, wear a scarf, or get turtleneck sweaters.
- Night time is the one time it might be beneficial to wear as much makeup on your face as you like in order to block the blue.

Choosing the healthiest light bulbs

All artificial light is bad for you because it's different from the real, full-spectrum sunlight that powers and controls our biology. Man-made light contains too much blue, and not enough of the other colors. You also tend to get it at the wrong times. The more mismatched it is to real sunlight at that time of day, the worse it is for you. And you tend to receive it indoors, behind glass windows, which block UV and a lot of the IR.

Make this your mantra from now on: Real sunlight is good for you, and artificial light after dark is bad for you. However, there's a little more to it than that. You may be surprised to learn how remarkable the differences are when it comes to composition of light produced by the different bulb types, and the health effects they invite into your life.

Ranking light bulb types, from least healthy to most healthy

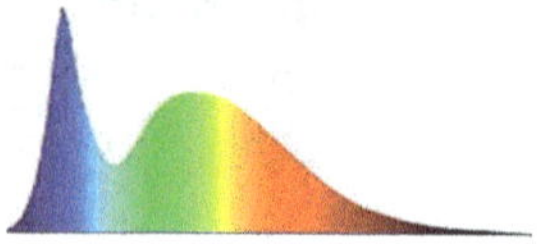

LED is by far the worst for you, because **cool white LEDs** (~5700°K) put out a big spike of blue light, reduced amounts of the rest of the visible spectrum, no UV, and almost no IR. They're made by encasing a blue LED in yellow phosphors to turn the light whiter. The bad news is, blue light between 435 to 465 nanometers destroys melatonin, mitochondrial function and dopamine – making light from cool white LEDs the most unfriendly type of light you can buy.

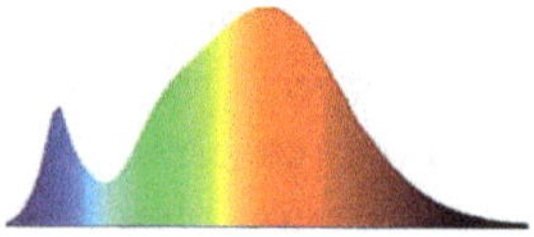

Warm white LEDs (~2700° Kelvin) are a little better for you. They peak in the green-to-orange range, depending on design, and have a smaller, but still sizable, spike of blue, with no UV and very little IR, if any. That makes warm white LEDs one grade better for you than cool white LEDs, but still detrimental. Both flicker and mess with your circadian cycle more than you'd think.

Fluorescent. Compact fluorescent bulbs are worse than worthless. At least you can say LEDs are the most energy efficient, and longest lasting, of all bulb types.

But fluorescent lights:

- use more power than LEDs;
- don't last nearly as long as LEDs;
- flicker at twice the rate of the power grid, which stresses the brain smoothing out the strobe effect;
- produce huge amounts of dirty electricity (compact fluorescents);
- produce huge spikes of mostly blue, green and orange, with no UV and IR to heal you;
- electrify mercury to produce UV, which hits the bulb's phosphorescent coating, making it glow; this homeopathically embeds the vibrational frequency of mercury onto the wavelengths emitted (somewhat like dirty electricity does), which radiates all your cells with destructive frequencies (particularly brain cells), while mobilizing toxins in the body such as mercury;
- release large amounts of mercury when broken – accidentally, or when disposed of improperly.

Candlelight-style OLED. By comparison, candlelight-style OLEDs are not terrible. They do flicker, but they produce a full spectrum of visible colors that looks like a mountain with orange in the middle, no UV or IR. Most important, they don't have much blue, so they avoid the big spike that makes cool white LEDs so destructive to melatonin, dopamine and mitochondria.

Halogen. Halogen lights are a brighter, more energy-efficient type of incandescent light. So they don't flicker very much. They make nearly full-spectrum light weighted toward the middle (yellow), with good IR, and no UV. That means they're modestly disruptive to our circadian biology due to their altered light spectrum, and when you tend to receive that exposure. Overall rating: above average.

Incandescent. Traditional incandescent is the best type of light bulb for you, in terms of health impact. Their light comes from a hot tungsten filament, so they produce full-spectrum light weighted toward the red/infrared. That makes their light minimally-disruptive to circadian rhythms from a color standpoint, and potentially disruptive when used hours after nightfall. They are the most expensive to run in terms of electricity, since they produce a lot of heat that goes to waste. And they don't last as long as the others. However, higher wattage bulbs (>60 watts) don't flicker like LEDs and fluorescents. And their color spectrum is best of the artificial bunch – next best thing to fire.

Fire and candle light. The best type of light for human biology, after sunlight, is fire. That's because it contains the full spectrum of colors, and a healthy dose of IR (42%). That means its mixture of colors and intensities is generally not harmful. But the time of day/night when you get them can potentially upset your circadian system. Too much red, too often, long after sundown, can lower your melatonin level. And it can give you cataracts. However, most people have got much bigger EMF concerns to worry about. Verdict: cavemen were mitochondriacs before it was cool. Fire's still best.

Flicker

The way LED and fluorescent lights flicker is more disturbing to the nervous system than people think. In order to be energy efficient, LED bulbs turn on and off 100–120 times per second, which is faster than the eye and brain can perceive. They may be more economical in terms of electricity consumption, but they're not better for you and me. That's because the brain consumes energy knitting still images into smooth-motion video. So what starts out as merely energy-consumptive can become draining after hours of exposure. As a result, flicker is a major source of eye strain and mental fatigue as the brain converts the digital on-off signal (i.e., square waveform) into a continuous analog one.

But even worse are dimmer switches on LED bulbs. They alter a bulb's waveform so the gaps are both wider (requiring more interpolation) and more unnaturally shaped (requiring more transformation). Dimmer switches make the flicker much worse for you by increasing workload on the brain.

Excellent video by Dr. Alexander Wunsch explaining the hazards of artificial lighting: vimeopro.com/alexanderwunsch/alexander-wunsch-in-english/video/270946568.

The thinking man's way to choose light bulbs

It is possible to save money on electricity with energy-saving light bulbs. But you're going to end up spending far more on medical bills over time. So save up that sick time, make sure you have good health insurance, and get used to taking pills. We call that being pennywise and pound-foolish. Don't be fooled: LED and fluorescent light bulbs are more economical to buy and operate. But you're going to be sicker, dumber, and more controllable as your reward. Instead, invest in the single easiest lifestyle choice you can make for better health: getting your light environment right. Just spend the extra hundred or two on electricity each year to buy yourself more wellness.

Lighting prescription: Use candle light, a patio fire pit, or oil-filled lamps as your preferred light source at night, if you can. If that's not doable, go with incandescent or halogen in the places where you spend the most time. And use LEDs in places where the energy savings outweighs any erosion of health – such as outdoor security lights, laundry rooms and attics.

Wireless nnEMF mitigation strategies

Start your EMF remediation effort by expanding your knowledge and awareness of nnEMFs. Learn when and where you're exposed to EMFs, what dangers they present, and how to disentangle yourself from their influence. For example, instead of streaming music or videos to your smartphone in real time, download them to your device while it's away from your body, or to your (wired) home computer. Then play them back from your phone's memory. Simple strategies like this dramatically cut down on EMF exposure from your cell phone – probably by a factor of 10 to 100, or more.

Next, try simple assessments like the following to get your head and heart on the same page: Pay attention to the way cordless phone radiation makes your brain feel – particularly on that side of your head on which you rest your phone. If you feel warm, tingly sensations – or any odd effects – realize that foreign frequencies are impairing ATP production and charge on the cell wall, which routes more blood flow and electrical flow to the area. You need to notice when damage is being done, and understand it for what it is. Keep in mind, this type of stimulation is different from Nature's own electromagnetic waves. Our bodies know what sunlight and heat from a fire are supposed to feel like. But we're not wired to know what non-native frequencies feel like. This tends to make us misread our senses.

In other words, *programming* short-circuits a person's own critical thinking so they lose independence and become reactive. When critical thinking is not there to protect you, *medical experts* and *media stories* become your default truth, as a result of conformist thinking. Lately, people who don't think for themselves have been dubbed "normies" to poke fun at just how fast asleep some people are to what's really going on. On the other hand, *independent* thinkers take responsibility for what they believe, and the outcomes they get in life.

Most important, mitochondriacs treat all information coming from the mainstream as suspect when separating truth from fictional narratives. Unfortunately, none of us are a given an accurate B.S. detector at-birth. Instead, open-minded individuals develop a sense of skepticism by seeing false narratives crumble to pieces, one after another. Over time, we question centralized sources after catching them in lie after lie. However, it's human nature to want to believe others. So a survival skill we all need to practice in a wireless world is the power of discernment.

To illustrate bias in action, try this thought experiment: Say you just gave your Chihuahua a bubble bath. It's shivering, so now you want to dry it off. Why don't you just put it in the microwave oven to warm it up, and dry it off, pronto? How about for only a minute? What... not even on low power? I don't know about you, but this sounds like an incredibly bad idea – even for one second on low power. And, yet, we do almost the same thing each time we put a cell phone against our head.

Why do we deliberately shoot microwaves into our brain, many times a day, if we wouldn't do the same thing to our pet? Answer: Because those we trust *said it was okay*. That's it. Experts have told us cell phone signals are safe AND everyone is doing it. So we assume *'every product I buy is safe because I trust "the experts" to protect me.'* Sadly, this is why most people set aside their own reasoning and fall back to groupthink, 99% of the time. You see how societal bias is ingrained into all of us? We're all prejudiced towards trusting authority figures, instead of exercising our own judgment. Like good consumers, we just think and do as we're told. Unfortunately, that's not going to go well for you anymore. I advise you to take the red pill and see the world around you the way it truly is.

Red pill: A rude awakening when a person's false beliefs are dispelled after they were programmed to think in another way.

General guidelines to reduce your nnEMF exposure

1. **Increase distance (the Inverse Square Law):** Distance is your friend when it comes to reducing nnEMF exposure. When you can't turn off the source, and you can't shield yourself from it, your best strategy to reduce exposure is through the Inverse Square Law. It states the total amount of nnEMF exposure goes down in proportion to the square of the distance. That means if you double the distance between you and an EMF source, your exposure level isn't just cut in half. It's a quarter of what it was. So if you move from 10 feet away to 20 feet away, you're getting 25% of the exposure. Of course, the opposite happens when you halve the distance to an EMF source: you quadruple the dose/exposure. Unfortunately, these days you may have no other choice but to move to a new location in order to put more distance between yourself and hordes of obedient idiots irradiating themselves like addicts.

2. **Decrease exposure time.** Perhaps easier said than done, because "they" make it easy to get hooked, and darn difficult to unplug. That's because blue light and social media are addictive by nature. Technology is getting harder and harder to turn off – literally, and in terms of dependency. The tech companies just keep you coming back for more, and spending what you will, consequences be damned. Fortunately, getting natural light, at the appropriate times of day, rebuilds your body's endogenous supply of pleasure chemicals so you don't need external sources as much in order to feel good. Find those 'off' buttons and give the real world a try for a while. You can break the cycle by not supporting the industry that is intentionally harming you.

3. **Population density.** Along the same lines, your biggest exposure concern should be how many people live and work around you. Irksome to the informed, the average person today uses between 7–17 wireless devices. Most have no idea what they're doing to themselves and others. That number will only increase as people saturate their lives with smart devices. Which means if *you* experience serious mitochondrial issues from *other people's* nnEMFs, you may have no choice but to move to an area with less population density to get relief.

In other words, don't suffer through chronic illness, or even become a statistic on a chart, just because lifestyle changes are inconvenient, or cost more. It's been proven time and time again that people with severe cases of mitochondrial dysfunction, circadian disruption, and EMF sensitivity can feel like a brand-new person in a matter of days when they get away from the exposures that made them sick. Just go camping in the woods for several days, or take a trip to Mexico, to try it before you buy into the belief.

4. **Nighttime exposure is much worse than daytime.** Your stress level is supposed to be higher during day so you can function better. Conversely, you're designed to rest and recuperate at night. So activating

your stress response at night with microwaves messes up your hormones, metabolism, and regeneration substantially more than during the day. That means you'll get the most bang for your mitigation buck focusing on your nighttime exposure first, and daytime second. Accordingly, if you choose to shield your home from nnEMFs, shield the room(s) you spend the most time in, which is usually the bedroom. I remind you: Shielding your sleeping quarters gives you passive and automatic benefits requiring no effort, ongoing expense, or extra time to achieve.

5. DHA, oxygen and water. The more blue light and nnEMFs you're exposed to, the more DHA, oxygen, and water you need to offset their effects – **DHA** because you need more DC electricity to offset the heteroplasmy damage of blue light; **oxygen** because your mitochondria make more ATP and fewer free radicals with O_2; and **water** because it stores energy and maintains the architecture of proteins, which are otherwise lost in dehydration.

Targeted tips to reduce nnEMF exposure

- Filter dirty electricity out of your living spaces. Bedrooms and kitchens are the worst for dirty electricity and stray electrical fields.
- Reduce your exposure to electricity. Move your bed away from walls that have electrical wires running through them. Turn devices off when not in use. Power strips are great for this. Install kill switches at the breaker box, if necessary. Or unplug devices entirely.
- Choose devices that use less power. Battery power is preferable. Direct current is usually better for you than 120 or 220 volt power that oscillates.
- Check for bipolar magnetic fields where you spend the most time, and correct the ones presenting the highest exposure.
- Turn off your home computer network's Wi-Fi. Hardwire your router with CAT6–8 cable. Or put a kill switch on the router that turns it off when you go to sleep. Consider a timer. Some Wi-Fi routers also have a setting that lets you turn down its signal strength.
- Shield your house from nnEMFs coming in. Unfortunately, it can be fairly involving, but totally worth it. Start with your sleeping location. Shielding paint is a pitch-black paint containing graphite fiber, carbon fiber, and conductive particles that traffick nnEMF pollution out through a ground connection.
- Shielding clothing can potentially help, but it's not easy to achieve full protection. Don't get a false sense of security just because part of your body is covered. Specifically, your head and neck may still be exposed. And, don't forget, most devices transmit and receive in all directions. Rule of thumb when it comes to ambient electrosmog: Total

amount of body exposure is more important than where nnEMFs hit you. As a result, larger people tend to be more electro-sensitive simply because they have more surface area acting as an antenna.

nnEMF Rx: If you've got the financial wherewithal, get a whole home assessment from a good nnEMF remediation specialist. They have advanced tools and training to evaluate your home or apartment (whether owned or rented). They can recommend devices and techniques to reduce the adverse effects of nnEMFs around you. But bear in mind: There's an art and a science to nnEMF remediation as with any professional trade. And you're likely to get better results, at a better price, over the long term, all things considered. It could be the best money you ever spend.

Strategies to reduce exposure to cell phone radiation

The main goal in reducing your exposure level is to get the device away from your body, and minimize its 'on' time. And, let's not forget, most of the people around you have a cell phone in their pocket too.

- **Activate airplane mode** when you aren't using your phone, so it's not pinging cell towers every few minutes.
- **Disable other modes of connectivity when not in use (of the 5 or 6 different kinds).** Smartphones have up to six different microwave antennas operating at any one time – including call service (e.g., 3G, 4G, 5G), a data connection, Bluetooth, Wi-Fi, a geo-locator, and a hotspot. You might need to disable some of them separately.
- **Turn your phone off at night.** Nighttime exposure is much worse for you than daytime, because you need to be as stress-free as possible when you sleep.
- **Charge your phone away from your body.**
- **Hold it away from your ear.** Speakerphone is best. An air-tube headset is second best. A low-power wired headset is third best. If you can't do any of the three, just holding it away from your head is fourth best. Even an inch or two can cut your exposure dramatically. A general rule of thumb is to hold your phone at least 5 mm away from your head – the more, the better – at a 15 degree angle.
- **Avoid using laptop computers and tablets against your lap.** Even if you don't mind being infertile, no one wants sexual dysfunction and hormone imbalances. Not attractive. A radiation shield can help.
- **Avoid using your phone when you have poor reception.** Cell phones crank up their signal strength when reception is weak in order to maximize call quality. So you could get 1,000 times as much exposure when your phone is struggling to stay connected. Even a loss of one bar can translate into 10–100 times the radiation beamed directly into your brain.

- **Watch out for poorly-shielded cases.** Shielding cases for smartphones have become popular among the wise and proactive. But beware of breaches in their shielding that make your phone increase its signal strength.
- **Impact-protection cases.** Some impact protection cases increase your absorption of radiation by 20–70%, according to the Environmental Working Group.
- **Avoid using cell phones in moving vehicles.** Not only do nnEMFs bounce around inside metal boxes, but signal strength goes way up when your phone switches from one cell tower to another.
- **Switch to a previous generation signal for lower power.** Some smartphones allow you to switch to older communication protocols (e.g., 3G signal instead of 4G) in the settings. You may not notice any difference in call quality, but using 4G instead of 5G can reduce your nnEMF exposure by more than 80%, because each successive generation radically increases the power density of its signal.
- **Be wary of frequency harmonizers.** Harmonizing low frequency microwaves with stickers and pendants reduces some forms of damage and not others. It can make you feel less electro-hypersensitive, for example. But cancer and neurological injury could remain, or even increase.

PART 4

The New Way to Wellness

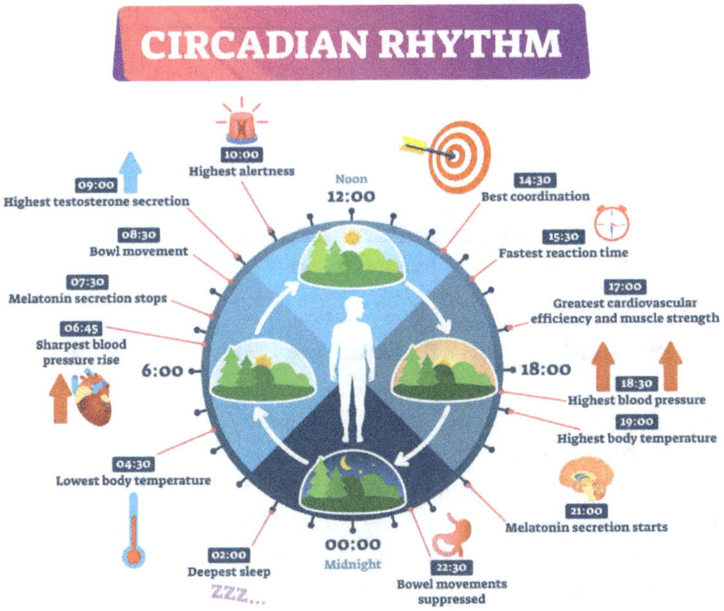

16

RED, BLUE AND O2

My two cents on the "Red, Blue and O2" protocol from Dr. John Lieurance, Dr Jason Sonners and Brian Richards

Sleep and circadian disruption (here, blue is bad)

Sleep is regulated primarily by two biochemicals: adenosine and melatonin. Conversely, cortisol, adrenaline, dopamine and stimulants such as caffeine inhibit sleep. On the promoter side, the *need* to sleep is driven by an abundance of adenosine and a temporary scarcity of ATP (they're inversely proportional). While the *ability* to go through the necessary stages of sleep is enabled primarily by melatonin. That right there is the basic formula for effective sleep: **(Melatonin + adenosine) – (cortisol, caffeine and adrenaline) = healthy, restorative sleep.**

Let's dig further into the details to find out why sleep is dysregulated in millions of Americans and how simple changes to daily routines can get your circadian rhythms cycling properly.

"Sleep pressure" is controlled by adenosine (AMP)

ATP supports wakeful activity, while adenosine and melatonin run sleep and recovery programs. An abundance of adenosine and (temporary) lack of ATP compels you to sleep. It's what sleep scientists call "sleep pressure." Sleep pressure makes you feel mentally and physically tired, with an urge to close your eyes and shut down conscious brain activity.

How adenosine creates sleep pressure: While you're awake, mental and physical activity breaks apart adenosine *tri*phosphate (ATP) into adenosine *di*phosphate, then one of the molecule's component groups, adenosine *mono*phosphate (AMP). The more work you do during the day, and the more brain power you use, the more ATP is consumed. As ATP gets used up, adenosine is produced. Thus, ATP and adenosine are functional opposites of each other. For this reason, the body uses adenosine to coax you into a sleeping state when less ATP is around to keep you active. Pet owners will recognize this process at work in how sleepy their dog gets after running around at a dog park (i.e., after consuming lots of ATP).

On the other hand, caffeine does the opposite. It counteracts adenosine. Caffeine heightens your mental and physical state by binding to adenosine receptor sites which tell your brain and body you're tired and/or sleepy. In effect, caffeine winds up the sympathetic nervous system by blocking sleepiness receptors.

Dr. Andrew Huberman explains the midday crash after morning coffee

"Caffeine is an adenosine antagonist. The longer you're awake, adenosine builds up in your bloodstream. It's what makes you feel fatigued. Caffeine essentially blocks the adenosine receptor. But then, when caffeine wears off, the adenosine that's still around [in your bloodstream] binds to that receptor and you crash. You feel really sleepy. So, one thing that you can do [to avoid a midday crash]: when you wake up in the morning, don't ingest caffeine for the first 90 minutes or so. Really push that off so the adenosine and adenosine receptor interactions can all take place and dissipate. Then you drink caffeine, and what you'll find is that if normally you would crash around 2 or 3 in the afternoon, you don't experience that crash anymore."

—Dr. Andrew Huberman.

From other interviews, Dr. Huberman says adenosine continues to buildup in your bloodstream while your first cup of coffee is still onboard. So, when that coffee wears off, you're hit with a double-dose of adenosine – from both the unprocessed backlog and the amount newly added. I call this an adenosine debt, having to be repaid… or else deferred with more coffee. He goes on to say that caffeine makes the adrenals release adrenaline. It makes the locus coeruleus release norepinephrine. And it increases the sensitivity of dopamine receptors. Put all these influences together and now you know why stimulants such as caffeine postpone tiredness/sleepiness through adenosine debt – without increasing your sum total energy reserves.

My own health crisis related to stress, sleep and repair

Near the end of 2023, I experienced health problems that could have become debilitating had I not known what caused them and how to relieve their true source. Financial pressure was the final straw that exposed my abuses of light and sleep/wake cycles. Psychological stress basically exhausted my body's ability to tolerate artificial light at night and lack of real sun in the morning. But the thing that really struck me was the speed with which these difficulties overtook me. In a nutshell, my disconnection from Nature culminated in a health crisis that needed to be dealt with promptly or there would be a biologic toll to pay.

These were my symptoms: I couldn't fall asleep, I would wake during the night for no reason, my heart felt like it was pounding, my muscles were sore, my hips hurt, I felt sluggish, and my brain function was off. Now, for context, I don't drink caffeine, I don't get stressed out, I've always been a good sleeper, I work out every day, and I use incandescent light bulbs, so there's no reason to suspect adrenal exhaustion.

On the other hand, my laptop screen (which I use long hours) can go very bright, with a high color temperature (bluish-white). And I was going to sleep and waking very late. I would say my EMF environment is average for a modern human: bad by virtual of neighbors' cell phones and Wi-Fi, but no cell phone signal or smart devices of my own. I've always had low stress levels, excellent health, and unremarkable sensitivity to nnEMFs. So circadian influences were my only risk factors. Poor sleep was the main symptom, but I knew there was more to the situation than that because timing issues were throwing my hormones out of whack.

My (self) diagnosis: Mitochondriacs will recognize that my symptoms are signs of circadian disruption. The crisis was caused by inappropriate light signals to my suprachiasmatic nucleus desynchronizing my circadian rhythms. Biochemically, this results in improper levels of cortisol, melatonin, adrenaline, dopamine and tryptophan, relative to time of day.

On a side note, when the body isn't working properly, biochemical imbalances are often in the middle of a malfunction, but they are not the real source. As a general rule, circadian dysregulation due to light mismanagement is the #1 cause of hormone and neurotransmitter imbalances. And this was what caused my situation too. Psychological stress depleted my adaptive capacity into dysfunction. However, like most people, it took a personal health crisis to make me change my daily routine. Personally, I think the Universe made me unwell so I would live the lesson and pass on the dangers of light mismanagement to others.

But the lesson within the lesson: Sleep problems get our attention because we feel the dysregulation and dysfunction. But the thing that Average Joes miss is that the root cause of poor sleep – melatonin deficiency – derails the body's maintenance and repair efforts. You see, melatonin does more than just start or stop circadian programming. Vital to our long-term wellness, it heals and detoxifies the body, and it recycles our mitochondria. Specifically, melatonin increases antioxidant enzymes, suppresses pro-oxidant enzymes, directly scavenges free radicals, and it renews mitochondria through mitophagy. That's why some call melatonin the Swiss Army knife of antioxidants. Through multiple means, it helps clean up waste products and repairs cellular damage from daily activities.

Melatonin is the master antioxidant

Melatonin regulates sleep cycles. That's its claim to fame. But it has two special skills we can't live without: it repairs oxidation damage and invigorates mitochondria. Of all the antioxidants the body makes or uses, melatonin is said to be the strongest – not by acting alone, but by starting a cascade of antioxidant/anti-inflammatory effects. Melatonin doesn't simply donate its electrons to neutralize free radicals the way ordinary antioxidants do. Instead, the melatonin molecule rearranges itself to form three breakdown products: 3-OHM, AFMK, AMK, which are powerful antioxidants in their own right. So, when all's said and done, one melatonin molecule can scavenge up to ten free radical species, thereby combating oxidative stress like no other antioxidant. More simply, completing the redox cycle of oxidation and reduction eases inflammation.

Melatonin also keeps your mitochondria in shape through mitophagy. Mitochondrial recycling is imperative in keeping ATP production as strong as can be. So sleep isn't the only program that collapses when you're deficient in melatonin. Equally important, but often overlooked, the body's renewal efforts suffer when circadian disruption upsets melatonin production. Cells, organs and tissues then repair oxidation damage more slowly, resulting in oxidative stress and chronic inflammation. Toxins and cellular waste linger longer. And mitochondria lose efficiency. That's the seriousness of the situation when your sleep is impaired from circadian disruption. Melatonin deficiency (usually from light mismanagement) is a secret saboteur of healing and recovery programs – particularly mitochondrial renewal.

Sleep is regulated by melatonin

Functional, restorative sleep is fueled by melatonin. For sleep to be productive – including REM sleep, brain cleansing and memory storage – you need to have enough melatonin to fall asleep and stay asleep through its 4–5 stages. Just being unconscious is not enough. As we've all learned (or lived), sleep is complicated and sensitive to disruption. So, if you were to knock yourself out with a sedative, you might miss dream cycles, glymphatic detoxification, or growth hormone release. Take melatonin exogenously and, over time, your body gets lazy making its own. Or, if you were to wake prematurely from too much caffeine/not enough melatonin interrupting your sleep cycles, your immune system may postpone healing programs to another day. Consistent with its job of regulating sleep, melatonin is the body's strongest parasympathetic activator. It calms mind and body.

How brain cells put you to sleep and wake you up

When the sun goes down (less blue light, more red), melatonin starts to be released. Melatonin acts as a major magnetic force in your mitochondria to condense brain cells and transport biochemicals to replenish stores that were used up during the day. Melatonin level continues to rise into the evening, and exerts its calming influence to counteract cortisol's stimulating effect. This makes you fall asleep at night. The body then uses the condensed state of brain cells, along with the regenerative effects of melatonin, to recover and regenerate while you sleep.

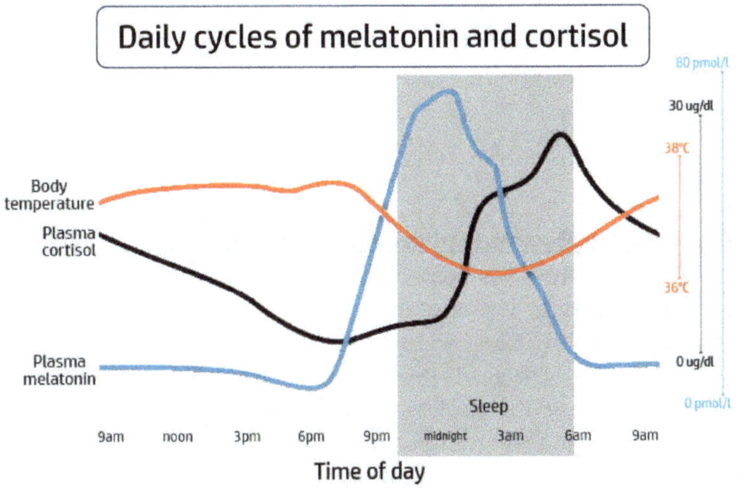

Around 4am (or a couple hours before waking), the circadian system begins maximizing your cortisol level. Cortisol unwinds the triple helix of collagen, which releases water. This water expands the brain's neurons, glial cells, and intracellular space (in the cell). This wakes you up. To reverse the process, photons in morning sunlight start re-zipping collagen, which recaptures the water and re-condenses brain cells. Light activates the switch that turns off melatonin production to start the whole sleep-wake process over again. This contraction process continues throughout the day until the sun goes down. This is the rhythmic, pulsating nature of circadian biology. It's unfolding by day and contraction at night. Uptime followed by downtime. It's but one aspect of the circadian breath of life.

Note: In the above graph, cortisol level is shown as a predefined line when, in reality, it fluctuates in response to needs in the moment.

Other biochemicals released cyclically along with cortisol: ACTH, adrenaline, noradrenaline, α-MSH, β-endorphin, thyroid stimulating hormone, estrogen, and prolactin.

Melatonin cycling

Melatonin in the eye is recycled through morning sun exposure. UV and IR stimulate regeneration of ocular melatonin through the day. That means sun exposure during the day helps you sleep at night. Since blue light is always present in terrestrial sunlight, its absence after dusk signifies to our internal clocks that (1) the sun has gone down, (2) it's now nighttime, (3) we need to turn on sleep mode by releasing melatonin. Cold exposure is another cue for the circadian system to raise melatonin. So if you don't feel any coldness before bedtime (as you would in the wild), your sleep can suffer. Or, in extreme cases, you can get sleep apnea. After four hours of darkness (around midnight), melatonin offloads to your semiconductors the photonic power it absorbed during the day and the cycle starts over.

As the graphic above illustrates, circadian rhythms are 24-hour expend-and-recover cycles requiring an absence 400–550 nm light after dark. That's all of the blue spectrum and part of the green. Unfortunately, blue light in modern bulbs and tech devices stops your tide of melatonin from rising at night. Break the cycle and you experience melatonin insufficiency along with inflammation in the eye, brain, and blood plasma. To make matters worse, cortisol and dopamine stay switched on in the presence of blue light. Combined, these factors inhibit sleep, recovery, and energy production. Therefore, you need to get real sun exposure during the day, and block blue-green light at night, in order to have enough melatonin for a full night of productive sleep. It's that simple (but not convenient). Unfortunately, this is a rule of Nature modern humans break more than any other, and our circadian rhythms crash from this photoreceptor abuse, as we may call it.

Melatonin is made in the gut when tryptophan mixes with serotonin

Another thing you can do to increase melatonin is help your gut make more serotonin. It's a precursor for melatonin. Remember, most serotonin is made in the gut, so if it's busy digesting food late at night it will be distracted from making serotonin for melatonin synthesis. Whereas, resting your digestion in the evening benefits metabolism in many ways. That's why you want to fast several hours before going to bed. Shift your carbohydrate consumption in particular to earlier in the day so your circadian system metabolizes them more efficiently – storing less as fat.

The solution

As philosophers and scientists have said, once you understand a problem completely, it's halfway solved. In my case, recovery was simple. For the nighttime, I ordered switchable-spectrum LED bulbs and blue-light blocking glasses, got the f.lux app that reduces blue light emissions from my laptop, turned off my house lights after sundown, reduced my TV time, and went to bed at 9:00–10:00. I avoided food containing stimulants such as chocolate.

My new morning routine: I forced myself to wake up early and watch the sun rise for several days until it became habit. The more often you resynch your SCN by watching sunrise, the better your internal pharmacy will be at releasing hormones and neurotransmitters on time. For instance, seeing sunrise boosts cortisol by 50%. Metabolism also works better when cell clocks are synchronized because cellular signals are time-based.

I later added Cold Thermogenesis. I go out underdressed for morning sungazing. To my amazement, experiencing coldness after sleeping in a warm bed is the most exhilarating part of watching sunrise because it really wakes you up. I can feel my biochemistry being activated, including a blast of dopamine five minutes in. To exercise cold flexibility even more, I finish my showers with fully cold water for a minute or two. This makes mitochondria and cells release V-UV to IR light, which is outstanding for you because endogenous light supplies so many things the body lacks from modern living. Combined, these changes corrected daily biochemistry fluctuations to match that of the day and season.

Blue light glasses

Clear lenses. Blue-blocking glasses with clear lenses can block up to 30% of blue light. Often sold for computers use, they're good for relieving headache or eye strain from excessive blue relative to other frequencies, including red, IR and UV. So if you stare at a computer screen for many hours a day, and/or your workplace is lit with LED bulbs, blue-light filtering glasses can help relieve light stress during the day. However, blue-light *filtering* glasses don't help at all with melatonin, sleep and restoration, because even miniscule amounts of blue light in the eye tells the SCN it's daytime – keep releasing cortisol and not melatonin.

Yellow lenses can block up to 50% of the blue. Often used for color therapy, consider yellow glasses if you're depressed, have seasonal affective disorder, or just want to elevate your mood because it's cloudy out.

Dark amber lenses (good ones) can block all of the blue-green light. That's what you need to turn on melatonin and turn off cortisol. You need to block all of the blue and a majority of the green.

But, be forewarned, there are a lot of cheap glasses being sold online that don't do what they claim. If you want to improve the quality of your sleep, healing and mitochondrial health, you absolutely must get good quality blue-light blocking glasses. That means blocking 100% of light from 400–550 nanometers. Clear or yellow lenses can't do that. They have to be dark amber/orange or red in color to block 100% of the blue. Though, lenses that are dark amber in color don't necessarily block all the wavelengths they need to. Ask for optical-grade lenses too.

Blue light spikes before sundown (the "de-crescendo effect")

Sunlight contains a surge of blue light twenty minutes before sunset. This tells your suprachiasmatic nucleus that the sun will be setting soon, turn off daytime chronology, and start nighttime programming. Most important for this discussion, the extra blue light tells your body to turn off cortisol and other stress hormones, and start releasing melatonin and tryptophan to prepare for sleep. This gets your body ready to clean up free radical damage incurred during the day, replenish biochemicals consumed through activity, and refresh your mitochondria.

I admit, I didn't know about this spike of blue when I wrote the First Edition. I just heard sunsets were beneficial to catch. But now I see how incorporating sunsets into your circadian system locks in the timing and length of daylight. Sunrise ends the night and starts the day. Then, sunset ends daylight programs and starts nighttime routines. Sunrise is definitely more important that sunset in correcting hormone and neurotransmitter levels. But sunset and darkness do offer additional benefits when fixing major issues. Do both if you want maximum response from your biology.

Three influences control circadian rhythms

Uncle Jack says light exposure is the 'hour hand' of the body's clock timing system. Light has the greatest influence over our daily rhythms because countless operations such as melatonin and vitamin D response are driven by narrow bands of light, if not single frequencies (81×10^{36} frequencies around the visible spectrum). Since white light is a mixture of many colors, the individual wavelengths contained in white light trigger and/or power countless processes in us.

Next most influential is darkness. It's like the minute hand of our clock system. A lack of light enables certain activities to take place, such as melatonin only being made or released in the absence of blue light. Just think: photoreceptors in the eye are so sensitive they can detect a single photon of light hitting them. As a result, you need your room to be as dark as possible to sleep as well as you can, because even indicator lights on electronic devices can give a false signal to your SCN. Even "on/off" lights hitting your closed eyes can disturb sleep programming.

Third most influential is the cold. It's like the second hand of the body's circadian system. Through hibernation programming we've adapted for nightly use, cold temperature (particularly on the face) facilitates sleep and ketosis. Our circadian systems use cold temperature as an environmental signal to tell the body when to switch to burning ketones for energy, when to get busy detoxing, and when pigments like melanin should offload the light they captured during the day. This is why most people sleep best in a room 5–10 degrees cooler than their living spaces, and why we sleep so poorly in hot, humid conditions: coldness controls operations that take place when we sleep.

One consequence of circadian disruption, not already mentioned, is that enterocyte cells lining the gut wall don't turn over in ~48 hours like they're supposed to. They then stick around past their prime and: (1) let toxins and undigested particles through the gut wall, (2) digest food poorly, and (3) delay deuterium removal, as they're sloughed off.

Chrono-nutrition

In 2018, researchers discovered that breastmilk pumped during the day has high levels of stress hormones such as cortisol, and low levels of sleep hormones such as melatonin and tryptophan. On the other hand, milk pumped at night contains high levels of sleep/recovery hormones melatonin and tryptophan, and low levels of cortisol. Breastmilk is, after all, a filter product of the blood. It's basically whole blood with the red blood cells removed and immune factors added.

Not having a circadian system of its own, this is how baby gets the biochemicals it needs to regulate its sleep-wake cycles. And, as you may predict from this knowledge, this is one reason babies wake up during the night: stress hormones in milk pumped during the day are unknowingly given to baby at night (or vice-versa)... because who labels the hour of day at which milk is pumped? No wonder some babies can't sleep through the night when they get milk pumped during daylight hours.

Babies borrow their circadian rhythm from their mother. But what else are they picking up?

Babies have a circadian system encoded in their DNA, but it's inactive. So their timing mechanisms need to be regulated by mother's circadian signaling until theirs matures. Circadian entrainment happens first through cord blood in the womb, then hormones in breastmilk. That's how Nature designed the maturation process to unfold. Unfortunately, when a mother's circadian system is corrupted by faulty light exposure, toxins or acute stress, baby's daily cycles get messed up too.

This means baby's hormonal system in the womb will be exposed to whatever chemical influences are present in mom's bloodstream – good or bad. Stress chemicals released by blue light, social media addiction, or psychological stress all affect baby the same way alcohol, drugs or smoking corrupt their biochemistry – possibly their growth and development. Anything in mother's bloodstream finds its way into baby, and carries with it the potential to upset their biochemical baseline indefinitely. This becomes painfully obvious when you think about it.

Striking example: Psychological disorders were extremely common in children born to mothers traumatized by World War II atrocities. Orphanages in Europe were filled with kids whose mothers went crazy from unspeakable pain and suffering in occupied countries. In these cases, I'd say stress hormones in the mothers not only preconditioned the babies' systems to normalize aberrant biochemistry, thus perpetuating it, but chronically-elevated cortisol, adrenaline and dopamine probably interfered with physical and psychological development as well.

I would assume these biochemical imbalances in times of war are happening to a lesser extent in modern humans who subject themselves to blue light, nnEMFs and social media pretty much non-stop. The sources and situations may be different, but stress in any form releases similar sets of biochemicals into the body. This must be a primary pathway through which children acquire traits from their parents – particularly psychological and personality disorders.

Genes are supposed to make a person who they are. But I say the hormonal and neurotransmitter cocktail a baby's brain is exposed to in early development could conspire with learned behaviors in shaping a person's psychology and personality (disorders) – even more than genetics or epigenetics. I propose that fetal and infant exposure to the biochemicals associated with stimulation, stress and anxiety may hardwire a person's brain to be like that into adulthood. This predisposes individuals to the disorders of modern society – including ADD, sleep problems, chemical addictions, chronic stress, anxiety, depression, anger management issues, hormone imbalances, and a whole lot more.

Food for thought: Nature designed the circadian entrainment process to occur through biochemicals passed from mother to baby. Now that we know the placenta is more of a sieve than a barrier due to this process, medical science needs to rethink the role that a mother's biochemistry plays in the development of a child's psychology, personality, stress level, sleep patterns, anxiety, depression, addictions, propensity for disease, energy level, and pretty much everything else. Simply put, baby inherits mom's biochemistry. Knowing what we know, how could it not?

Skin has its own circadian rhythm

Skin is designed to be active by day and healing by night. But artificial light shuts down its repair mode. Short wavelengths of blue-to-ultraviolet contain more energy than longer wavelengths. Higher energy causes wear and tear, so Nature designed blue light to defer healing activities until skin has an opportunity to rest and recover. That would be nighttime. Hence, complete darkness is the circadian signal that activates repair processes. These keep skin looking as young as can be. On the other hand, blue light inhibits restoration of skin. Bottom line: blue light ages the skin, while red light slows it. Modern living then, with hot blue everywhere, is a driving force behind the appearance of aging.

Infrared sauna and photobiomodulation

The benefits you get from sauna, as well as your overall experience, come from the type of heat source it uses. Of course, all saunas get you hot to make you sweat. But wouldn't you want as many benefits as an activity can give you, in the shortest possible timeframe? I know when I'm doing an activity on a regular basis, I want the time and place to be as enjoyable as possible so I'm less tempted to skip days, or stop altogether. I insist that benefits be easy to understand and noticeable when cycled on and off.

Clearly, Brian Richards and his team at SaunaSpace think the same way as they develop their products. Now in their second decade of study and iteration, they've created the ultimate sauna experience. But to fully appreciate what they've put together, you must realize how far sauna technology has come through the ages. You have to know where predecessors reached their design limits, while the benefits of SaunaSpace continue to grow the more we toxins and technology we face in our lives.

Origins

Sauna use spans thousands of years, dozens of cultures, multiple formats, and going by many names such as sweat house, Turkish bath and steam room. Take wet saunas inspired by geothermal vents and pools: a heat source such as hot rocks vaporize water to create steam. That steam heats the skin surface, which then transfers heat into the body. The transference of heat between neighboring materials is called conductance. Conductance is a wet sauna's main way of getting heat into the body. Unfortunately, it's also their biggest drawback because it's uncomfortable and uncomplimentary to how the body works. Just think: The room has to be oppressively hot and humid to raise core body temperature. At the same time, hot air fights the body's efforts to cool down with sweat.

Then there are dry saunas, aka Finnish saunas. Classic styles use a stove or open fire for heat. Some have hot rocks to add humidity, while others keep humidity to a minimum. Like wet saunas, dry saunas must first heat the air, which then heats the surface of the skin to raise core temperature. This is why dry saunas typically need to operate at 160–200°F. Bottom line, Finnish saunas offer good benefits, but most people find the heat too intense and claustrophobic.

In 1891, John Harvey Kellogg invented the incandescent electric light bath using light bulbs with an emission spectrum similar to that of the sun. This was the first use of electrically-powered light and heat therapy to solve maladies of modern living. He proceeded to treat thousands of people in his wellness center called a "sanitarium," culminating in a book about their experiences, called *Light Therapeutics* (1910).

Wavelengths of red: Visible red=600-700 nanometers. Near infrared=700-1500 nm. Mid infrared=1500-3000 nm. Far infrared=3000-1,000,000 nm.

More recently, manufacturers came out with saunas that heat you up via infrared light. Using non-visible infrared, their innovation was to warm the body using radiant heat. Instead of heating air or water molecules and then the body, most of the infrared heats tissues beneath the skin surface. That way, you don't need the air in the enclosure to be extremely hot. It only has to be a little hotter than body temperature to overshoot the cooling effect of sweat. This was their big advantage in terms of comfort, convenience and effectiveness. However, the kind of infrared most-commonly used – far infrared – has a huge disadvantage: water absorbs it quite well. Recall that water absorbs red light from 600 to 3000 nanometers. That means IR light stops at (1) water in the blood, (2) water in cells, or (3) perspiration on the surface. Some still gets past, but this left the door open for a technology that reaches deeper tissues (where vital organs reside).

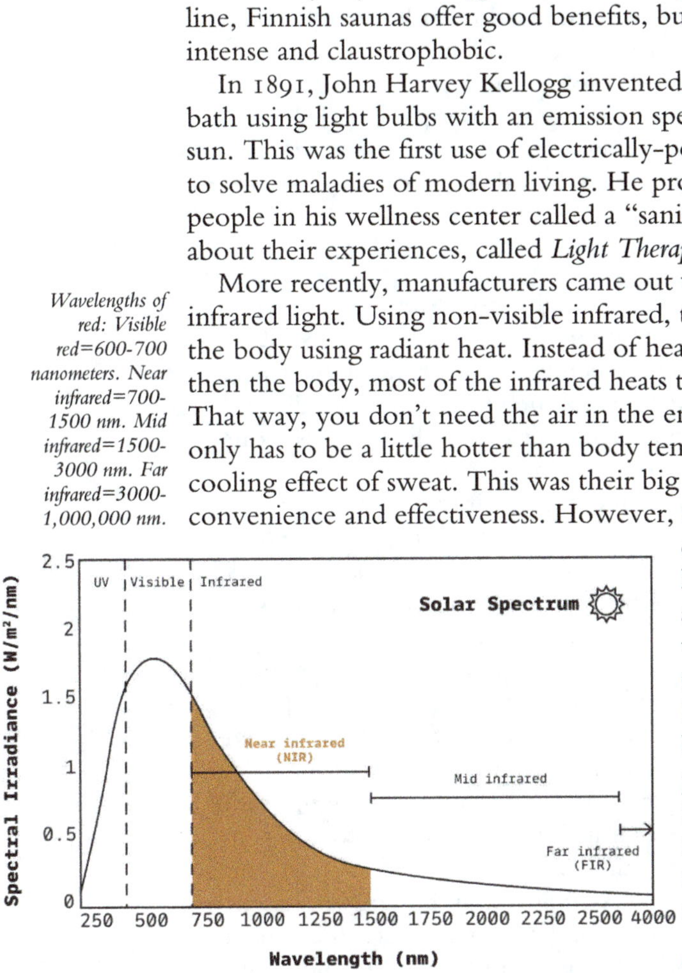

Graphic (above) is the copyrighted property of SaunaSpace. Used with permission.

Welcome SaunaSpace

Near-infrared sauna is one of the best biohacks a health enthusiast can do for themselves. Specifically, the near-infrared saunas from SaunaSpace have earned a spot on many a mitochondriac's list of therapeutics to do on a regular basis because they combine multiple modalities into one practice with spectacular benefits and next to no downside. Their signature feature is incandescent red-light exposure containing (1) a lot of near-infrared and mid-infrared, (2) some visible red, and (3) a minority of far infrared. (SaunaSpace's other modalities – earthing and shielding – are discussed in more detail elsewhere.)

Red portions of the spectrum, in general, penetrate tissues deeper than any other. This enables sauna owners to use red/infrared light to accomplish desperately needed work around the body that it didn't otherwise have the energy and opportunity to do. Within that band of frequencies, near-infrared and mid-infrared penetrate human tissue the deepest, whereas most far infrared is absorbed within a few millimeters of the skin's surface, wherever water is present. So, right out of the gate, SaunaSpace has the advantage of heating the body from the inside out. In contrast, traditional Finnish saunas heat the body from the outside in, and far-infrared saunas heat you near to the surface, inward.

Heat therapy

All saunas are designed to heat you – though some do it more efficiently than others. Raising core body temperature by 2–3°F gives users the same healing, antimicrobial, and restorative benefits as fever. Just look at what fever does as it fights infection: it activates the immune system, dilates blood vessels, increases membrane permeability, mobilizes toxins, beats up germs, and increases protein production to speed recovery.

A nasty but necessary function of fever is that it releases toxins from storage. For what purpose? Perhaps Nature knows pathogens love to harbor toxins – and either or both promote sickness. Or, maybe illness is just a convenient time to expel toxins since you'll be phlegmy, inflamed, and teeming with white cells. Whatever the case, fever has a deep-cleaning effect. And, I would argue, heat therapy has the same set of effects, minus the white cells and inflammation. When heat therapy is employed, you get many of the benefits of fever – I think by accident – simply because cleansing and restoration programs are built into our biology as a kind of virtual spring-cleaning. By chance or design, heat is one of the original biohacks by which many elements of restoration programming can be harnessed on purpose.

Heat shock proteins are the workers employed in both situations. In addition to the benefits mentioned above, heat shock proteins can have sweeping effects on the body's operational performance – including stem cell proliferation and differentiation, gene transcription in repairing DNA damage, hormonal response such as insulin sensitivity, and cellular life-cycle optimizing, such as anti-apoptotic/pro-survival signaling.

Detoxification

The primary purpose of sweating in a sauna is to open a new channel for toxins to leave the body. Busiest pathways include (1) the kidneys and urine and (2) the liver and bowels. But, of all the avenues of elimination, skin is the largest organ of excretion. Yet most people hardly use it at all to remove toxins. This represents a huge opportunity to relieve disorder or add resiliency through toxin reduction – especially since most people's detox organs are already operating near full capacity. Which is to say we all need more detoxification, yet sweat is a superhighway seldom traveled.

Why's it so productive, besides its surface area? Sweating sidesteps the kidneys and urine – which are being kept busy by all the mineral salts, sugar, food additives and toxins we ingest. It's an alternate path to purge toxin once they've exited the cell. Toxins do leave through the lymph, blood and urine. But sweating has the sensational ability to release toxins from adipose tissue directly into sweat glands. Hence, avoiding the bloodstream offers less opportunity to redeposit toxins elsewhere. Conclusion: Sauna does work the kidneys and liver harder. However, you're simultaneously engaging more resources to offset the increased workload. This is one of the balancing acts – workload *vs.* detox capacity – that dictate how quickly you're able to acclimate to full sauna sessions.

Circulation. The body uses the presence or absence of red light to manage systemic nitric oxide and thus vasodilation, blood flow and respiratory rate. When wavelengths of red hit photoreceptors in the mitochondria, they liberate NO (from Complex IV, many people believe). NO then enters circulation, which dilates blood vessels. Hence, when red light isn't available, keeping nitric oxide in confinement suppresses Complex IV activity. Conversely, the presence of red raises respiration. This is one way that IR light/heat raises mitochondrial respiration.

Summation: As a major benefit of sun exposure or red-light therapy, nitric oxide dilates blood vessels, which brings more blood flow, oxygen, electrolytes and nutrients to cells and mitochondria. Waste is carried away. Not only does this set of reactions increase ATP, intracellular water, carbon dioxide and mitochondrial membrane potential – all aiding detoxification. But these are the main mechanisms by which red light increases heat, blood flow, sweating and detoxification capacity. This is how SaunaSpace saunas "stir the pot" and transport the toxins out.

Time-efficiency. The speed with which you start sweating profusely is essential in making your sauna-time as productive as possible. This is a phenomenal feature worth a premium because it pays you back in time. Without the need to preheat, SaunaSpace saunas get you sweating vigorously in just a couple of minutes, resulting in a session time of about 17–25 minutes with 4 bulbs, or 15 minutes with a 7-bulb setup. Compare that to *far-infrared* sauna's 30–120 minutes of preheating, 15–20 minutes to start sweating, and 45–60 minutes per session. Near infrared is, hands down, far more time-efficient than any other kind of sauna.

Stressed *vs.* unstressed. Another requirement for efficient detoxification is that you need to be in a parasympathetic resting state. You can't be exerting yourself. When you're active, the body diverts energy away from detoxification. Detox organs then can't focus on processing toxins in sympathetic dominance. For instance, toxins tend to stay put when you're running. And, let's not forget, stressors such as microwaves, electric fields, bipolar magnetic fields, toxins, emotional stress, noise pollution and artificial light all increase sympathetic activation, thereby reducing excretion rate. Thankfully, SaunaSpace's faraday model is specially-designed to mitigate modern influences like these for a truly relaxing, sanctuary experience.

Your ticket to tranquility

The radiant warmth and electromagnetic quietness of SaunaSpace saunas transport body and mind to a place of serenity. Their biophysical peacefulness is an unexpected pleasure you get on top of deep detoxification – particularly for those engulfed in sympathetic stressors all day, every day. To say they're relaxing and de-stressing would not do them justice. On the other hand, sympathetic activation is associated with most modern disease because it drains the body's resources, including antioxidant reserves, biochemicals, immunity and renewal capacity. Conclusion: Owners feel the centering effect of SaunaSpace makes it a rare health measure they consistently look forward to "doing" 3–4 times per week – ironically, through an absence of effort.

Photobiomodulation

Photobiomodulation (PBM) – a phenomenal feature of SaunaSpace saunas – is the purposeful use of light to activate renewal mechanisms. Of all the colors, IR/red-light therapy offers the most therapeutic benefit, and is therefore most popular and effective. The body responds to blue and green as well. But most devices use invisible infrared and visible red because they're better at reaching deep tissues. Within the red range, near infrared activates rejuvenation programs most powerfully, followed - closely by the shorter half of mid-infrared, then visible red, while far infrared offers no PBM benefits apart from heat therapy. These effects make logical sense because they match the proportions of light our ancestors received throughout history on earth (see graphic, page 265).

Immediate effects of red/infrared photobiomodulation:

- **Builds e-zone.** Charge-separated water powers work all over the body, including protein folding, blood flow and chemical reactions.
- **Vasodilation.** Red light liberates nitric oxide from the ETC, which goes into circulation. NO then dilates blood vessels to enhance circulation, oxygenation and detoxification.
- **Oxygenation.** More oxygenation means increased respiration, oxidation and reduction (i.e., fitness and healing capacity).

- **Light- and heat-gated ion channels.** Light-gated ion channels increase activity of cells, including respiration and detoxification.
- **Mitochondrial activation/ATP production.** Red light is absorbed up to several inches deep in the mitochondria. This makes the ATP synthase spin faster, making more ATP.
- **Lymphatic drainage.** Zeta potential of lymph fluid keeps particles in colloidal suspension, instead of separating into fluids and solids. Therefore, red light promotes lymph movement through charge-separation of water, while its absence encourages stagnation of toxins from depolarization.
- **Melatonin.** Red light enhances cellular melatonin production. Melatonin has the strongest antioxidant effect, enables productive sleep and healing, and tunes up mitochondria through mitophagy.

Lasting effects of red/infrared PBM:

- **Enhances muscle recovery and wound healing.**
- **Anti-inflammatory, immune modulation.** Red light shifts macrophage and microglial immune cells from a pro-inflammatory M1 state to anti-inflammatory M2.
- **Activates stem cells.** Red wavelengths shift energy production of stem cells from glycolysis (low ATP output) to oxidative phosphorylation (high output). This wakes them up. When inactive, stem cells use glycolysis for energy, which does not require oxygen. But when their mitochondria get fired up, stem cells are forced to leave their hypoxic niche to find oxygen to proliferate and differentiate. This makes them migrate to where they're needed.
- **Angiogenesis.** Red and near-IR grow new blood vessels by enhancing ROS signaling, DNA transcription and protein synthesis.

Conclusion: Photobiomodulation was first used in recovery of high-performance athletes, then mainstream orthopedics, and now general wellness for consumers. Its use and acceptance have exploded over the last few years because those in the know realize it is highly effective and significantly faster than traditional treatments. PMB, as it's abbreviated, has proven itself to be a surprisingly strong in healing many problem areas.

Red-light LED therapy, in particular, is being used quite successfully to treat joint and muscle pain, degenerative conditions and, most intriguing, sport injuries. In my estimation, red light therapy can heal major injuries – whether surgically-treated or not – about 25–40% faster than conventional treatments, or give the body enough resources to heal injuries that would otherwise not heal on their own. Minor injuries such as muscle soreness from overexertion might heal twice as fast. Why? It gives the body far more resources with which to restore and rebalance.

Heat shock proteins. Red light does much of its PBM work by activating heat shock proteins (HSP). Some mediating factors include protein refolding, DNA transcription/gene expression, growth factors, stem cell proliferation and motility, and cellular life-cycle signaling such as anti-apoptotic/pro-survival.

Protein refolding. The primary benefit of photobiomodulation we need to talk about is protein repair. On average, 20–30% of proteins in a person are misfolded. Since *form* enables a protein's *function*, the more deformed their shape due to shoddy transcription, oxidative stress or redox potential, the more glitchy their performance. PBM repairs proteins by calling heat shock proteins into action. These are repairmen of the immune system. HSPs refold proteins so they can do their job in the following categories: (1) structure (e.g., collagen) (2) communication (insulin receptors), (3) transport (calcium ion pumps), (4) enzymes (cytochromes), (5) defense (antibodies), or (6) storage (ferritin). As this list suggests, that's a lot of duties around the body. By activating genes that tune up proteins, you bolster maintenance and repair abilities all over the body. For instance, remodeling collagen for younger-looking skin is an end result of upregulating gene transcription to improve protein function.

See how a protein's structure enables its function in YouTube video entitled "What is a Protein" youtube.com/watch?v=wvTv8TqWC48.

Blood contains mitochondria. A study published in 2020 found functional mitochondria in the blood. Some would disagree, but if true, these mitochondria would be a fourth vehicle besides water, hemoglobin and melanin that captures red light and carries it around the body.

Improved healing capacity (through hormesis). Red light upregulates repair processes such as protein synthesis, and it downregulates inflammation by reducing pro-inflammatory cytokines. Here's how: When IR light boosts heat, nitric oxide and blood flow, more electrons flow through the electron transport chain. This results in a sudden rise in reactive oxygen species, which sets off distress signals in the nucleus. As the transport chain processes this surge of electrons (or any), more will naturally escape into the matrix (0.2–2% on average). These merge with oxygen to form superoxide, which creates more ROS.

This bump in ROS migrates out of the mitochondria into the nucleus of the cell and, through hormetic stress, activates the enzyme system abbreviated "NF kappa B or NF-kB." When activated, NF-kB triggers expression of over 100 genes which make proteins used in repair processes, while lowering inflammatory cytokine levels. As one crucial example, this process raises brain derived neurotrophic factor (BDNF), which repairs misfolded proteins in the brain. Some say BDNF can break down amyloid plaque, which is associated with Alzheimer's and dementia.

NF-kB regulates immune response and cell survival mechanisms by controlling DNA transcription and inflammatory cytokine production.

In summary, the red wavelengths in near-infrared sauna boost circulation, exercise mitochondria to make more ATP, and detoxify the body. In addition, the bump in ROS sounds alarm bells in the nucleus for several

Graphic (right) is the copyrighted property of SaunaSpace. Used with permission.

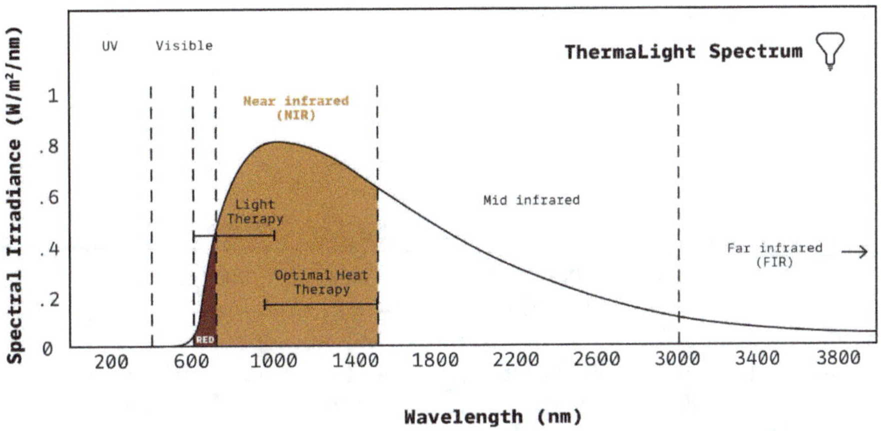

weeks that correct vulnerabilities and imbalances that turn into chronic inflammation, misfolded proteins and disease over time. These bursts of ROS, followed by corrective actions, tune up the body in wonderful ways. This is the magic behind the enduring benefits of near-infrared sauna.

How their light/heat is made

To our distinct advantage – though it's really expensive to do so – SaunaSpace is the only company that's gone to the trouble of custom-engineering incandescent bulbs that replicate both the types of infrared found in sunlight, as well as sunlight's wide distribution curve. Incandescence is the only light source that can even approximate that of the sun. You can call this type of light *analog*, as opposed to *digital* (referring to a light's broad-spectrum, non-flickering qualities).

Incandescence: Heating a material so it glows. In the case of traditional light bulbs, a tungsten filament is heated to ~2000°F, which makes the bulb emit light.

To learn how SaunaSpace saunas work, visit: https://sauna.sp ace/pages/how-

You see, normal infrared heat bulbs, like those found in bathrooms, emit mostly far infrared. Whereas, SaunaSpace thermal bulbs (FireLight®) use handmade stained glass and a special filament that operates at a higher color temperature than regular heat bulbs. Peaking in the middle of near-IR, FireLight® bulbs emit 40% near-IR, 40% mid, and 15% far-IR. As for power density (photons/second), the bulbs' total emissions are 70% near infrared, which is 2–4 times that of far infrared emitters. And the light is flicker-free. Conversely, LEDs used for photobiomodulation pulse on and off 120 times a second, which stresses the nervous system.

Titration

SaunaSpace saunas come with four-bulb panels, with switches to turn individual bulbs on/off. This lets you to fine-tune the dose and application of light you're getting. You can also open the front curtain to let your unexposed side cool down when you rotate. They also sell 3– or 4–bulb panels separately to add more coverage for faster sessions. Or you can use the panels to build a custom setup in a shower, closet, or full-body cabinet-style sauna from another maker.

A "full" sauna session includes the following: (1) raise core body temperature 2–3°F, (2) sweat out a pound of water, and (3) arrive at a "subjective feeling of exhaustion." This feeling of fatigue happens when your body has released enough toxins for the moment – it wants a break. Once acclimated, it may take 15–30 minutes to reach these markers. Session time varies from person to person, based on your toxin load and detox capacity. Start as low and slow as you need to and work your way up to higher intensity, longer session times if you're heat sensitive or intolerant in some way. Or add another panel to cut session time almost in half. Warning: This may be too intense for some people. When you're done, be sure to rinse off excreted toxins and replete the minerals and vitamins you sweated out. It's that easy and enjoyable to get extraordinary benefits from a modest time commitment.

Differences between types of light (incandescent *vs.* LED or laser)

The biggest difference between light from a sauna and light produced by LEDs or laser is that LEDs and laser are better at targeting treatment in deeper tissues due to their higher power density. They make light in only one or two wavelengths to focus their effect. Specific wavelengths are better at targeting and restoring tissues because they're good at avoiding absorption in outer layers to reach inner tissues of interest, such as brain or tendons. And LEDs have become cheaper to make than other sources. For these reasons, PBM devices offer the most targeted therapy at the lowest price, in terms of specific benefit. For instance, LED arrays can be configured to address one specific area such as the scalp, to serve one purpose such as hair regrowth. On the other hand, near-infrared sauna tends to produce systemic effects involving improved circulation and heat shock proteins, instead of one site on the body or a single issue.

Other applications for PBM (independent of sauna)

Photobiomodulation with LED arrays or laser is skyrocketing. Over a thousand new studies are published every year proving its efficacy in restoration and anti-aging. Clinics and practitioners are popping up all over the place. Almost every major sports team now uses PBM to enhance recovery of its athletes faster and better. While the remaining naysayers are being buried under a mountain of peer-reviewed evidence demonstrating excellent results treating challenging disorders, including

- Neurodegeneration, such as Alzheimer's, Parkinson's and dementia
- Joints and tendons like tennis elbow, tendonitis and carpal tunnel
- Traumatic brain injury, such as sports injury and car accidents
- Diabetic peripheral neuropathy and foot ulcers
- Psychiatric disorders, such as depression, anxiety, ADD, autism, insomnia and drug addiction
- Hair regrowth.

Final verdict: Photobiomodulation is absolutely here to stay. At this point, no one can plausibly deny that PBM is a top-notch treatment for many conditions perplexing Big Pharma. The only problem is, it doesn't fit Western medicine's model for making money. Therefore, most consumers only hear about it if/when they have a specific need PBM

To learn more about the science behind near IR sauna (PBM), read the paper "Mitochondria as a target for neuroprotection: role of methylene blue and photobiomodulation" by Yang et al. in Translational Neurodegeneration (2020).

solves particularly well, and they didn't like anything mainstream medicine has to offer so they go looking for a solution on their own.

Methylene blue: Panacea or poison?

Methylene blue (abbreviated MB, proper name: methylthioninium chloride) has achieved almost mythical status in the wellness community because it can do things no other compound can do, natural or synthetic. In fact, the term "magic bullet" was first used to describe methylene blue's amazing power to save people from serious or life-threatening conditions. Indeed, its potency and versatility even inspired industrialists to create both synthetic drugs and the medical system as we know them, based on petroleum derivatives.

Important point: Most descriptions of methylene blue, unless otherwise stated, are in reference to its use in low doses/concentrations. As conscientious practitioners are keen to point out, low-dose MB can have the opposite effect of high-dose MB.

From fabric dye to drug of first choice

Because of its intense blue color and propensity to stain, methylene blue was originally used in the 1870s to dye jeans blue. Then, in 1891, it became the first synthetic drug after Paul Ehrlich discovered in the course of developing laboratory stains that it also killed the bacterium he was studying in malaria. For many decades, it was THE antimicrobial treatment healers used. Notably, it was given prophylactically to soldiers in WWII to fight parasites, fungal infections and trench foot in jungles of the Pacific. Shortly thereafter, penicillin took its place in the late 1940s.

To unsung standards of care

Between the 1950s and ~2020, no one cared about MB as markets grew for health conditions as *problems that new drugs could solve* – including antibiotics, antidepressants, antianxiety, antipsychotics and stimulants. MB was virtually forgotten until biohackers rediscovered its amazing attributes quite recently. However, despite shifting interests, methylene blue has never lost its status as the solution of first choice in obscure niches.

- Emergency rooms keep methylene blue on hand to treat carbon monoxide (CO) or cyanide poisoning due to its ability to support crashed mitochondrial function. **Methemoglobinemia** – a condition which can be caused by CO poisoning – is when hemoglobin in red blood cells loses its ability to carry oxygen so mitochondria don't work.
- It's used regularly as a **stain** in microscopy work because various tissues take it up at different rates, which increases contrast.
- It's used to **disinfect donated blood products** – of hepatis and HIV especially. So if you've ever gotten a blood transfusion, you've unknowingly received a little methylene blue.

- It's been used for generations to **treat fish tanks**, and thereby the fish themselves, of fungal, bacterial or parasitic infections.
- **Hydroxychloroquine** (the drug that's great at fighting symptoms of COVID-19) is derived from **chloroquine** (standard treatment for malaria), which is derived from MB.

Hence, methylene blue is on the World Health Organization's list of 100 essential medicines. And yet, it's only garnered serious interest from biohackers starting around 2020. So, it's both the oldest, and one of the newest, biohacks in use today (perhaps a divinely crafted one at that). Which is to say, despite its lengthy track record, we're all still learning about it. And not everyone agrees on what it's doing in/for the body.

At the heart of all the mystery and confusion is methylene blue's ability to transition from one state – blue and oxidized – to another – reduced and colorless – and back again. People aren't used to thinking about one compound transitioning between two states acting in opposition to each other. It's both weird and wonderful… and why so many mitochondriacs are evaluating it today to improve redox reactions in people.

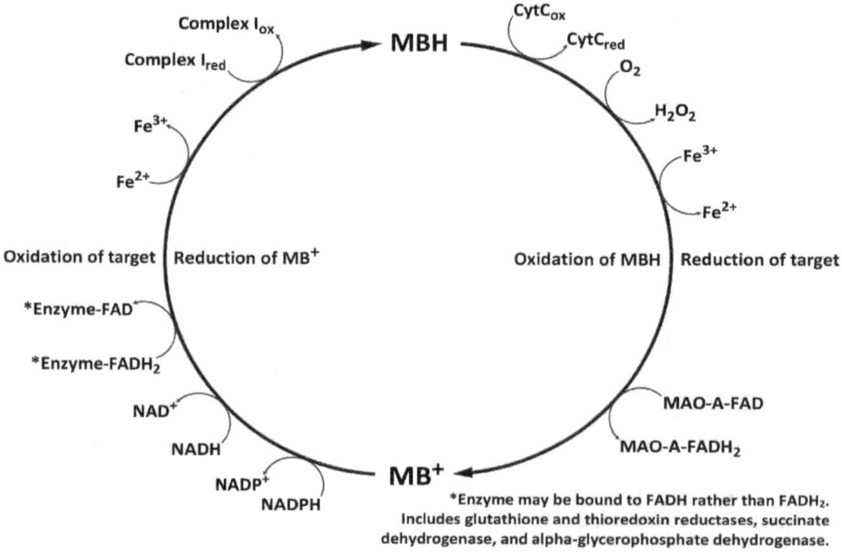

Methylene blue is capable of donating electrons or taking them, repeatedly, as this diagram of redox reactions illustrates. Hence, it's called an electron cycler. In low concentrations, it does both, slightly favoring donation. In high concentrations, it increasingly steals electrons (oxidation).

Key features

Far and away its best feature is the effortlessness with which MB switches from giving electrons to taking them. Adding two electrons creates its clear form: leucomethylene blue (MBH). Losing two electrons creates its blue form: methylene blue (MB^+). This is its natural, preferred, and most stable state – though, it easily transitions between states when the opportunity presents itself. To illustrate, a colorless mixture of (1) methylene blue, (2) sodium hydroxide and (3) glucose turns blue a few seconds after shaking it, then back to clear again after twenty seconds of rest.

The mixture can do this repeatedly through oxidation or reduction of the glucose. Its signature superpower – "switchable" electron exchange – gives MB the unique ability to provide resources the body lacks in times of deficit or quiescence. Its high redox reactivity enables MB to save lives in emergency situations; reduce oxidation damage, inflammation and skin aging on a day-to-day basis; as well as enhance healing capacity and mitochondrial fitness over the long-term.

For context. When the body's major means of energy production – oxidative phosphorylation – breaks down, cells resort to much less efficient backup methods such as glycolysis to eke by. Unfortunately, glycolysis produces about 1/10 as much power as ox/phos. This difference in energy output represents methylene blue's greatest opportunity to restore power supply and organ function. It's fundamental to why many observers call it a miracle molecule.

Gets around. Methylene blue has the unusual property of being both water soluble and fat soluble. And it's a small molecule. As such, it doesn't require agents like insulin or structures like voltage-gated ion channels to enter cells. It easily crosses membranes such as the blood-brain barrier and lipid bilayers to get into cells and mitochondria.

Reconnects faulty cycles. Methylene blue takes energy production cycles like Krebs that have stopped turning from lack of substrates which drive them and gives them a backup supply to restart production. This gets the interconnected gears of ox/phos, Krebs cycle, beta oxidation and glycolysis turning again that were missing teeth in their gears (steps in the cycle) – most often from lack of oxygen or NAD^+.

Electron donor. Methylene blue is famous for its ability to donate electrons into the middle portion of the electron transport chain. No need for electrons from food to enter at cytochrome I or II and travel the chain, or even have oxygen present to carry electrons away from the ETC. MB enables mitochondria to continue functioning in the absence of metabolic resources such as circulation, oxygen, ROS reduction or CO_2 removal. It just bypasses respiratory deficits as if it were designed to do so. That means you can prevent brain cells or heart cells from dying when tissues aren't getting any blood flow as a result of heart attack, stroke, brain injury or surgical procedure, as examples.

Rescues part of the respiratory chain. NADH normally feeds its electrons to Complex I, which then tunnel through the ETC. But MB can take electrons directly from NADH or the matrix and deliver them somewhere into the ETC – it's not known exactly where – so that they end up at Cytochrome IV. Complex IV is the most likely spot for electrons to enter via MB, followed by Complex II and Cytochrome C.

Whatever the case, this route skips Complex I and its proton pumping, and possibly Complex III as well, in the event that portions of the electron transport chain are malfunctioning from toxins, injury or age, as examples. Thus, you lose that/those energy input(s) into ATP production, compared to normal operations. However, it's better than nothing, were the ETC to shut down completely. More precisely, when MB is used to bypass faults in the respiratory chain, you lose between 33 and 67% of the ETC's ATP production through this workaround (recall that three complexes pump protons: I, III and IV).

To learn how methylene blue rescues distressed mitochondria and brain cells watch Dr. Been's YouTube video entitled "Methylene Blue and Near Infrared Light for Neuroprotection."

Prevents shutdown of glycolysis and TCA cycle. You then have ⅓ to ⅔ the normal amount of NAD^+ coming out of the ETC to feed glycolysis and the TCA cycles. Whereas these partner processes would shut down without NAD^+ coming out of ox/phos to run them.

Regulates vasodilation and metabolic rate. Methylene blue takes electrons that leak out of the ETC – whether smooth running or distressed – and it combines them with oxygen and hydrogen, to make hydrogen peroxide (H_2O_2) and maybe H_2O, depending on who you talk to. As oxygen drops in the matrix, mitochondria notice the hypoxia and tell Complex IV to release nitric oxide (NO) it's holding. This dilates blood vessels, which brings more blood flow, oxygen and nutrients to increase respiration. And it carries toxins away. As cells and mitochondria get the supplies they need, stress and injuries are reversed. Vasodilation and metabolism upregulation are major hormetic benefits of MB. On the other hand, the oxidative stress of high-dose MB inhibits nitric oxide synthase, which makes NO, resulting in respiratory inhibition.

Antioxidant effect. Electrons typically escape from 11 different places in the ETC, but mostly Complexes I and III, to form reactive oxygen species. Some call this "electron slippage." MB keeps free radical formation at normal, healthy levels (0.2–2%) by taking those orphaned electrons and giving them a home within more stable, productive molecules. One outcome: MB creates water on its own (without Complex IV) by bringing together: (1) one oxygen, (2) a free-floating electron, and (3) two hydrogens. Another: MB can keep electrons it absorbs from the matrix so they don't bind to oxygen and create more superoxide than the ETC already does. This is how methylene blue (oxidized) turns into leucomethylene blue (reduced).

In contrast, antioxidants such as vitamins C or E can grab hold of stray electrons in the matrix, but they don't release them to form water like MB does. The antioxidants we've all heard of certainly help, but a dead end as far as electron cycling is concerned. And that's where methylene blue shines. It's just as eager to donate electrons as it is accepting them. To illustrate, methylene blue's antioxidant properties help fight electrosmog and radiation damage – on an airplane, for example. Harmful frequencies knock an electron out of place. No problem. MB gives molecules a replacement electron to make them whole again. Similarly, you would expect free radicals to go up when respiration increases. But leuco-MB deactivates them with its antioxidant abilities.

Antimicrobial. Methylene blue excels as an antifungal, antiparasitic, antiviral, antibacterial agent because of its oxidative effects on prokaryotes. Starting at low concentrations, methylene blue increasingly steals electrons from their cell walls through the same process that our immune system uses to fight infections: oxidation. Our cells resist oxidation better because we evolved as oxygen breathers. But prokaryotes just perish as most give up their electrons more easily. Remember, methylene blue self-concentrates, according to a tissue's energy needs.

MB builds resiliency by exercising redox systems (hormesis)

MB has the rare ability to recover lost mitochondrial function by beneficially stressing metabolic pathways a variable amount via dosing. Through hormesis, you can increase your antioxidant reserves which, in a word, is resiliency. In this case, dose makes the difference between improving function through a "fitness response" (overload followed by increased capacity) *vs.* a large dose causing harm through net toxicity.

One way that MB triggers hormesis is by raising hydrogen peroxide level. Using its talent for trading electrons, MB makes hydrogen peroxide all by itself. A normal byproduct of metabolism, the free radical hydrogen peroxide (H_2O_2) is a pro-oxidant. It deconstructs cell walls and DNA by stealing electrons. But as long as you have enough antioxidant capacity to neutralize the cleanup done by H_2O_2, you regenerate stronger. Stress the system with H_2O_2, and it gains power and flexibility. This is how you build resiliency. Simply dial the amount of beneficial stress up or down by changing your dosage of methylene blue.

Once symptoms of mitochondrial deficiency get straightened out – including membrane electrical potential, oxygen delivery, water production, mitophagy, magnetic flux, DNA repair and ROS neutralization – irregularities and imbalances go away, normal function is restored, and the whole machine just works better. Like physical exercise, challenging your metabolic systems with a small amount of ROS – in this case, H_2O_2 – tunes up your metabolic engines.

Methylene blue is a mild mood enhancer

It acts somewhat like a psycho-active drug by inhibiting the breakdown of chemicals that run your ability think and feel. These "monoamines" regulate nerve impulses that allow the brain to communicate with the body. They include the neurotransmitters serotonin, dopamine, adrenaline, noradrenaline, melatonin and acetylcholine. With more biochemicals to facilitate neuronal communication of all kinds, you just feel better. But, curiously enough, this makes it impossible to tell *why* you feel better. Did it genuinely increase your mitochondrial output and cell function, etc.? Did it actually put you in a healthier state? Or do you just feel better because neuro-active compounds create that experience?

Methylene blue improves brain function

MB improves cognition, learning and memory. But it does it differently than pharmaceuticals. Most drugs such as Ritalin or Adderall improve brain function by increasing availability of neurochemicals such as dopamine, serotonin and adrenaline. These neurotransmitters not only carry communication between nerve cells that enable you to think and act, but they also produce pleasure and feelings in reward centers which motivate you toward stimulation and away from boredom.

Without healthy neurotransmitter levels, you can lose cognitive performance such as reasoning and creativity but, just as important to quality of life, you can also lose emotional stability (anxiety *vs.* calmness), pleasure, and motivation. MB is mildly supportive of neurotransmitter levels. But its major kick comes from improving performance of brain cells themselves by increasing circulation and respiratory rate. It raises their activity level by correcting energy deficits, thus restoring both power and function. For these reasons, methylene blue is very neuroprotective.

Examples: MB is being used therapeutically to enhance executive function in ADD. Some say MB works just as well, if not better, than Adderall at improving focus and concentration. It's widely known MB protects against brain damage in stroke or surgery. And, preventively, MB increases resistance to brain degeneration. Neurological disorders such as Alzheimer's, Parkinson's and dementia are often associated with abnormalities such as neurofibrillary tangles, their tau proteins, and amyloid plaque. MB inhibits tau formation, is thought to resolve amyloid plaque, and is being investigated as a possible treatment for Alzheimer's.

Phototherapy with methylene blue

Methylene blue can be used as oxidative therapy by combining it with light between 550–700 nanometers (half the visible spectrum, from red to some green). When this light hits MB, the MB molecule and O2 exchange electrons in several ways to form powerful radicals such as singlet oxygen, superoxide, MB radicals, and hydrogen peroxide. Using the oxidative powers of these molecules therapeutically, the combination of MB and light becomes a killer treatment (literally) for pathogens and tumors.

Anti-aging product for skin

MB is receiving lots of attention from skin care companies as an anti-aging ingredient. By donating electrons and absorbing UV light, MB creates a virtual reservoir of antioxidant protection against light-induced

DNA damage. MB just keeps replacing all those electrons lost through age or photo-oxidation, until its buffering capacity is exhausted. Makes sense to me. But the application is so brand new, I think the best is yet to come.

Risks, warnings and downsides

- **Biphasic dose response.** As with most exposures, dose makes the toxin. And methylene blue can be the poster child for this concept due to its biphasic response – specifically, its ability to shift states. At low doses it acts one way. In this case, benefits increase as you raise the dose. But eventually the benefits top out, then decline, then drop below neutral and become deleterious.
- **It stains almost everything it touches.** Methylene blue is famous for turning tissues of the mouth blue for 6–8 hours after dissolving a troche of it at the gumline. Once it enters circulation, tissues with large energy demands (i.e., mitochondria) hoard it, including the brain and endocrine organs. Normal color returns after the tissues' mitochondria can't use it any longer. And because some tissues collect it more readily than others, the contrast is easier to see on a slide in microscopy work. Thus, it's perfect for tagging parathyroid glands among adipose tissue in surgery, as another example.

Troche: Small lozenge dissolved slowly between the cheek and gums or under the tongue.

- **Turns urine funky colors.** Methylene blue turns your urine blue to green, depending on how clear to yellow it would have been otherwise. Its half-life in the body is 5–7 hours when taken orally, or 24 hours when given intravenously. That means your urine will turn blue for the better part of a day when taken orally. Or you'll pee blue for days after receiving an IV of it. Even more peculiar, in high doses, it turns all body secretions blue.
- **Three purity grades.** Extremely important: Methylene blue is available in three grades: chemical grade, industrial grade, and pharmaceutical grade (USP). The chemical grade is allowed 20–30% impurities – commonly arsenic, aluminum, cadmium, mercury and lead. All not good for mitochondria and brain/nerve function. Industrial grade can have 8–11% impurities. Whereas US Pharmacopeia grade (USP) should be at least 99% pure. European pharmaceutical grade is slightly less stringent than USP. USP-grade MB is a controlled substance, available by prescription only.
- **Warning: Methylene blue amplifies psychotropic drugs – including MAOIs, SSRIs, SNRIs and opioids – potentially putting you at risk for overdosing.** Methylene blue acts as monoamine oxidase inhibitor (MAOi). Monoamine oxidase is an enzyme that breaks down monoamines – including the neurotransmitters serotonin, melatonin, epinephrine, norepinephrine, phenylalanine, tyramine and dopamine. Drugs that slow recycling of

these neurotransmitters raise their availability. The problem is, when you're taking a MAO-inhibitor drug (e.g., for depression) along with methylene blue, you can oversupply your system with serotonin, which can create a rare but serious side effect called serotonin syndrome. For the same reasons, avoid using methylene blue with selective serotonin reuptake inhibitors (SSRIs), selective norepinephrine reuptake inhibitors (SNRIs) and opioids.

- **Can harm microbiota at high doses.** MB can potentially kill microbes through oxidation. In particular, pathogens and probiotics that lack enzymatic defenses to break down hydrogen peroxide get overwhelmed by MB's proclivity to steal electrons. Therefore, MB may present an antibiotic threat to microbiota at high doses.
- **Can increase metabolic rate and appetite.** MB raises your metabolic rate… without the use of stimulants. This has been known to reduce the need for thyroid medication. But, more often, it increases appetite. Mitochondria realize that their electron supply has grown, and energy production is up. So it asks for more substrates to run glycolysis and electron transport by making you hungrier.
- **Contraindications.** MB can be harmful to a developing fetus. Pregnant or breastfeeding mothers should consult their healthcare professional before taking MB. Individuals with glucose-6-phosphate dehydrogenase (G6PD) deficiency should not take MB due to risk of hemolytic anemia.

Dosing

For over a century and millions of doses, MB has proven safe and very well tolerated. It has a very low side effect profile when taken responsibly. Though high doses have the opposite effect of low doses. In fact, many studies have shown the biggest benefits are achieved at the lowest doses, while adverse effects increase at high concentrations.

- A typical dose, which practitioners consider a "low" dose is 0.5–4 milligrams/kg of body weight in one day via IV. Oral dosing should be slightly higher. In this range, effects tend to increase with dose.
- A "medium" dose is 4–10mg/kg. In this range, effects generally decrease or stay the same with dose.
- A "high" dose is above 10mg/kg. In this range, effects can drop below baseline and become detrimental.
- Toxic effects can start above 5mg/kg. Conventional wisdom says this toxicity is due to switching from sort of net neutrality to oxidation dominance. However, one paper theorizes that toxic effects are not due exclusively to excess oxidation or failed hormesis.

Rather, impurities can never be removed entirely in manufacturing, so heavy metal toxicity is to blame.

- At medium to higher doses, it can build up in the tissues, putting you at risk for toxic effects. Remember, whatever dose you take, MB concentrates further in certain tissues/body compartments. Therefore, most practitioners recommend taking periodic breaks to purge the system.
- MB absorbs equally well through many routes of administration – a rare quality. It absorbs exceptionally well through the oral mucosa – some 80–90% gets into the bloodstream through troche delivery. It passes easily through the stomach and GI tract. It survives well its first pass through the liver to enter the bloodstream. Mitochondria then collect it and use it up slowly.
- MB is not habit forming. It does not create tolerance or withdrawal symptoms.
- Digestive and detox systems do not break down MB to any significant degree. Rather, it gets used up by mitochondria, then excreted via the kidneys and urine. It concentrates in the bladder virtually the same as it entered the body. This makes MB an excellent treatment for urinary tract infections in low doses. Though in high doses, it can cause burning sensation in the bladder due to oxidation.
- Titration (increasing dosage to suit). When MB gains two electrons, it loses its characteristic blue color (MB) and turns colorless (leuco-MB), and vice-versa. This dramatic difference in color is a convenient way to visualize how oxidative your internal environment is – specifically compared to dosage of MB. The more blue your urine is after taking it, the more inflammation is occurring inside you, thus leuco-MB is being used up and converted into MB. The less blue it is, the lower your inflammatory activity, thus more electrons stay on the leuco-MB molecule, because they aren't needed for reduction. With this information, you and your practitioner can understand in real time how much work your body is doing to overcome upsets and injury, relative to how much leuco-MB is available to neutralize oxidation.

Conclusions

Everyone agrees that methylene blue is indispensable in dire situations. The question is, whether it helps or hurts when taken long-term for performance-enhancement, maintenance or prevention. My assessment: It comes down to how much oxidative stress and chronic inflammation you have, compared to the dose you're taking – specifically, the concentration present in an area.

I think some commentators may not have completely wrapped their heads around the degree to which electron exchange changes, depending on concentration (technically, the methylene blue-leucomethylene blue ratio). At low-to-medium concentrations, redox reactions are highly fluid and fairly neutral in both directions, leaning slightly in favor of reduction. But, as concentration rises in relation to need, more oxidative pressure is exerted to steal electrons – particularly from the ETC. Don't forget, methylene blue, in its natural state, is oxidized. It wants to receive electrons and prevent them from turning oxygen into free radicals.

Think of it this way: MB, in low concentrations, is a casual participant in redox reactions – only stepping in when it sees an occasional electron ejected from the fray. Otherwise, it minds its own business. Whereas, in high concentrations, it's a troublemaker. It meddles in the mitochondria's business by stealing electrons directly from NADH, FADH2, and out of electron transport itself. It does give them back, but in an uncustomary location, thereby changing efficiency of energy production. It's better than nothing, like with stroke or heart attack, but it's suboptimal.

Bottom line: MB only becomes intrusive when the medium is so saturated it can't help but misplace electrons and disrupt standard operating procedures. To that end, mitochondria-dense tissues sequester MB, which raises concentration level, thereby increasing oxidation dominance (i.e., more MB, less leuco-MB). Concentration makes the difference between low-pressure oxidation *relieving deficits*, or high-pressure oxidation *butting in to processes such as ox/phos* whose efficiency can only go down, should the process be altered. In other words, more than three factors can confuse our impression of whether MB is helping or hurting mitochondrial function: (1) It concentrates in certain tissues and not others. (2) It switches from oxidative to reductive, and back again, based on concentration *vs.* redox needs in that environment. (3) It's eliminated slowly through the urine.

Therefore, MB might improve mitochondrial function in some tissues, while harming others. It can transition from hurting when first administered to then helping as concentration lessens. And redox needs of cells differ *and* change continually. All these factors affect many complex relationships we've discussed. Which brings us back to the ancient axiom: dose makes the toxin. Because, let's face it, when it comes down to it, almost every exposure that's beneficial in some way still exhibits a biphasic dose response. It's possible to overdose on almost anything you take or do for your health, including water and oxygen.

17

THE GREATEST STORY DR. JACK'S EVER TOLD

How POMC, melanin, UV biophotons, and biologic semiconductors fueled the rise of mammals... and our downfall today

When Dr. Jack says 'all life on earth is powered and controlled by light,' at first listeners can't believe the extent to which that statement could possibly be true. They think he's exaggerating for clicks or cachet. Heck, everyone knows he's super smart. But surely, he can't be serious, can he?

Prior to his current "POMC Phase," most of his assessments and predictions have come true. Yet, many felt there must be some pieces missing because none of us could see the whole picture. It wasn't until he sat down with Dr. Andrew Huberman and Rick Rubin, then Max Gulhane, that Uncle Jack began telling us the whole story about POMC, melanin, semiconductors, and non-linear optics guiding mammalian evolution.

Beginning in 2023, he laid out the mechanisms by which light directed life forms to evolve through revolutionary periods in earth's history such as endosymbiosis, the Cambrian Explosion and the K-T event. It's just within the last year and a half (as of summer 2024) that the process by which light turned early mammals into hominids became much clearer. Now mitochondriacs know how light sculpts our biochemistry, brain, and body to suit our surroundings. This is the story of life's inseparable relationship with light. This is precisely what makes us beings of light.

For 180 million years, dinosaurs ruled the earth

But all that changed in an instant when a giant asteroid, 65 million years ago, hit shallow water off the coast of Mexico's Yucatán Peninsula. On that fateful day, the environment for which dinosaurs were perfectly adapted changed so dramatically that only one class of dinosaur had the genetic advantages – flight and mitochondrial energy reserves – to survive the cataclysm. The existential crisis: Atmosphere-altering debris, and smoke from forest fires, darkened the skies for years, collapsing photosynthesis and most of the food chain along with it. In the aftermath, dinosaurs that could fly survived because they could go to where the food was. They evolved into birds. Other classes of creature weren't so lucky.

This period, called the K-T event, caused ¾ of all animal and plant species to go extinct. However, one species' tragedy is another one's triumph, as the K-T event set the stage for mammals to inherit the land after the dust settled. This is how a novelty of mammalian biology back then – the ability to capture and use light proficiently – became a central feature of our biology today. This chapter is about mammals' ability to use light to: make sustenance, evolve bigger brains and the biochemicals to wield greater intelligence, control circadian rhythms, and regulate cell cycles. This is how light sculpts life.

An environmental catastrophe gave dinosaurs an existential threat

Some paleontologists say the impact from the Chicxulub asteroid sent enough particulate matter into the atmosphere to blot out the sun for 10 to 1000 years. My guess is, most of the debris, forest fire smoke and climate disturbance would have died down after about 2–4 years. Whatever the case, animals had to have enough epigenetic versatility in their genes to survive years with photosynthesis being restricted in plants and plankton. They would have had less oxygen to breathe and a food chain that was barren from the bottom up – likely getting so bad it must have been like seeing ten to twenty winters in a row.

As we know, everything in the food chain links back to sunlight and photosynthesis. But many mammals hibernate, requiring long stretches without food. Many are nocturnal, which meant they could sustain themselves without sun exposure. So, the way they lived before the K-T event is the same way they subsisted through disruptions in the food chain: Their biology was replicating photosynthesis inside them using POMC products and light produced internally. Products of POMC literally make sugar and insulin that early mammals could live on through lean times. Indeed, you might call this process "animal photosynthesis."

When the environment changed, dinosaurs lost their competitive advantage, and mammals picked it up

When resources are plentiful, bigger, stronger animals have an evolutionary advantage. Size and strength help animals outcompete the smaller and weaker. Conversely, when a challenging environment is your biggest threat, *adaptability* gives you the advantage. This is one reason dinosaurs grew to be so large over millions of years, while early mammals stayed small and inconspicuous – basically living in the shadows of dinosaurs for the first 150 million years of their existence.

These furry critters probably did their best to avoid other animals. In all likelihood, they burrowed underground, hibernated, and were active at night. And for good reason: Fur contains melanin. And melanin, a breakdown product of POMC, fluoresces in sunlight. Since modern birds and insects can see into the UV range, dinosaurs likely could too. Thus, lighting up like a neon sign would make these shrew-like animals easy prey for carnivorous dinosaurs and prehistoric insects if they went out during the day.

Fluorescence: Absorption light of a shorter wavelength, such as UV or x-rays, and emission at a longer wavelength, often in the visible range.

But then an asteroid happened and, all of a sudden, hiding and sheltering became the little critters' competitive advantage because they were able to make their own light endogenously to control cell cycles, as well as sustain themselves without eating. POMC programs such as glucose and insulin production set up their biology to spend a lot of time underground, hibernate, and live a nocturnal life. POMC helped mammals survive in weak-sun conditions. However, it wasn't until dinosaurs left the scene that the melanin in their fur stopped being their biggest vulnerability. Dinosaurs had to go for mammals to receive the benefits of POMC in niches once occupied by giant reptiles.

Darwin did not say evolution comes from random mutation and natural selection

Evolutionary biologists like to use the work of Charles Darwin to explain how life evolves over time. They claim his writings say a series of incremental, adaptive changes over generations add up to gradual transformation of species. They say Darwin proclaimed evolution happens from random mutation and natural selection – the best genes win, basically. But that's not what Darwin's work really said – at least not in the first six editions. Truth be told, the notion of *survival of the fittest* is actually a misrepresentation of what Darwin originally wrote in his book *The Origin of Species* when it was first published in 1859.

What the book said before it was changed around 1871 was that *conditions of existence are far more important than natural selection.* Life purposely changes its genetic expression to suit the environment it finds itself in. It does not change by random accident, then survive to reproduce, or not. Meaning, the ability to adapt to your surroundings is no accident. It's built into the biology of all species through adaptability programming hardwired into our DNA and expressed through epigenetics. Through epigenetics, Nature gave every species some degree of flexibility in how its genes are expressed into physical traits and behavioral characteristics.

So Darwin, unaware of epigenetics, mitochondria, and biophysics in his day, was actually telling us that genes or accidental DNA alterations don't initiate change, which the environment then judges to be successful or unsuccessful. Instead, when your environment changes, life alters its characteristics as much as it can based on how much adaptability it has built into its genome – man being the most adaptable of all higher life forms.

As he suspected, and we now know, mitochondria change their state of expression based on environmental forces such as light and temperature. And that state of expression is what produces wellness or disease in a person through epigenetics, not your own genes. Mom's mitochondria then get passed on to her children in that state of vitality or fragility. That's how characteristics appear to evolve in an individual, and over generations. More than anything, mitophysical exposures drive biologic changes and adaptations, including metabolism, food compatibility, body shape and composition, skin color, constitution and longevity. Simply put, it's our mitophysical environment that makes us adapt, not genetic accidents that survive or fail the reproduction process.

Cephalopods are like prehistoric versions of the human brain

Cephalopods – including squid, octopus and cuttlefish – were one of the first kinds of complex life to appear on earth some 530 million years ago. Researchers study them because they appear to be us – humans and mammals in general – before UV light and wideband semiconduction upgraded our brains, bodies and biochemistry. They're like version 1.0 of the human brain, compared to us at version 1,000,000 or something.

Semiconductor: Put in one type of light, get one of a different frequency out.

Wideband semiconductor: Semiconductor that operates on high-energy, short wavelengths.

Take the cuttlefish, for example. They produce dazzling movement of colors across their bodies. But they forgot to evolve, as they're still not proficient at recapturing light and reusing it. Their cousins, squid, do make melanin. But they expel it in their ink to confuse their prey or escape predators, instead of using melanin for higher purposes. And then there's the octopus. Their intelligence is impressive. However, their aquatic environment never required them to adapt to light and temperature cycles of dry land, or skeletal/neurological complexities of a vertebrate.

Cephalopods, as a group, have the genes necessary to evolve into higher life forms. However, they never used light to drive their evolution. They did fine in the ocean, so adding complexity to their management of energy (i.e., use of light) was pointless. It would actually be wasteful. Bottom line: when you examine the evolution of organisms through history, you'll find their complexity is deeply connected to their ability to retain light and use it to run high-end endeavors such as language and creativity.

- **Bacteria and archaea** – 3.5 billion years ago. They don't retain light. Their lives are short, so they quickly spend all the light they get.
- **Cephalopods** – 530 million years ago. They expend light, but don't conserve it for environmental or intellectual adaptation.
- **Dinosaurs** – 252 million years ago. They started to figure it out, but not soon enough. Only flying dinosaurs survived a change in light.
- **Post K-T mammals** – 65 million years ago. Their ability to use and reuse light became their superpower.
- **Modern humans** – 300,000 to 50,000 years ago. Now, we rely on light to power our biology. Unfortunately, if that light is adulterated, so will our biology be corrupted.

How UV light made life on earth possible... then and now

Most primitive life forms such as bacteria don't like large doses of UV. It killed them 3.7 billion years ago when the earth was new, and shortwave UV still kills bacteria today. But starting 1.45 billion years ago, archaea teamed up with bacteria to create multi-cellular life. One specialized to make energy, the other focused on creating biologic complexity. The epoch upgrade being, both sides figured out how to make use of UV light to power and control its now-unified biology. Somehow, some way, mitochondria turned a once-toxic exposure, UV light, into a means for many cells to collaborate as one. Without ultraweak UV light, aka biophotons, advanced life on earth might not be possible. To illustrate, we go back the K-T period.

After the impact, debris in the atmosphere would have blocked UV frequencies more than any other because faster oscillation means more collisions with liquids, gasses and particles in the atmosphere. Since cells use ultraweak UV to trigger cell division (for growth and repair), melanocytes at the surface went looking for a better supply. Where did they go? They moved along the pathway they knew best: the ectoderm. Embryology teaches us that skin and nerve cells in embryos develop from the same germinal layer of tissue – the (neuro)ectoderm being one of them. This connection between related types of tissue is still maintained in the adult form. Hence, when melanocytes travel, they don't go stay at a neighbor's place; they move in (at least temporarily) with their closest *relatives* in the ectoderm family.

Melanocyte: cells that make melanin.

Deprived of UV from the outside, melanocytes in the skin wandered inward. The extra melanin, made by melanocytes, then built up the power and information network of the brain. This mass migration of melanin amplified mammals' ability to tan on the inside – i.e., collect light to put it to good use. The energy hog in our heads then had sufficient light, electrical charge, magnetic flux, chemicals and communication to increase capacity of the frontal lobes. Efficient energy management became the superpower of our species, effectively jumpstarting intellectual development of hominids and the biochemistry to run it.

During that time, our sun which is a G-class star entered the early-middle of its life. When G-class stars reach mid-life, their UV output goes up by 10%. That may not sound like a lot, but it has huge implications for life forms that use those frequencies to power and control adaptation in their biology. Remember, only a small percentage of sunlight is ultraviolet, and our biology responds to it in a non-linear fashion. So even small increases in UV add innumerable frequencies to stimulate amplified effects.

This set up a good/bad scenario. A larger prefrontal cortex gave hominids more brain power. But that also means our engines have the highest energy demands, so we're more sensitive to energy deficits and adverse exposures. Plus, more intelligence isn't better in all ways. Let's just say it presents a lot of temptation when technology, greed and arrogance grow faster than wisdom, caring and empathy.

Melanocytes can move toward any source of UV

The most reliable source of UV for melanocytes in the skin comes from their mitochondria, or at least it should. That's your endogenous production. The more powerful, but often inconsistent, supply comes from sun exposure. That's an exogenous supply. Both sources would have to be missing for melanocytes to pack up and leave the skin to go looking for UV inside. And that's exactly what happened following the K-T event. Darkened skies and food scarcity starved melanocytes in skin of UV, so they went looking for more on the interior.

On the flip side, melanocytes in the brain get their UV locally. That means they migrate only after their mitochondria and DNA stop making V-UV. In short, if a location has UV present, melanocytes stay put and replicate. But if there is none in the area, melanocytes mobilize. They may not know where they're going, but they know when they've arrived, because UV light is their signal to settle down and multiply.

V-UV is light between 100-200 nm. It's called "vacuum-UV" because air absorbs it well, so it's made (artificially) in a vacuum. You can think of it as UV-D, which is right next to x-ray.

But melanocyte mobility is not necessarily a bad thing

Melanocyte mobility operates under both healthful conditions and disease states. To illustrate healthful functionality, consider how white blood cells (WBC) travel to infection sites. Biophoton signaling must be turned off to enable WBC movement. WBCs are loaded with POMC for light to work its magic. The WBC activation process: The nucleus turns off V-UV light, which turns off mitosis. This stirs WBCs into action and they enter the bloodstream. When they run across bacteria or other microbes causing trouble in a tissue – sensed from their V-UV emissions – the WBCs stop and squeeze through the vessel wall to fight infection on the other side. On the other hand, they don't know when to leave, or where to go, unless/until V-UV is turned off.

Similarly, the disabling of mitosis in **neuroplasticity** is not pathologic either. Neuroplasticity (nerve tissue repair) is programmed deactivation of mitosis, followed by migration, cell division, and growth. Conclusion: The brain is way more plastic than once thought. But not every person's body has the energy available, and fluency of communication, to heal damaged/degenerated brain matter. Consider it a premium service that takes strong mitochondria to produce sufficient light, electron flow, magnetic flux, redox signaling and ATP availability – plus, lots more good exposures than bad.

All living cells use ultraviolet light to signal

In the 1920s, Alexander Gurwitsch showed that cells of all life forms emit light. The experiment: He cut onion roots in two and found they could be stopped from dividing and growing by separating them with glass. Or growth would resume when quartz was put between them. Since glass blocks short wavelengths and quartz does not, he knew non-visible, short wavelengths were triggering mitosis. What he didn't know at the time were the exact frequencies that were triggering cell division (spectroscopes and photomultipliers had yet to be invented), so he called the emissions "mitogenic radiation." He replicated the experiment on every cell type imaginable from single-celled to complex and found they all emitted this invisible energy.

Later, in the 1960s, Fritz-Albert Popp extended Gurwitsch's work by developing photomultipliers. With these, he discovered Gurwitsch's mitogenic radiation was actually ultraweak UV light. He called the individual particles of light "biophotons." More recently, scientists have found our mitochondria and DNA (when it unwinds) use UV-C or greater to tell cells when to divide or sacrifice themselves. That's light of a higher frequency than in midday sun!

Why biology chose ultraviolet for its signaling needs

First point: Dr. Robert O. Becker demonstrated the crucial first-step of the cell cycle is controlled by electricity. He found red blood cells dedifferentiated into generic stem cells when specific voltages were applied. Cells then re-differentiate into whatever cell types are needed. This starts the growth/healing process. As a partner principle, the formula to calculate band gap tells us that the stronger the light that goes into a semiconductor (i.e., frequency), the greater the electron voltage that comes out… and vice-versa. Therefore, the body has to use UV light to generate sufficient voltage from our semiconductors to bring new cells to life – voltage which other frequencies can't provide.

Formula to calculate a semiconductor's electrical input vs. *light frequency output (to jump the virtual band gap): 1240 ÷ electron volts = frequency of light emitted in nanometers. We use the start of photosynthesis to illustrate: 1240 ÷ 12.60 electron volts = 98 nm light (V-UV/soft x-ray range).*

Exhibit 2: Nature chose to control optical signaling with endogenously-produced far-UV because it needs to be able to extinguish the signal at certain times. It would not be able to stop the signal if sunlight had far-UV in it. The solution: the body makes its own UV light which is stronger than sunlight. And UV wavelengths don't penetrate deeply, so stray absorption is minimized. This comes into play when nitric oxide, for example, needs to act locally.

On the other hand, biology can't use infrared for signaling purposes because all matter emits infrared photons all the time, and it can penetrate inches deep. Hence, you'd get a lot of noise among the signal. In the middle, I would think visible-frequency photons would be both weird to

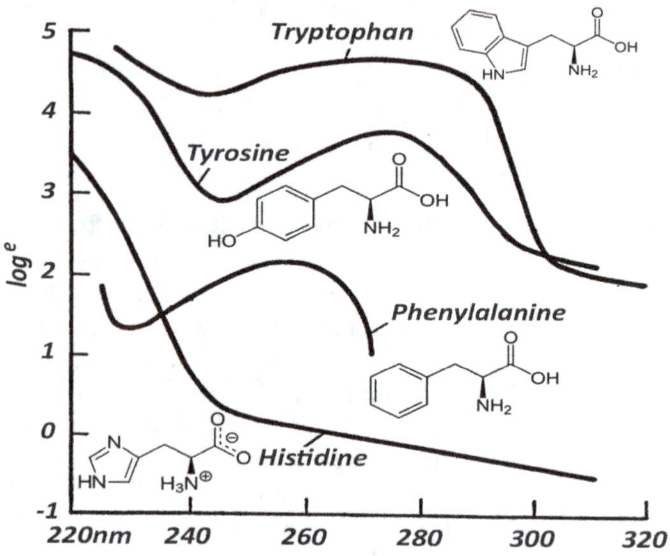

Aromatic amino acid absorption spectra

see escaping from a person at night and thermodynamically wasteful being stronger than infrared. Therefore, minute doses of UV was and is the only viable option.

Essential application: Our most important neuro-chemicals – including dopamine, serotonin and melatonin – are made from aromatic amino acids such as tyrosine, phenylalanine and tryptophan. These all have absorption spectra between 400 nm UV-A to 150 nm V-UV (see graphic right). That means most of the biochemical pathways that run human (and mammalian) biology operate on UV light. This gives Nature optical control over our neurochemistry, circadian rhythms, cell growth and metabolism.

Note: The action spectra of amino acid precursors all ramp up sharply in the far-UV range (under 300 nm). These make chemicals such as melanin, dopamine, melatonin and serotonin.

Graph is a reproduction of Dr. Alexander Wunsch's work.

Another example: mTOR (mammalian/mechanistic Target of Rapamycin) is a regulatory protein. Think of it as a switch that controls whether cells go into an anabolic state of survival, growth and proliferation, or a catabolic state of breakdown with autophagy and apoptosis. And what flips the switch to build cells up or tear them down? mTOR has an absorption and emission spectrum of 380 nm light. That tells you UV-A activates anabolic programming inside of a cell.

Based on myriad clues like these, Dr. Jack predicts UV photons (along with IR frequencies) will ultimately be shown to regulate not just the start of the cell cycle but also autophagy, apoptosis, chromosome separation, DNA replication, and instructions on how to run our semiconductor networks (encoded into our so-called "junk DNA").

We use UV light from outside the body and in

Mammals need UV light to control cellular division and life cycling. But external UV is only available at midday, part of the year, at higher latitudes. The way Nature compensates for a limited solar spectrum in fall and winter is by releasing heat and light from cells and mitochondria. Of course, cold exposure releases infrared to warm the body. But coldness also makes our semiconductors emit UV light from the inside. That's crucial, because if you never let your body get cold in winter, your mitochondria don't release the UV your cells need to make up for weak sunlight at high latitudes.

That's a primary reason modern humans are UV deficient. We can't bear to be cold, so the body fails to make its own UV internally. When that happens, you've lost UV from outside and inside. Hence, cell cycling breaks down, infradian rhythms get messed up, heteroplasmy rate rises, and inflammation persists. That's why cold exposure is such a potent tool for healing, regeneration, immune modulation, and anti-aging.

But I thought UV-B and above causes DNA damage and cancer?

X-ray: 0.01–10 nm, V-UV: 10–200 nm, UV-C: 200–280 nm, UV-B: 280–315 nm, UV-A: 315–400 nm

Ionization starts damaging DNA in the UV-B to UV-C range. So, why would Nature use frequencies that *can* hurt us, potentially? A: For signaling and energy-to-matter conversion, biology needed to use light frequencies that our semiconductive tissues react to. However, because we absorb and respond to UV in a non-linear fashion, we can easily overdose on UV and greater when we get a "linear dose." We do, in fact, suffer sunburn and DNA damage when our use/defense mechanisms such as melanin and vitamin D are down and we're overexposed. Conclusion: The body is designed to repair sun damage. But modern living raises our rate of injury with blue light, nnEMF and dehydration, while rate of recovery is lowered in the absence of red light, redox potential, vitamin D, strong mitochondria and melanin.

You can't get cancer when cells are receiving UV light

Metastasis: Cancerous cells that spread.

"Modern metastasis happens in humans because they no longer can make internal UV light from their melanin semiconductors. When the light signal is lost, mitosis stops and the cancerous cell can migrate to another tissue where the UV stimulus is present. That is what cancer really is at its core. It is cells looking for UV light."

—Dr. Jack Kruse.

Malignant melanoma might not be what we think either. As Dr. Kruse goes on to say, melanoma might be melanocytes from the inside, coming to the surface to find UV light because their semiconductors inside couldn't make it. Therefore, mitosis was halted, stalling the cell cycle.

We're the only primate to lose its coat of body hair

Other primates kept a full coat of body hair because they don't need as much sunlight to run their biochemistry. They make vitamin D endogenously, so they don't need as much external UV. They don't make melanin on demand. Their epigenetics are simpler. And they don't need as much energy for their brains and immune systems. Hence, they have a bigger margin for error, and their skin doesn't need to be bare.

We, on the other hand, lost our hair coats because we need more sun exposure to run our programs. Hugely important, we need UV-B to make vitamin D (we're the only primate that can't make vitamin D). And we need external light to enter our semiconductors so key frequencies and electrical charge will come out. Bottom line, we need more energy to run our biology than other primates, so most modern humans now live on a fine line between health and sickness. We consume more energy, so any new exposure that changes our collection or conversion of light can push us into improved health or quiet degradation.

Cross-species differences: Human babies are born fat because sunlight helps them metabolize subcutaneous fat into myelin to coat brain cells. In contrast, our closest relatives, chimps, don't need baby fat because they're born with fully functional brains. After our brains get to be fully myelinated in our mid-20s, we then need sunlight to run complex biochemistry such as cognition, metabolism, immune system, hormone-endocrine function and longevity. Unfortunately, when we don't get the sun exposure we need in adulthood, the body sometimes regresses to early-development programming. This evolutionary "atavism" adds unneeded/unwanted fat mass in preparation to myelinate the brain. Later, when subcutaneous fat depots are perceived to be full, atavistic programming shifts fat deposition to internal organs – aka visceral fat.

Another reason our babies need more fat than chimps is it contains leptin. As the nervous system, endocrine organs, and immune system grow rapidly, they need precise control over energy allocation. Leptin, the energy accountant, helps them sense how much light and DC electricity is available for crucial construction projects in pre-adolescents. Crucially, the leptin-melanocortin pathway put leptin in visceral fat to tell the hypothalamus to make more melanin to mature baby's brain postnatally.

UV light, in the eyes, tell the body to make vitamin D

UV light in the eyes is a signal that tells your circadian system it's daytime. We're out in the sun. It's time to make us some badly-needed vitamin D. By way of review, vitamin D supports immune function, bone strength, muscles, metabolism, the heart and the brain. It's widely needed in organ systems throughout the body. However, the vast majority of people are not just low in it, they show significant signs of deficiency, whether they and their doctors know it or not.

Unfortunately, sunglasses and standard prescription lenses block UV light so wearers don't get the signal to turn on vitamin D production. Hence, the modern practice of trying to protect the eyes from overexposure (or just look cool) undermines the body's ability to manage this key circadian function. This is one of many root causes of modern disease that's so simple and straightforward it's dismissed because it appears trivial. Don't be fooled by misdirection and false narratives. Instead, trust in the genius of Nature's design, and you can avoid the clutches of centralized medicine as often as possible, as long as possible.

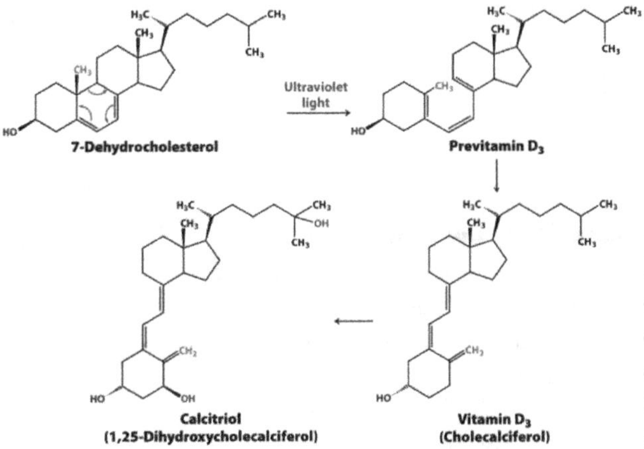

How vitamin D is made. Step 1: In the skin, UV-B converts 7-dehydrocholesterol into inactive forms of vitamin D, including D3 cholecalciferol and D2 ergocalciferol. Steps 2 and 3: The liver, then the kidneys, convert D3 and D2 into the active form of vitamin D: 1,25-dihydroxyvitamin D. Only then can vitamin D do the many jobs it does. Sadly, practitioners and public alike think vitamin D in supplement form works as well as that made by sunshine. They're mistaken. You see, vitamin D outside the body is fat soluble (e.g., in supplements). However, vitamin D needs to be water-soluble to circulate well in the blood. For that, the cholesterol that makes vitamin D needs to be sulfated. You need sunlight to do that.

When you ingest supplemental D3 and D2, the body ramps up production of LDL cholesterol to form water-soluble vitamin D. Yes, the body actually raises LDL cholesterol because it needs vitamin D to get to the liver and kidneys. On the other hand, vitamin D produced through UV-B exposure doesn't have this problem. When cholesterol converts to vitamin D naturally, it's sulfated and water soluble. Hence, you don't need to jack up your cholesterol level just to transport vitamin D around… then risk being put on statins by your doctor.

Bottom line, many folks can take supplements, and/or go out in the sun, and their vitamin D level hardly budges. That's how a person can look tan on the outside, yet their serum vitamin D remains low. Reasons are, most people's vitamin D production is broken today at multiple levels: (1) UV deficiency and/or (2) dehydration in the skin. (3) Previtamin D can't get to the liver and kidneys because its cholesterol wasn't sulfated. The (4) liver or (5) kidneys can't convert inactive D3 and D2 to active 1,25-dihydroxyvitamin D because they're missing melanin in those organs as a result of too many bad influences such as wireless radiation and blue light, and too few good exposures, including UV and cold. Another thing you can do to raise your vitamin D: Drink deuterium-depleted water and avoid nnEMFs to help hydrate the skin. This improves isomerization, which upregulates endogenous production of vitamin D.

Melanin

Before the K-T event, life forms assimilated light from their surroundings predominately through one of two semiconductors: chlorophyll in plants or hemoglobin in animal blood. After the K-T event, melanin became mammals' favorite transformer of light into biologic function. For context, white surfaces reflect light, while dark colors absorb them. And melanin is the darkest of pigments.

Melanin protects you from overexposure. But how does it do this?

When ordinary folks hear that melanin "protects" a person against sunburn, they assume that means melanin shields the skin so UV light doesn't cause premature aging or cancer. They think you need to avoid ionizing ultraviolet to stay youthful-looking and cancer-free. Most misleading, the very moniker *sunscreen* makes people believe: (1) UV light is bad for you, and (2) melanin's job is to prevent UV from getting inside the body. Therefore, it's discarded. But that's not what melanin or UV really do.

These are destructive half-truths infecting impressionable minds everywhere. Details matter when making your own health decisions, so here's the truth: Melanin does indeed protect the skin against sunburn from UV, but not how most consumers think. Melanin's real job is to absorb light and transduce it into energy and information. Simply put, melanin harvests light; it doesn't throw it away. At the heart of the issue, melanin is designed to conserve light and convey it to other tissues. In fact, it's the darkest pigment on earth because it absorbs all frequencies of light. This is how it "protects" the skin against sun damage and cancer.

On the other hand, when there's not enough melanin around to carry out this function (lack of morning sun would be one reason), cells that are poorly-equipped to handle UV light can absorb it in an unregulated fashion. These cells don't have the right photoreceptors to capture and use UV light, so they can easily get too much of it, too fast. Ionizing radiation from midday sun then increases oxidation and inflammation in the skin at best, while DNA gets damaged on the way to melanoma or accelerated aging when cells decay faster than they're repaired. In most cases, that results in skin losing healing capacity from too much blue and not enough red/IR. It's yet another organ system sucking our antioxidant/anti-inflammatory reserves dry. And that is the recipe for accelerated skin aging and cancer, over time.

Melanin's #1 job: charge-separating water

When light hits melanin, this changes the bond angle of hydrogen in surrounding water. This reconfigures H_2O (water) into H_3O_2 (e-zone, aka electron-zone water). This produces free electrons, hydrogen protons and oxygen. And what do mitochondria need to run their engines? They need all three of these raw materials. That's why melanin is always located around mitochondria and water. Now, light does have the ability to charge-separate water on its own, but melanin makes the process more efficient by spitting out four electrons of its own.

Next, melanin takes electromagnetic frequencies from the environment and converts them into electrical signals for sensor-regulator organs. Crucial in connecting us to the outside world, melanin sheets are used as sensor-transducers in all five of our senses. These biologic solar panels receive inputs of light, sound, smell, taste or touch and translate them into electrical/light signals for the brain to decipher.

Melanin mediates our five senses

Melanosome: Organelle in melanocyte that actually makes the melanin.

- **Skin.** Melanocytes are cells that (normally) live between the outermost layer of skin (epidermis) and middle layer (dermis). They host organelles, called melanosomes, that produce melanin. This melanin absorbs all electromagnetic frequencies – harvesting the useful ones and degrading the bad.
- **Eye.** The retinal pigment epithelium at the back of the eye uses melanin to transduce light into DC electricity. DHA carries that electrical signal to the suprachiasmatic nucleus to tell the hypothalamus which light frequencies are around you.
- **Ear.** The cochlea uses melanin to hear light and transduce it into sound.
- **Taste.** Melanin in taste buds enables taste perception.
- **Gut.** The gut lining (villi) is made of enterochromaffin cells (ECCs). They're best-known for making serotonin. But, more important for this discussion, ECCs help regulate digestion by sensing what's happening in the GI tract and informing the brain along the vagal nerve. The ECCs' melanin sheets basically turn light from microbiota into electrical signals by which the brain modulates secretions and gut motility, while controlling microbiome populations.

Melanin aids magneto-chemistry

As we've learned, free radicals have a net positive effect in healthy individuals. They just become pathologic when they break down tissue faster than the body can repair or replace it. It's a two-sided equation. In short, the body runs and repairs itself through free radical signaling. Melanin supports free radical signaling by retaining ROS and RNS. It holds onto free radicals to help mitochondria control metabolism locally.

For example, beta oxidation and urea cycles are enhanced or suppressed by ROS and RNS signaling and light produced by semiconductors. Unfortunately, when you lose melanin in an organ or tissue, power production goes off the rails, as does biophoton emissions and coherence of communications between the tissues and brain. State of tune in mitochondria – from ROS and RNS influence – is a big reason why melanin matters tremendously in manifesting wellness or disease.

Melanin detoxifies

Other mammals use melanin on their insides like we do, but they don't use it in their skin. Instead, they put melanin in their fur and feathers. For them, melanin detoxifies heavy metals by grabbing hold of them and keeping them out of sensitive tissues like the brain and heart. As you may guess, fur and feathers are safe places to put toxins because they're metabolically inert. Thus, the metals just hang out, not causing injury until they're eventually shed.

Similarly, melanin captures deuterium in all animals so it can be sweated out or sloughed off in dead skin cells. Unfortunately, when melanin systems malfunctions, you lose this detox pathway. Or the opposite can happen. Probably the #2 way melanin gets depleted inside us is through detoxification because melanin is consumed in capturing and escorting toxins out of the body. That's a root cause of numerous conditions. Outlier example: Melanocytes can collect in the gut and dump heavy metal accumulations directly into the colon when primary detox channels are backed up.

Finally, melanin's common claim to fame is perhaps its least useful

Melanin is a natural sunscreen and nnEMF-shielding material. However, it operates differently than artificial sunscreens. Melanin absorbs all frequencies of light, visible and invisible; incorporates beneficial wavelengths around the visible range; and attenuates frequencies that can damage DNA and cause cancer. More simply, melanin's job is to integrate light productively, rather than destructively. In summary, we use melanin in our skin, eyes, hair, and sensory organs like the ear. Melanocytes and melanin production are stimulated by UV and inhibited by blue light and free vitamin A (retinal). Equally problematic, microwave radiation trashes melanin wherever it is. Oxygen and nitric oxide renovate melanin after it's used.

Creativity may be an evolutionary regression

Dr. Kruse thinks of creativity as a kind of genetic throwback – like an epigenetic reversal, which I might contextualized in this way: Advanced cognitive functions such planning, abstraction, aesthetics and creativity are survival adaptations that are not needed when your species is doing fine without them. If you can eat, sleep, live and reproduce without using creativity, what difference would it make in your success? Remember, there's a cost to every use of mind and body, including intelligence. So, sometimes high-performance is irrelevant. Think about it: You don't need to be creative, or even think too hard, when you're winning. You're only required to employ brain power when you're facing challenges.

Whatever the case may be, creativity is powered primarily by dopamine fueling frontal lobe function. Dopamine is made when melanin production breaks down. This happens in an altered light spectrum – be it blue-light LED, incandescent, or even fire. That means rarified brain function is unlocked by bad light – specifically, an absence of UV. A lack

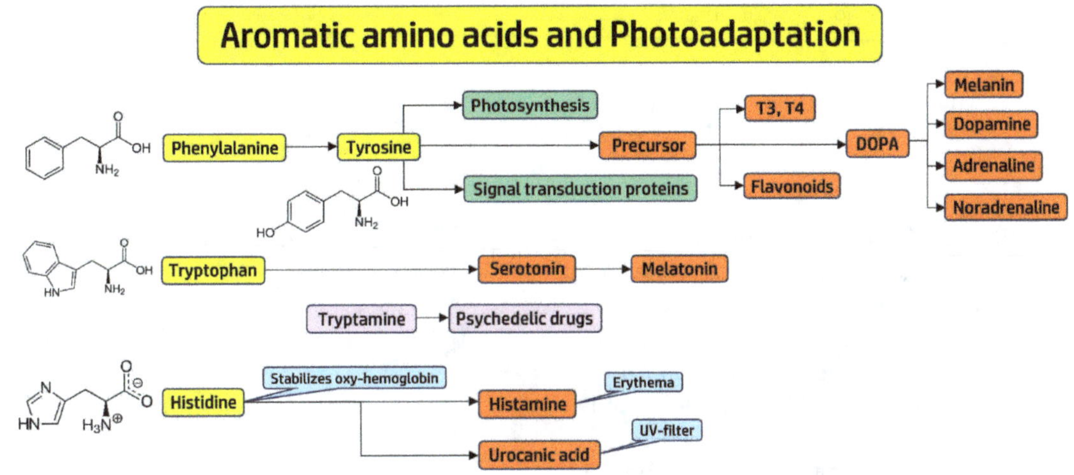

Melanin synthesis starts with UV light converting phenylalanine into tyrosine. Two step later, vitamin C is required as a co-factor to turn DOPA into melanin, dopamine and noradrenaline. Note: T3 and T4, which run the thyroid, are also made (or messed up) along this pathway.

Diagram is a reproduction of Dr. Alexander Wunsch's work.

of UV makes dopamine, instead of melanin and β-endorphin. Just think: how many great songs were written and performed in dark places – whether emotional, in-studio, or concert venue?

I never knew why this might be. But now it seems being indoors, under artificial light, may play a role. Without a doubt, music and darkness go hand in hand. Conversely, happiness and joy are virtually synonymous with sunlight. So, lowering serotonin in depression, while raising dopamine with blue light, produces out-of-the-box thinking needed for artistic expression. Case in point, the first evidence we have of abstract thinking, cave paintings, were produced under (artificial) fire light with skin covered. Similarly, how many people say they're most productive late at night? I used to think it was because the environment is quieter, including the thoughts of others. But now that I think about it, artificial light in combination with fewer distractions is what makes more dopamine to maximize frontal lobe function.

Conclusion: High latitude countries consistently outperform stronger sun regions in terms of art, music, architecture and innovation. Part of the reason may be that serotonin (a feel-good neurotransmitter) and dopamine (a concentration/reward chemical) tend to work in opposition to each other. When you have lots of serotonin in your system, you chill. You're okay *being*. But when you're running high on dopamine, you're motivated to create. You feel like *doing*. This state of dopamine arousal is a knock-on effect of melanin degradation. Hence, artificial light, with its lack of UV, may play an unsuspected role in artistic output because it encourages creativity and productivity, rather than rest and relaxation… that is, until you overdose on blue light and run out of dopamine. That's when you need Ritalin and Adderall.

POMC sculpted human biology

Linchpin of the leptin-melanocortin pathway, POMC is an intricate, compound molecule made in the pituitary of mammals.
"Proopiomelanocortin" doesn't do anything by itself. Rather, it's a precursor polypeptide (piece of a protein) that's kind of like crude oil in that it needs to be split into useable products by heat – in this case, light – before each breakdown product can do its jobs. Mammalian biology

chose to amplify it because it's an energy-efficient way for one gene to make many molecules the body uses, all in one go. Most important, specific frequencies of light snip off pieces of the POMC molecule which are independent biochemicals that once guided mammalian evolution long ago, and which control human biology today. These include

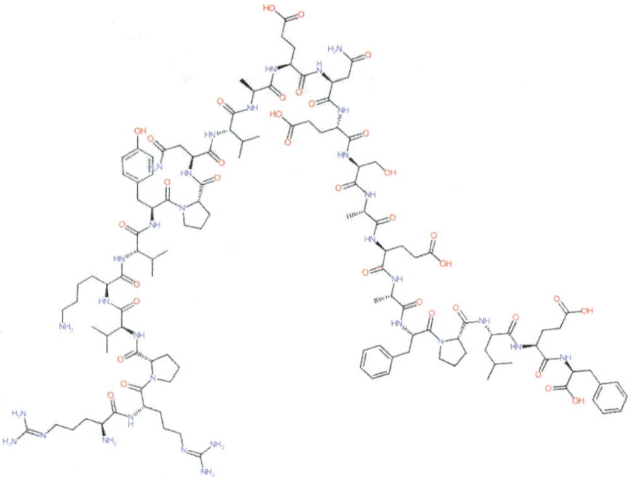

The POMC molecule. Ridiculously ornate, don't you think?

- **β(eta)-endorphin** is an endogenous feel-good chemical and pain killer. It's a drug the body makes when you're out in the sun.
- **ACTH.** Blue light cleaves adrenocorticotropic hormone from POMC, and so does UV-A to a lesser extent. ACTH raises cortisol in response to blue light. Cortisol raises glucose, especially in the absence of UV.
- **CLIP.** Corticotropin-like intermediate peptide raises insulin.
- Neuropeptides (aka peptide hormones, melanotropins or intermedins) in the melanocortin family stimulate melanocytes in response to UV light, which then make melanin. These include
  - **α-MSH.** Alpha-melanocyte stimulating hormone. UV-A cleaves alpha-MSH from POMC. α-MSH suppresses appetite.
  - **β-MSH.** Beta-melanocyte stimulating hormone controls energy balance and body weight (via appetite).
  - **γ-MSH.** Gamma-melanocyte stimulating hormone regulates production of a large number of steroid hormones (called steroidogenesis), energy, appetite, blood pressure and sodium.
- **Met-enkephalin** (aka opioid growth factor) is an opioid peptide that blocks pain signals and makes you feel good.
- **Lipotropin.** Stimulates melanocytes, mobilizes fat stores in fasting or exercise, produces steroid hormones, affects mood and behavior.

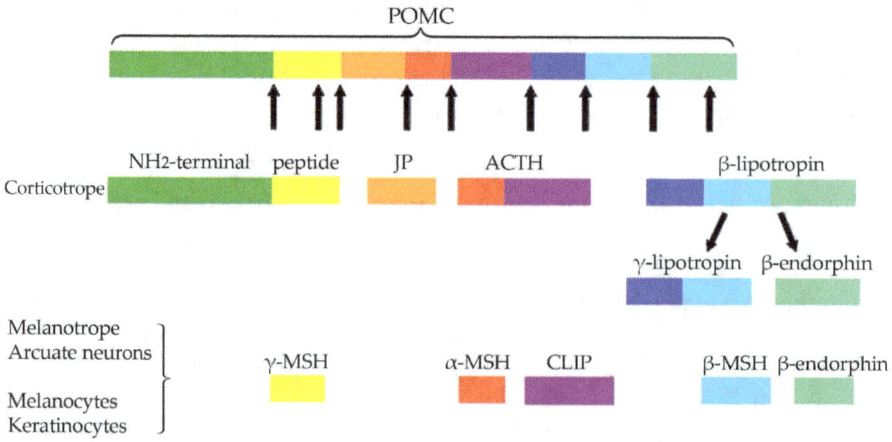

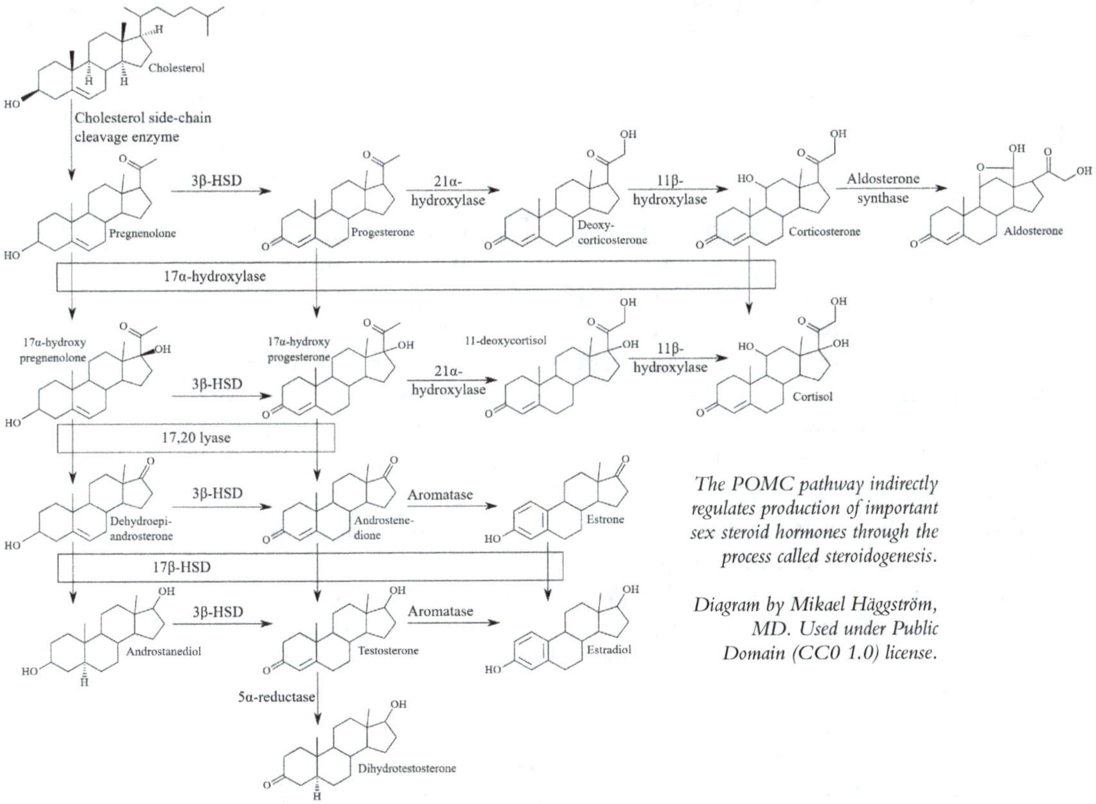

The POMC pathway indirectly regulates production of important sex steroid hormones through the process called steroidogenesis.

Diagram by Mikael Häggström, MD. Used under Public Domain (CC0 1.0) license.

To unpack the word proopiomelanocortin, "pro" means pre, as in precursor, "opio" stands for opioid, "melano" gives you more melanin, and "cortin" tells you it raises cortisol. These eight products promote the production of others, including

- **Melanin.** As you may have guessed by their names, the three varieties of MSH stimulate melanocytes to make melanin.
- **Sex steroid hormones.** ACTH promotes adrenal androgens such as testosterone, estrogen and DHEA (diagram above).

Androgen: Sex steroid hormone that regulates development and maintenance of male characteristics.

UV light and oxygen increase POMC expression, while high blood sugar suppresses it. Cold temperature also raises POMC and its products because it prompts mitochondria and cells to produce endogenous V-UV to IR-A light, while also increasing receptor sensitivity of biochemicals such as dopamine and β-endorphin.

To learn more about POMC, watch the YouTube video entitled "Pro-opio-melano-cortin (POMC), Melanocytes Stimulating Hormone (MSH), Skin Pigmentation – Endocrine". https://youtu.be/U0QeS7yhA-k?si=2faeMpiKP5BAHM3-

We're now being taken out by our own frontal lobes

Prior to the K-T event, POMC helped early mammals live a reclusive, niche existence. Through the photosynthetic brownout, POMC gave the little critters a way to sidestep a broken food chain. Then, after the K-T event, POMC helped the survivors adapt to a changing environment, expand their intellect, and take over the earth. POMC was their savior, turned secret weapon.

But now, our POMC pathways have turned against us. The way it reacts to the exposures we give it today silently imbalance our systems, or flat out steals health away from us. Indeed, many feel a mass extinction is well underway and our addictions to comfort, convenience and consumerism are to blame. How ironic: The cranial capacity that enabled us to rise as a species and dominate the planet now degenerates our bodies faster than centralized medicine is able to fix the damage our cleverness causes. At the same time, we're arrogant enough to not see that our intellect is the very root of our evolutionary regression. We are our own worst enemy, because when you don't go out in the sun, you effectively de-evolve to an earlier era, before POMC empowered higher brain function and advantageous gene expression.

Without UV and IR exposure charging up your battery packs, the epigenetic programming of your brain and body default to simpler settings to make best use of available power. Our bodies simply don't have the juice to run all the advanced programs modern humans want to run, at least not with all the setbacks they must endure. The result: basic functions such as metabolism lose their coherence, while high-end operations such as reasoning and reproduction deteriorate.

In a dreadful environment, the POMC pathway punishes you because flawed signaling messes up our contingency programming. Given stifling adversity without countermeasures, modern living inevitably creates grave dysfunction. This is the primary reason people are getting sick with so many diseases today. By design or by chance, the POMC gene has become humanity's self-destruct button. And our intelligence, combined with poor choices, is each person pressing his/her own button dozens of times per day through intentional exposures or passive acquiescence.

Here's what those POMC problems look like in real life:

POMC-related problems explain most diseases of civilization

Weight gain and obesity. Obesity is caused by breakdowns in the leptin-melanocortin pathway. POMC plays a starring role in both function and dysfunction of this pathway. For instance, α-MSH suppresses appetite, while UV light causes melanin to increase satiety through the hypothalamus. UV exposure is also supposed to make vitamin D. It binds to the vitamin D receptor in Cytochrome III to slow electron transport. This downregulation of the ETC reduces your desire to eat, while red light makes ATP without food. Note how many processes falter when you forgo UV and IR. Meanwhile, concentrating blue destroys POMC and α-MSH. Blue light alone increases blood glucose and insulin through ACTH and CLIP, respectively. Conclusion: Nature designed sun exposure and tanning to suppress appetite and weight. It's infradian programming, built into our biology, which we're comprehensively circumventing with our technology addictions.

To put that more potently, you may be able to lose weight by starving yourself (or any unnatural diet such as carnivore, low-carb, vegan/vegetarian or keto). But if you don't employ natural calorie restriction with α-MSH, β-MSH and γ-MSH, your basal metabolic rate doesn't stay elevated from exercise, diet, and lifestyle changes the way it should. Instead, thyroid speed dips to bring you back to baseline and you put the weight back on. **That's the mechanism by which pounds come back with a vengeance after significant weight loss.**

You employed a hack to lose some weight, but the real problem remains: infradian and metabolic dysregulation causing weight gain. You didn't raise satiety and lower appetite through α-MSH, β-MSH and γ-MSH. You didn't change your metabolic set point. No doubt, this is the gigantic missing piece – you could call an epic fail – of dieting and weight loss programs as food gurus have long proffer them. So, what happens? A diet works for a while. But as the months and quarters go by, the diet has to be more and more restrictive to keep the weight off because you didn't reset your thyroid and leptin systems to that of sun-power seasonal programming, instead of food for energy. Solution: You need to leverage UV A/B/C/V and cold exposure to manage your weight the way Nature does it (the healthy, sustainable way). These control POMC, nitric oxide, the vitamin D receptor, and the hormones T3 and T4 which help the thyroid regulate your metabolic rate. Conversely, blue light stymies weight management chemicals and systems.

Diabetes, metabolic syndrome. ACTH stimulates cortisol production. Cortisol makes glucose in the liver through the process of gluconeogenesis. Another cleavage product of POMC is CLIP. CLIP raises insulin. ACTH and CLIP are both stimulated by blue light. This means that every mammal on the planet creates sugar and insulin when they're exposed to blue light. You don't have to eat anything to increase propensity for hyperglycemia and hyperinsulinemia.

What follows is this: Melanin holds onto free radicals. The body's magneto-chemistry (signaling pathways for metabolism and repair) runs on free radicals from their unpaired electrons. Unfortunately, glucose quenches free radicals. That means melanin deficiency (from blue light) impairs the magneto-chemistry of the pancreas because repair signals go missing. Or glucose itself, also from blue light, deactivates superoxide so it's not available to trigger repair of insulin-producing beta cells in the pancreas. Two related mechanisms of action, same result.

High glucose and insulin also turn off the other part of POMC: alpha-MSH, beta-MSH, gamma-MSH. α-MSH reduces your appetite. This is one way Nature balances energy intake *vs.* consumption. When you get energy directly from the sun, the red part of the spectrum stimulates ATP production, so you should eat less. Unfortunately, blue light in isolation leaves your appetite switch in the 'on' position. Elevated glucose and insulin also speed up epigenetics and fertility, causing **precocious**

puberty, male and female **hormone problems, premature aging** and **hair loss.** I mean, we're seeing young girls starting to have periods at 6 years old due to blue light from tech screens. How crazy is that?

Conclusion: You don't have to eat carbs and sugar to get diabetes when blue light elevates your glucose chronically. Red light of 650 nm and longer can help reverse those effects. Studies show 670 nm light reduces blood-sugar by 27% after eating. But first, stay away from artificial light, particularly blue in isolation and at night.

The primary function of modern **sunscreen** is to reflect or absorb UV light in its mineral compounds. But sunscreens also contain ingredients such as PABA, salicylic acid and benzoic acid that inhibit the enzyme which converts tyrosine into melanin and dopamine, called tyrosinase. To our detriment, tyrosinase inhibitors reduce melanin production in the skin (causing pale white skin tone), gut (faulty vitamin D receptor), eye (RPE dysfunction), ear (hearing loss, tinnitus), and basal ganglia (dopamine production, Parkinson's, depression, poor executive function, brain fog). Without melanin in your skin, you can't make vitamin D3, you burn more easily, and POMC production decreases. At the same time, the tyrosinase inhibitors in sunscreen reduce dopamine (see tyrosine pathway pg. 291), thereby inhibiting photoreceptor repair.

Similarly, **sunglasses** block α-MSH. They block melanin creation. That's a big reason why modern humans don't **tan** well, can't make **vitamin D**, and have **atrophic skin**. Lack of full-spectrum sun atrophies your eyes and skin. That creates havoc in most of your biochemical pathways because skin is a solar panel for the brain. No wonder millions are dying from brain and heart diseases.

Hypothyroidism. Phenylalanine converts into tyrosine. Tyrosine forms thyroid hormones T3 and T4 using tyrosinase. The thing is, sunscreen inhibits tyrosinase activity, thereby upsetting melanin production. So, if you shut down tyrosinase activity with sunscreen, you can't make melanin, T3 or T4, which run the thyroid. You're effectively turning off the whole mechanism of POMC, while setting yourself up for melanoma. This is why melanoma and hypothyroidism often go together.

Cancer. Normal cell division (mitosis) is triggered by ultraweak UV light from the nucleus and mitochondria, which is captured on melanin. But, when cells can't divide due to melanin dysfunction, they migrate. That means metastasis (cancer) is caused by the inability of cells to divide, which forces them to spread. At the same time, autophagy and apoptosis are broken from lack of UV, IR and melatonin. For reasons that are unclear, cellular metabolism also shifts to runaway aerobic glycolysis (Warburg), instead of the normal oxidative phosphorylation. Consequently, cancer is supported by acidic, low-oxygen conditions, and it's subdued by oxygen and alkalinity. For their part, Big Pharma is busy trying to find drugs that block mitosis in cancer cells, which is exactly the wrong thing to do because it accelerates proliferation of malignant cells. In other words, their thinking is completely backwards and utterly destructive... as usual. (Cancer story continues on page 323.)

Autoimmunity is caused by dysfunctional CLIP, ACTH, α-MSH, β-MSH and γ-MSH. These POMC products modulate both arms of the immune system (innate and adaptive) and inflammation. Under normal circumstances, α-MSH controls immune cells such as white cell macrophages and dendric cells which initiate antigen-related (adaptive) immune responses. α-MSH also keeps inflammation under control by downregulating proinflammatory cytokines such as IL-1, IL-6 and TNF-α. But, under excessive blue light, the immune system malfunctions when its signals get messed up. Incoherence results in overzealous immune cells attacking your own cells, which we call autoimmunity.

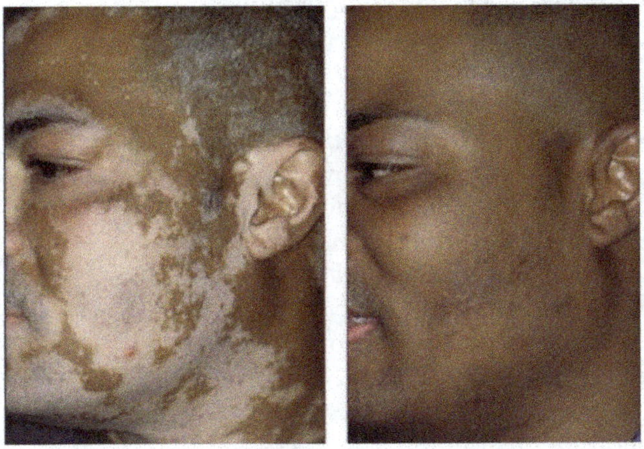

Note: This is not Dr. Jack Kruse's patient that he talks about. Image used courtesy of Tareen Dermatology in Minnesota (www.tareendermatology.com).

Vitiligo is caused by melanocytes abandoning the skin surface in search of UV deeper inside. In this condition, melanocytes travel inward in bunches (hence, the blotchiness). This often includes pigment-producing melanocytes in hair follicles, resulting in gray hair. The solution is UV light exposure on the skin and eyes. Dr. Kruse once achieved complete remission of vitiligo in six weeks using therapeutic doses of UV and natural light. The hair's natural color also returned. He believes others can achieve similar results in that timeframe.

What he didn't anticipate was the patient's diabetes to worsen. But, after giving it some thought, the regression did made sense. Melanocytes left the interior to soak up new-found UV at the surface. That depleted melanocyte populations and POMC inside. Result: He convinced her to stay the course and, in a few more weeks, melanocytes were back at full strength both outside and in. Her type 2 diabetes went away, and she lost 80 pounds. This proves, in Uncle Jack's mind, the neuroectoderm is way more plastic than medical science thinks. It turns out, melanin is super, duper inducible in quantity and location. Now we know the recipe.

Pregnancy melasma (aka "pregnancy mask"), hyperpigmentation. When ACTH is released, the pituitary can't help but release β-MSH at the same time because they have a common precursor: POMC. That means when cortisol is released due to blue light or nnEMF stress, you're at greater risk for hyperpigmentation. Normally, the amount of β-MSH released through this pathway is not enough to cause any changes. But, under pathologic conditions such as a doubling or tripling of provoking factors, melanocytes have been known to run amok and over-produce melanin in patches. This is called pregnancy melasma. Similarly, **melanoma** is melanocytes migrating to the surface to get UV light because they were being starved of it inside from breakdowns at the melanin level, or UV-C not being made from mitochondria or DNA.

Infertility. β-MSH and γ-MSH control fertility through leptin. The leptin-melanocortin pathway is responsible for picking an immature egg to ripen and release. Leptin basically counts electrons to make sure you have enough energy to carry a baby to term. But when the pathway isn't working due to (1) blue light stress, (2) circadian disruption, (3) poor redox, (4) inflammation and/or (5) DHA-deficiency, you get infertility. These are primary reasons for the infertility crisis in the US.

Psychological disorders (aka GAPS conditions). I used to think gut dysbiosis and neurotransmitter imbalances were responsible for causing psychological disorders. Threats to the microbiome include antibiotics, glyphosate and other pesticides, vaccines and heavy metals. However, now I place more of the blame on POMC and melanin dysfunction, resulting from deficits in full-spectrum sun during the day, and not enough darkness at night. Specifically, too much blue light and too little UV inhibit conversion of the precursors phenylalanine, tyrosine, tryptophan and others into the neurotransmitters that run cognitive functions – including dopamine, serotonin, noradrenaline, GABA and acetylcholine.

Gut dysbiosis and junk light both have the ability to damage neurotransmitter pathways. But biochemicals that run the brain pretty much all have action spectra in the far-UV range. And we know from Jeff Leach's work that real sun exposure protects the microbiome from harm. Therefore, poor light hygiene and nnEMF exposure have overtaken antibiotics and other tight junction hazards as the primary threats to cognitive function, in my mind.

Gut dysbiosis. When the POMC pathway is broken in the gut, the microbiome malfunctions along with it. That's because melanin sheets in the gut are enterochromaffin cells. They work with the light being released by gut bacteria to regulate the gastrointestinal tract – particularly intestinal motility and secretion of signaling molecules such as serotonin that control the gut's nervous system (enteric nervous system or ENS). So, in the gut, the melanin in enterochromaffin cells are the screen and Bacteria and Company are the projector. When both are working correctly, melanin sheets in the gut lining transduce light from the microbiome into an electrical signal for the ENS and brain to read. Together, they make sure food is digested properly and the brains are up to date – the one in your tummy and the one in your head.

Solution: You need to fix POMC over the gut. Get out in the sun and you can suck melanocytes that had migrated to the surface back into the gut, where they belong. This fixes a lot of problems that we blame on food intolerance, bad bacteria, poor digestion, irritable bowel syndrome, etc. In fact, the skin over the abdomen contains more POMC than any other, just so its cleavage product melanin can help regulate the microbiome.

Autism. Uncle Jack believes autism is an evolutionary regression in which the individual's genetics fall back to an earlier, simpler design. In this paradigm, melanin dysfunction causes de-evolution of the central and peripheral nervous system. POMC problems disturb the melanopsin system, which ruins neural migration. In other words, sensory processing tracts of the brain fail to develop properly because problems with POMC, melanin and melanopsin build the brain defectively. And what controls neural migration? It's vitamin A and melanin.

"I have a sense that we will find out in the future that autistic children have wiring defects in their chiasms and in their sensory relay pathways because of a broken Vitamin A cycle in the melanopsin system along with a lack of melanin and POMC activity in the germ cells… Vitamin A derivatives, called retinoids, are involved in a complex signaling pathway that regulates gene expression and, in the central nervous system, controls neuronal differentiation and neural tube patterning."
—Dr. Jack Kruse.

Alzheimer's and Parkinson's. These are caused by neuromelanin degradation in dopaminergic neurons of the substantia nigra. Sitting atop the brainstem, this area of the midbrain helps control movement and reward through dopamine production. Unfortunately, melanin breaks down in the area from the usual suspects: blue light in the eye (destroying melatonin that repairs photoreceptors), nnEMFs (such as wireless headphones), circadian dysregulation (at the SCN), and lack of light from unproductive mitochondria (namely, UV). Dopamine production then suffers and you lose memory or muscle control. At the same time, blue light raises glucose and insulin, which shrinks the brain. Result: blue light subjects modern humans to evolutionary regression, where brain centers malfunction and can't talk to each other.

Multiple sclerosis (MS). Multiple sclerosis is a neurologic condition in which the insulation around nerves is damaged so they can't carry a signal properly. This can result in loss of vision, numbness, tremors, impaired coordination, pain and fatigue. Centralized medicine thinks it's caused by the immune system attacking the myelin sheath around nerves for unknown reasons. We can narrow that down some more. MS is caused by an inability of mitochondria in the central nervous system to make V-UV light. Melanin then migrates away from nerves to go look for UV near the surface. That leaves nerves with weak mitochondria and semiconductor activity, which results in low redox, incoherent communications, and persistent inflammation.

Similarly, **fibromyalgia** occurs when you lose melanin around nerves, causing them to misfire. Melanin deficiency contributes to **chronic fatigue** syndrome. Red light can help by improving mitochondrial function. UV can help by making more β-endorphin. **Gout** (a form of arthritis) can occur from local loss of melanin. **Non-alcoholic fatty liver** is the liver storing extra fat so its mitochondria can make deuterium-depleted water to renew itself after melanin destruction.

You can prevent or reverse **eye diseases** by optimizing dopamine, melatonin and very long chain polyunsaturated fats called "elovanoids." They all help regenerate photoreceptors. Dopamine is increased by optimizing your light environment and avoiding stressors which deplete dopamine. Note: Dopamine precursors phenylalanine and tyrosine act in the UV range. Melatonin is increased by minding your circadian exposures (natural light during the day, no artificial light at night). Melatonin's precursor, tryptophan, also acts in the UV range. And the latest research by Nicolas Bazan, MD, PhD, tells us that breakdown products of DHA, called "elovanoids," are instrumental in photoreceptor regeneration. In the UV spectrum, frequencies from 340 nm to 380 nm appear to be where photoreceptors like operating, because lots of ROS and RNS are made in that range which control metabolism. Whereas it appears frequencies in the 380–400 nm range are designed for photoreceptor recovery, because free radical production drops sharply above 380 nm.

DHA breaks down into long chain polyunsaturated fats known as elovanoids. These, with dopamine and melatonin, help repair non-visual and visual photoreceptors damaged by free vitamin A (retinal).

Pain management in bone healing. People that have more melanin are more resistant to pain. Here's why – using a metric that's part therapy, part proxy: Dr. Kruse noted that those with a high vitamin D level needed far less pain medication after bone fractures or failed lumbar fusions. When a patient's α-MSH, β-MSH and γ-MSH are high – from UV exposure – they didn't need whopping doses of pain meds.

My interpretation: When you make α-, β- and γ-MSH, β-endorphin is produced in the process (or the other way around). β-endorphin is a natural pain-killer. At the same time, UV-B makes vitamin D, which helps repair bone. Conclusion: This is one of the rare instances where Dr. Jack may approve of vitamin D supplementation in certain situations to reduce intractable pain, because vitamin D aids bone healing while MSH (from UV) throws off β-endorphin. On the other hand, non-healing fractures can devastate quality of life.

My wish list: Many people, including myself, would love to see Uncle Jack publish a comprehensive list of diseases – both their complete mechanisms of causation as he sees each condition and what he would do to fix them. For mitochondriacs, that would be the most interesting document ever created because he explains more about how biology really works than anyone, or any source, ever. It would be like asking the Universe to explain the advantages or disadvantages of each evolutionary decision ever made. I wish. We wish. From what I gather (just a guess), he has personally been able to reverse or substantially improve about ⅔ to ¾ of medical conditions he discusses on podcasts. Point being, there are real cases in his experience to support his claims.

The POMC gene made humans what they are

As the preceding list should make abundantly clear, POMC connects your biochemistry and epigenetics to your biophysical environment. POMC takes environmental influences – including light and electromagnetic fields, as well as temperature and water quality – and uses those inputs to regulate your metabolism and energy production, appetite and body weight, seasonal changes, mood and behavior, sensory perception, immunity and inflammation, brain function and mental health, sexual response and reproduction. That means POMC pathways directly or indirectly run most aspects of human (and animal) biology. We just didn't realize it before, because no one had put all the pieces together so we could see the whole picture of what POMC was really doing inside us.

"Possibilities and probabilities for life is made tunable just by changing incident light photons. That is how these chemicals all operate. This implies that your mitochondrial metabolism creates an adaptable light spectrum during metabolism and it is this light that tunes and controls the chemicals in you to act in different ways."
—Dr. Jack Kruse.

Every mammal has adaptation programs built into it. But humans operate on the slimmest margins because our brains consume the most power. To compensate, our epigenetics are the best-equipped of all species to tolerate imbalance and dysfunction. But, as modern living depletes our healing capacity and we live closer to the edge, that just means there are more ways for our biology to degenerate from blue light and electropollution. We're able to live in more disorder longer.

Dr. Robert O. Becker

In 1941, Albert Szent-Györgyi gave a talk in which he said proteins in the body look like they have an electronic structure with semiconducting properties. Turns out, he was right. Our bodies are made of semiconductors with densities ranging from bone to collagen (in cartilage, ligaments and tendons) to respiratory proteins to melanin. This casual comment at the end of his talk inspired Becker, then a medical student, to pursue a lifelong journey of proving Szent-Györgyi right. And he did just that.

Over his career as an orthopedic surgeon, Dr. Becker studied the effects that internal and external electromagnetic fields have on growth, healing and regeneration. His findings explained for the first time the electromagnetic nature of life. However, centralized medicine didn't like his message (not enough chemicals and surgery for their pocketbooks). In short, he discovered that minute electric currents – produced by semiconductive tissues and carried on nerve myelin – instruct new cells where to go and what type of tissue to turn into (e.g., bone cells, skin cells, muscle and nerve cells).

He basically figured out the mechanisms by which embryos grow into their adult form, injuries heal throughout the animal kingdom, and animals regenerate missing appendages that are still able to do so, including salamanders and flatworms. As for human regeneration today, Becker's work offers tantalizing clues as to how clinicians may one day heal sports injuries faster, fix worn out joints such as hips and knees without replacement, and restore function to damaged organs such as kidneys and heart. Published around the mid-to-late '80s, his work is still cutting-edge science that many health professionals know nothing about.

We're made of semiconducting proteins

Materials can be divided into three categories: (1) Those that carry electrons/electricity easily – called "conductors." Copper, gold and silver are excellent conductors. (2) Those that carry electricity poorly – called "insulators." Rubber, plastic, wood and fat make good insulators. And (3) those that conduct electricity under certain conditions – called "semiconductors." Earth's most common semiconducting material, silicon, goes into making the chips that run our computers and devices. Whereas, the semiconductors in us go by the names collagen, bone, melanin, hemoglobin, cytochrome C oxidase and others.

Using the Bohr model of atoms (concentric rings), the outermost orbital ring containing electrons is an atom's valence shell, aka valence band. The next higher shell/band/orbital is the conduction band where electrons can leave the atom entirely and go to another. Electrons jump to a higher shell when a specific light frequency or voltage is added. This mobilizes electrons to jump to neighboring atoms, thereby moving through a material. Otherwise, electrons stay in their "home" orbital.

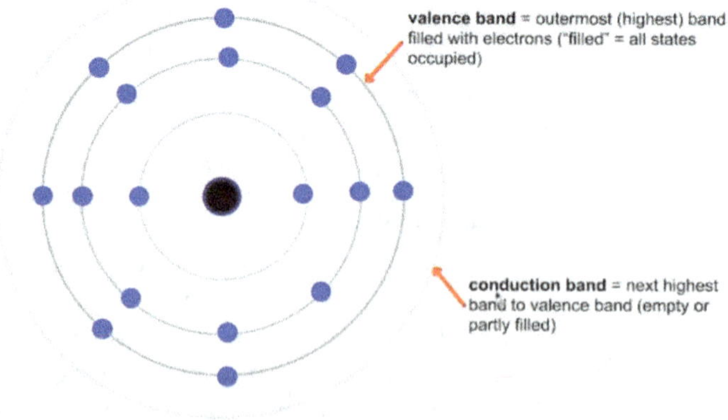

Image of valence band and conduction band used courtesy of Paul (YouTube channel: youtube.com/@PhysicsHigh/videos) and (www.physicshigh.com).

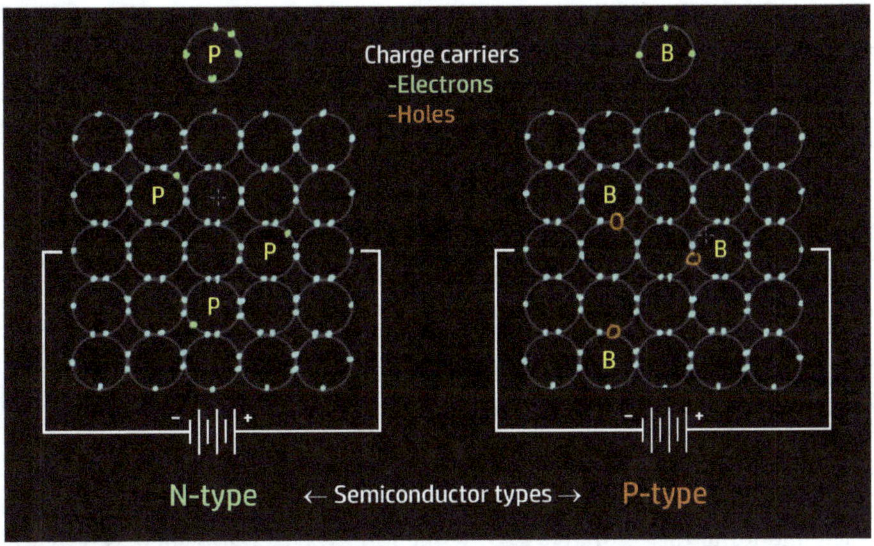

P- and N-type semiconductors (man-made, inorganic)
The atoms of N-type semiconductors have an extra, unpaired valence electron that can jump from atom to atom, given sufficient voltage, light or heat. These surplus electrons in N-type semiconductors hop from one atom to the next because their outermost orbitals can't hold any more electrons. Therefore, electrons have nowhere to rest; they have to keep moving – generally forward. Whereas, P-type semiconductors have a void that can move around. This migrating hole (unpaired electron) sucks electrons through the material because electrons are attracted to the void in orbitals. To say the preceding another way, the thing that carries electrical charge around – aka the "force carrier" – in N-type semiconductors is electrons and repulsion, while the attraction of a void carries positive charge in P-type semiconductors.

Image of band gaps used courtesy of Professor Dave (YouTube channel: www.youtube.com /@ProfessorDave Explains).

Band gap
You have to add a specific wavelength, voltage or heat to get a semiconductor to conduct electrons – too much or too little and valance electrons stay in their orbital. This extra energy needed to mobilize electrons is called a semiconductor's "band gap." In biologic systems, band gap represents the material's absorption and emission spectrum.

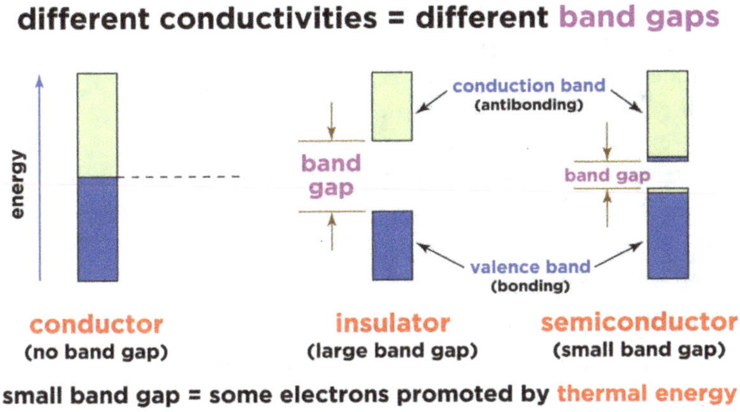

When its excited electrons shed the higher orbital energy – that is, when electrons fall back to the ground state of a lower orbital – the semiconductor releases light, typically of a different frequency. More simply, put light in, get some other light out. The question mitochondriacs are most concerned with is, *what kind of light is being emitted?* That determines what will happen to semiconductors down the line, including melanin and circadian sensor-controllers such as the suprachiasmatic nucleus.

A: When mitochondriacs say 'wide bandgap,' they're referring to semiconductors that act at high energies, as opposed to low-frequencies in the red or middle of the spectrum. Wide bandgap semiconductors absorb above visible violet or 3.0 mv, and/or they emit high frequency wavelengths, such as V-UV or UV-C biophotons. UV is high energy because it adds enough photonic power to elevate an electron's orbital shell and jump the figurative gap, while long wavelengths such as reds are low energy, narrow bandgap. Semiconductors with wide bandgap absorption and emission – together, called "action spectra" – run our "non-linear optics" (i.e., put one light in, get an amplified response).

To illustrate non-linear optics

When 312–320 nm UV light hits cholesterol, vitamin D is formed. Through vitamin D synthesis, a thousand other genes are turned on that code for biochemicals to be made. What most don't know is that sun exposure triggers this biochemical cascade. And fewer still realize UV light, made internally, can have the same effect. Miniscule numbers of biophotons can turn on or turn off hundreds or thousands of genes. Now can you see how massive amplifications like these exemplify non-linear optics? Just as peculiar to think, UV biophotons, made inside us, literally give us an internal tan, as melanocytes make melanin.

Biologic semiconductors work better when the body is cooled

Coldness dampens the vibration of proteins and water so their motion doesn't disturb light propagation through the body. In the non-uniform environment of living systems, cold reduces photon/phonon scattering and absorption. This is a little-known reason we sweat, as cooling carbon-based semiconductors allows light to pass through living tissue more easily. Example: Sweating cools semiconductors embedded in subcutaneous cholesterol, which improves vitamin D production. At the other end of our comfort zone, Cold Thermogenesis improves performance of our semiconductors, which then make more UV from within. This is particularly valuable when you aren't getting sun exposure with UV in it. Conclusion: Heat mobilizes electrons through *inorganic* semiconductors for better electrical conductivity, while coldness increases performance of *biologic* semiconductors because there's less molecular vibrational and thermal noise of intervening tissues hindering light absorption.

There's a lightshow going on around our semiconductors

The photoreceptors in our mitochondria absorb and emit a kaleidoscope of colors, aka their action spectra. Cytochrome I (NAD+) is activated by

UV-A. Cytochrome II acts on purple to blue. And Cytochrome IV responds to infrared. Every time an electron transits a cytochrome, its redox centers strip light frequencies off the electron to refine molecules such as phenylalanine and tyrosine into biochemicals such as melanin and dopamine. Hormones and neurotransmitters like these then shape our mood, behavior and endocrine functions.

Depending on what light gets liberated through mitochondrial respiration, our DNA gets methylated in different ways. This controls all kinds of epigenetic expression – including embryo development, detoxification, birth defects and disease progression in general. Altering (i.e., corrupting) light exposure also changes the free radical signal produced by mitochondrial metabolism. When you make different free radicals, the action spectra of your proteins change so they no longer do what they're supposed to. They might not make a biochemical you need or fail to turn off a gene that prevents disease, as examples.

Light spectra control metabolism

In the 1st Edition, I was under the impression that the liberation of light from electrons tunneling the ETC was incidental – like it was an effect with benefits. I've learned since then that light is *at cause*. It's both cause and effect. Light drives steps of metabolism – including those in glycolysis, TCA/krebs/citric acid cycle, beta oxidation, urea cycle, PPP and the electron transport chain. Light is then emitted from some of those chemical transitions and is recaptured on photoreceptors such as melanin to run other operations around the body.

Light drives metabolism. But, in other cases, it applies the brakes. For example, Dr. Jack's Leptin Prescription has you eat soon after waking. The reason is, you want to get those food constituents processed through the ETC early in the morning, before UV light shows up later to limit metabolism via nitric oxide and vitamin D. Recall that NO and vitamin D slow electron transport via cytochrome IV and the vitamin D receptor in cytochrome III, respectively. The result: While electron transport is restrained, it's detrimental to overeat (reductive stress?). Plus, UV exposure raises α-, β- and γ-MSH, which increase satiety/reduce appetite. Combined, these are levers and switches by which midday sun enacts natural calorie restriction, or "partial afternoon fast," as we might call it.

Conclusion: Just look at how many steps are involved in metabolism (some depicted in Chapter 3: Mitochondria), and the precise timing with which they all need to happen. You'll soon realize light is the only influence that can possibly control countless interactions happening simultaneously, and at the tiniest time scales. And that's just metabolism. To put light in perspective, we all know fiber optics and microwave frequencies carry massive streams of data to your smartphone. So why wouldn't biology use light to control its 100,000 biochemical reactions? Think about it. It's the only medium capable of carrying such bandwidth.

Biology uses light for everything

"The light that's created from metabolism is truly… that's actually the sun inside of you that is driving everything. It's driving animal photosynthesis through the non-visual photoreceptors. It's affecting water chemistry [e-zone creation]. It's affecting polarization in different proteins. It's affecting binding of different minerals. It's affecting how proteins can bend. It's affecting how protons spin through the ATPase. It also controls [distance] between cytochrome proteins [heteroplasmy]. It controls endocytosis… and ubiquitination. It controls when the decision is made to pass through certain parts of the cell cycle. Light and different frequencies control all that."

—*Dr. Jack Kruse.*

How semiconductor dysfunction botches biochemistry

Hypoxia chokes mitochondria. This upsets calcium distribution and signaling. Calcium is a wide band semiconductor. It, like other semiconductors inside us, receives light of one frequency so it can emit another. This means hypoxia reduces the power and range of the light that mitochondria produce. So, instead of making ultraweak UV light from 200–400 nm, mitochondria might only produce biophotons at 330 nm.

This affects performance of all the neurochemicals made from phenylalanine, tyrosine, tryptophan and histidine. Production of melatonin, serotonin, dopamine, epinephrine, melanin, nicotine and metabolism (TCA, ox/phos, glycolysis) might then be affected. That's about as broad an impact as any influence can have. **Contact lenses** are another influence that can lower oxygen tension. This too effects calcium efflux and mitochondrial function, which alters light emissions.

Exposing your semiconductors to **artificial light** does the same thing as hypoxia: it feeds them an alien spectrum which becomes aberrant signals to regulatory centers in the brain. Namely, one surefire way to create neurodegeneration is to damage photoreceptors in the peripheral part of the retina. This disturbs transmission of light from melanopsin in the back of the eye to the frontal and temporal lobes in the case of Alzheimer's, or to melanin in the basal ganglia in Parkinson's.

Change the way proteins bend in the eye – from **blue light or even improper focal length** – and you change the optics of the system, often altering circadian regulation. Other things that change light exposure: **sunglasses, contact lenses and cataracts.** Junk light basically makes you much more likely to suffer neurotransmitter imbalances that show up as attention deficit, anxiety, depression, schizophrenia, bipolar or OCD. Most important, blue light ravages melanin, and a cornucopia of problems shows up chronically and acutely as a result. The thing they all have in common: faulty signaling in the leptin-melanocortin pathway.

Any time you distort the light going through the eye to the CRP, you mess with biochemical synthesis and signaling to the SCN. You might then lack frequencies you need to control executive function or mood. Extreme example: People have been known to become severely depressed and kill themselves two weeks after getting **laser eye surgery.** The mechanism: The laser wrecked the photoreceptors neuropsin and melanopsin, which disrupted optical signaling to that part of the brain regulating emotions and behaviors, called the habenular nucleus. Similarly, **fixing retinal detachments with laser photocoagulation** can seriously dysregulate your central retinal pathway.

So how can we reverse POMC problems? The first thing you can do to fix dysfunction and disease is stop exposing yourself to the influence that causes more damage than any other: unnatural blue light. Closely related, eliminate as much wireless radiation from your life as possible. Then, get real sun exposure on your (naked) eyes and skin, UV and IR in particular. To stack the deck in your favor, indulge in coldness. Coldness improves transmission of light to your semiconductors, which increases their emission spectra. Coldness is also proven to raise dopamine and noradrenaline (improving cognitive function) for the same reason.

Let's recap semiconductors

The property that determines which type of material you're dealing with – be it conductor, semiconductor or insulator – is called bandgap. Bandgap is the amount of energy needed to get electrons/electricity moving through a material. Insulators require extremely-high energy inputs to move electrons through it. Conductors move electrons effortlessly, without any assistance. While semiconductors require precise inputs of energy before they conduct electrons. And/or you can add stuff to a material to help mobilize its electrons and improve conductivity.

Computer chip makers, for example, help semiconductors move electrons by deliberately adding an impurity in manufacturing, called a "dopant." Small amounts of dopants such as boron, phosphorus, germanium or arsenic make it easier for electrons to flow through silicon in microchips. Second, application of an electric current makes it even easier for electrons to flow through a semiconductor. Combine these technologies and you have key elements of a microprocessor. That is, intersperse a dopant throughout a semiconducting material, add circuitry such as resistors and capacitors, miniaturize them, apply voltage, and you have mind-blowing processing power to run our tech devices.

But, as important as microprocessors are to society, the semiconductors inside us are the juicer story because biologic systems like homo sapien run on semiconductors. Semiconduction is an enabling technology for mammalian biology. Specifically, carbon-based semiconductors (rather than silicon) inside us turn one form of photonic energy into those we can't get from sunlight. We literally have light emitting diodes within us that power and control our cell cycles and biochemistry. Believe it or not, semiconductors in our mitochondria turn visible light frequencies into UV-C+ photons approaching the power of x-rays!

So when you think about the ecosystem of *sun, earth and biology* as a big circuit, you begin to understand what Uncle Jack means when he says the sun is a **cathode ray** and everything on earth with water in it is an anode. The sun shines many frequencies of light upon us. The water in us converts that solar energy into DC electricity, while our collagen, melanin, hemoglobin, cytochromes and other semiconductors transform that energy and shuffle it amongst themselves. Yes, there's a substantial distance between the sun and our semiconductors, but the connection forms an electrical circuit nonetheless.

Dissipative structures, entropy and time keeping of cells

Dissipative structures are self-organizing systems such as a cell, organism or tornado that aggregate energy or something else from its surroundings and directs it. This is opposed to entropy, which is energy or matter leaving a condensed state and spreading out until it reaches equilibrium with its environment, or maximum disorder. Conceptualized by Ilya Prigogine, dissipative structures explain how order emerges from chaos, and why the universe favors complexity over entropy – life itself being the ultimate example. Basically, someone had to explain why living things, by nature, coalesce rather than falling apart. However, the term "dissipative structure" doesn't make sense to me, because it doesn't articulate the most important side of the equation: clustering, organization and increasing complexity.

Dissipative structures come into play when Uncle Jack says that 'time is a flowmeter for entropy.' That means time doesn't flow the same for all things, in all situations. Instead, time is relative. And change should be perceived like a video showing some sort of aggregation decreasing in complexity over time, increasing in disorder. When you fast-forward or rewind the "tape," time determines state of entropy/randomness of the heat, particles, or whatever you're measuring. In this model, and in life, time regulates entropy. But, if each cell, organelle or molecule experiences time differently, you see how that can be a problem keeping them all synchronized ('cause that's how it actually works).

To function as one organism (think evolutionarily), each subcellular "zip code" of the body needs to be in synchronicity for circadian rhythms and metabolism to work properly. On the other hand, when the time-keeping mechanism of cells (e.g., their clock genes) is not perfectly synchronized, the collection and usage of energy becomes chaotic. Control centers of the brain then can't talk to cells of the body and energy gets mismanaged, your biochemistry loses coherence, biocycles get dysregulated, and inflammation doesn't stop. For example, glycolysis or beta oxidation might misfire and produce excess ROS.

Our interiors need peace and quiet

Biologic systems go to great lengths to maintain quantum coherence inside them. But scientists don't know how this would possibly be accomplished. Therefore, physicists, for many decades, laughed at the idea of quantum processes happening in warm, wet environments (i.e., in biology). They couldn't believe light-controlled processes could happen in living tissue because it was just too chaotic an environment to maintain fidelity of signal. This section answers that.

With lots of ultra-fine EMF communication happening around cells (often involving individual photons), the internal environment needs to be quiet. Cells need electromagnetic silence to maintain quantum coherence – i.e., accurate light-based signaling. That's because electromagnetic interference creates chaos. Electropollution confuses cells about what they should be doing, programming gets disrupted, and corrective actions may be taken such as inflammation. This is where topologic insulators/insulation comes into play.

Electromagnetic isolation is established using melanin, deuterium and vitamin D. They shield quantum processes such as entanglement from external signals. Melanin does this by absorbing all frequencies from radio waves to gamma rays. Alongside melanin, Uncle Jack speculates that deuterium is welcomed in the circulatory system, but not cells and mitochondria, because deuterium's atomic mass – thus, its magnetic moment – absorbs so much energy it acts like a force field around tissues, protecting them against nnEMFs. Vitamin D does the same thing. It acts like a faraday cage for mitochondria by safeguarding the nuclear spin of electrons and protons inside them.

Atomic, molecular, optical physics OR atomic, molecular organization – collectively called "AMO physics"

Def: Field of science exploring how the arrangement of atoms affect the way light is handled by the body. Mitochondriacs study AMO physics because atomic-molecular organization helps regulate the body's clock/timing mechanisms by the distinct ways in which molecular arrangements resonate when hit by certain light frequencies. Ultimate expression of atomic-molecular organization: The "optical lattice clock" in the brain's SCN controls daily and seasonal cycles, while "peripheral cell clocks" in the rest of the body keep track of time in a location, which controls entropy, which manages energy flows.

Oddly enough, each cell runs on its own, independent time, which some call a "space-time domain." Time regulates energy collection and usage in a cell – in metabolic processes such as beta oxidation, for example. The passage of time speeds up or slows down entropy. That's why peripheral cell clocks are flowmeters for entropy. Interesting example: **Pharmaceuticals work on clock genes.** 75% of pharmaceuticals target circadian clock genes as their method of action.

You may have heard Uncle Jack refer to "atomic lattice" or "optical lattice clock"

These refer to the fact that many molecules in the body have repeating, uniform geometry at an atomic level which help our brain and cellular clocks tell time of day, season or location. In mineral structures such as diamond, quartz or ice, we call repeating geometry a *crystal.* But in you and me, smaller, simpler "mini crystals" are incorporated into larger (bio)molecules to help us run our circadian/infradian programs. Which is to say, molecules in the body are not crystalline throughout their entire structure the way diamond, quartz and ice are. Instead, mini crystals are the centerpiece – the business portion – of larger molecules. And the reason is, when Nature found simple structures that do something special, it plugged these "molecular modules," if you will, into all sorts of molecules and biochemical pathways.

Benzene rings trap photons

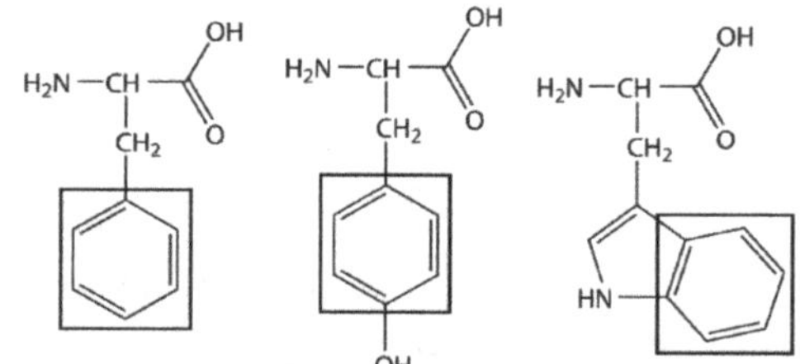
Phenylalanine Tyrosine Tryptophan

This is the mechanism by which molecular organization incorporates light into living systems. It's how $E=MC^2$ turns light into life. Quintessential examples: Tyrosine, tryptophan and phenylalanine molecules are built around 6-carbon, double-bonded rings that enable these aromatic amino acids to capture light. Benzene rings, as they're called, trap UV and IR photons to help regulate all aspects of our biology… with light we can't see. For example, the leptin molecule contains multiple benzene rings to capture lots of light. And DNA itself is filled with crystalline structures that hold onto UV biophotons. Some biophotons are freed when DNA is opened up (bisected) to be read and replicated.

Time crystals

Def: Repeating atomic structure (i.e., a crystal) whose electrons and protons perpetually oscillate in spin state and magnetic moment (think tick-tock of a metronome). When undisturbed by external influences, they flip-flop at regular intervals between two different states, alternating their response. In other words, they have a pre-set "periodicity" to their pulsation. This makes them crystalline in four dimensions (3 in space, 1 in time) and therefore a distinct phase of matter distinguishable from a liquid, solid, gas or plasma. However, influences such as light can make them favor one state over the other and thus their effect. Moreover, their fixed oscillation rate may be used for time-keeping purposes.

How light affects a time crystal: *Photons* in a beam of light polarize a time crystal's *electrons* in angular momentum and magnetic moment by aligning their spin states. Acting like tiny magnets, the spin state of *electrons* then partially-align the crystal's *protons* (or neutrons) in the nucleus. The thing is, the magnetic force of the protons push back against that of the electrons and you get wobbling, or precession of spin. This repeating *action-reaction* results in an extremely delicate balance between flip-flopping states. Thus, different frequencies of light, at different times of the year, bias a time crystal's precessional oscillation into one configuration or the other. This is how tiny inputs of light produce amplified effects in the SCN and elsewhere.

To see the specifics, watch the YouTube video entitled, "Impossible Time Crystal Breakthrough – Explained" by Dr. Ben Miles. youtube.com/watch?v=Zr8zwfm1Ub8

To illustrate, tryptophan and methionine are time crystals. When light hits them, they resonate differently at different times of the year. This helps regulate seasonal programming. Case in point, tryptophan can be broken down into either a ketone for fat burning or a glucogenic amino acid, which makes glucose. Important in adjusting to seasons, the spectral composition of sunlight on the eyes and skin (plus cold) tells your infradian programming to burn fat in winter or make glucose in summer. What controls how tryptophan is metabolized? The energy signal that comes out is a function of how the light frequencies resonate with tryptophan. That light signal to the SCN tells the body to make glucose to burn immediately, or ketones to make heat and conserve fuel.

ROS and RNS are protagonists in magnetic reactions

Magnetism controls timing. Timing controls entropy. Entropy controls the flow of energy in cells. More precisely, unpaired electrons make free radicals magnetic. Through free radicals, magnetism controls the absorption and emission spectra of semiconductors. The emission spectra of semiconductors in mitochondria controls metabolism. Generally speaking, free radicals are like force carriers of magnetism, except they don't initiate. Rather, they respond, then influence. Free radicals modify the way atomic arrangements turn light into other forms of energy.

Oxygen and free radicals control light production

Light (as biophotons) is produced through mitochondrial metabolism. As electrons tunnel through the ETC, respiratory proteins liberate light. Light-producing processes requires oxygen to both accept electrons at Cytochrome IV, and it requires ROS/RNS to magneto-chemically regulate metabolism. Metabolic pathways are thus turned up or down, based on that tissue's needs in the moment. Unfortunately, problems commonly happen when light is improperly received or translated into other forms – in flawed melanin or semiconductor function, for example.

DHA

In order to carry signals with high fidelity, leptin's central retinal pathway (which controls circadian rhythms) needs to have more electrons than any other part of the body. That's why DHA is so important. DHA's electrons are like the highway upon which photoelectric signals are carried from the retinal pigment epithelium to the suprachiasmatic nucleus. Leptin and melanin function allow our circadian system to maintain flawless control over timing-keeping mechanisms that run the body.

The language of creation

The memory of water

As we touched on in Chapter 8: Circulation, water retains energetic imprints from surrounding chemical and biologic exposures. Water records states of consciousness such as thoughts and feelings, movements of waterways, and energetic residues of chemicals. It carries these memories around indefinitely until they're overwritten by new exposures.

This information is stored in the hydrogen bonds of water, not unlike a computer hard drive. Hard drives store information as ones and zeros on a magnetic surface. Of course, you need lots of memory locations on these disks to store megabytes and gigabytes of data, but how does biology do it? A: Water molecules are extremely small. Each drop of water contains 3 sextillion (3×10^{21}) hydrogen atoms. That offers lots of bits and bytes of data-carrying capacity. Each hydrogen atom changes its bond angle with oxygen when its electron captures a photon. Thus, the bond angle of each hydrogen represents a 1 or 0 in the binary code of water's memory storage.

Luc Montagnier proved this water memory experimentally. He put DNA in water, removed the DNA, then transmitted the energetic imprint left in the water to another lab, over the Internet. Colleagues then reconstituted the DNA perfectly at the other end. Montagnier showed that water absorbs DNA's code, and that program can be read by scientists, transmitted, and written into nucleic acid sequences to form DNA. He transmitted the code for life over a fiber optic cable.

Biophysically speaking, the DNA emits UV biophotons which are captured on hydrogen's electrons, which changes water's "hydrogen-bonding network." That's how exposures/experiences such as love, hate, food cravings, swirling or falling movements of a natural waterway, or crop chemicals are recorded into water. Well-known example: Dr. Masaru Emoto demonstrated that water is affected by people's thoughts, words and emotions. When people directed loving thoughts into water, it would freeze into beautifully-patterned crystals. But when people directed negative thoughts at water, it froze into ugly, chaotic shapes. Both are a result of water recording states of consciousness in their hydrogen-bonding networks. Finally, let's not forget hemoglobin and chlorophyll. They're effectively liquid crystals that take light and transfer its energy and information to/through water.

Now we know from Marcel Vogel's work that quartz crystal can infuse UV light into water. He cut quartz into distinct shapes, now called "Vogel cut" crystals. By rotating water around a Vogel cut crystal, it absorbs UV photons. For example, Vogel took mediocre wine and vortexed it around a Vogel crystal. This dramatically enhanced the wine's taste and aroma. Reason was, the wine absorbed photons of light that replicated superior growing conditions. He basically augmented the finished product by performing post-processing photosynthesis. What's more, the effect was lasting. The wine had been permanently improved.

Equally impressive, Vogel crystals have the mind-boggling capacity to create coherence (i.e., unidirectional and in-phase) in the light passing through them. That is, unorganized light goes in and laser light comes out. The implications are staggering. This confirms Montagnier's findings: (1) Water does retain UV biophotons. (2) Crystals such as DNA hold biophotons in their molecular structure, and later release some of them when they unwind. And (3) consciousness is infused into water through this phenomenon. In fact, this is how light gets loaded onto electrons throughout nature. In plants, the liquid crystal structure of chlorophyll infuses electrons with photons. In animals, hemoglobin acts as a liquid crystal to encode these energies into water.

Organ recipients inherit food cravings and "emotional baggage" (good or bad) from donors. Dr. Kruse explains that one of his patients, after receiving a heart transplant, developed intense cravings for McDonald's French fries. Turns out, the donor loved McDonald's fries and was in the middle of enjoying some when he was killed in a car accident. This craving, combined with trauma from the event, was recorded as an energetic imprint on the hydrogen bonds of water in the heart.

The recipient then found herself so absorbed by this new craving she was compelled to track down Dr. Kruse and ask him why this might be happening (all she knew was that he was involved in the donor's case somehow). Dr. Kruse had three other cases confirming this method of transference. And he's not alone. The literature is filled with stories of energetic memories finding their way to organ recipients. Recipients of heart transplants in particular have been known to acquire cravings for mustard, as a memorable example. Now you know why.

Conclusion: Nature uses crystals such as chlorophyll, hemoglobin and quartz to load (predominately) UV biophotons onto water molecules so they can pass that information (and energy) from plant to animal, prey to predator, object to object, place to place, and/or one point in time to another. It's a primary mechanism by which light is captured, stored and distributed throughout the material world. When Nature needs to communicate information and energy asynchronously, hydrogen bonds (of water) are the medium, biophotons are the data, and crystals are the recording device. This is the secret language of creation.

Why older couples often look alike

In light of what we just learned, we must consider the possibility that biophoton exchange is the main mechanism by which long-time life partners often grow to look like each other over a course of decades. Biophoton emissions from the hands, face and body – in addition to sharing of body fluids and the entangled particles they host – must play a role in harmonizing gene expression among couples (as well as pets). When you're right next to a person eight hours a day or more, as you are in sleep, you can't help but share lots of biophotons containing guidance for each other's epigenetic phenotype.

Phenotype: Observable characteristics of an individual resulting from interaction of genes and environment.

Just think: communications of light are happening all the time inside us to tell distant cells and organs what to do. So why wouldn't exogenous instructions get received by another individual's cells and be followed? When you understand the extent to which light sculpts biology, the way a person looks is most certainly influenced by external direction of stray light signals. Now you know why the longer a couple is married, and the closer they are, the more they tend to look like each other. Without a doubt, they're sharing light.

Consciousness

Consciousness or intuition (Uncle Jack uses the terms almost interchangeably) is a function of mitochondrial redox – in other words, electrons. Why? Because you need electrons to be able to interact with light. To illustrate, anesthesia blocks DC electric current through nerve pathways. In the brain, it stops electron flow to deactivate your consciousness. The anesthesiologist then restores consciousness by allowing the anesthesia level to drop in your system, which reestablishes electron flow, which restarts your brain's ability to create electricity from light, which revives you from an unconscious state. That's how anesthesia works. It stops the flow of DC electricity to interrupt consciousness.

Correspondingly, variation in redox causes people to awaken from anesthesia at dramatically different rates. Teens who have ample redox often recover quite quickly, while an older person who's known to have low redox (e.g., in cases of Alzheimer's) might take 6, 8, even 24 hours to completely return to normal. The time needed to get the brain back to full functionality depends on getting DC electric current flowing again.

God

An interesting implication related to consciousness is that God, Nature, the Universe, and the Creator – all the same here – are not just bearers of light, or delivered on light. Those are two ways we conceptualize a deity. But, more precisely, God and light are one and the same. The intelligence of Nature is delivered on light. So God is light. God is Nature – or the Universe, if you prefer. Light is the Creator, because everything in the universe is simultaneously created, delivered, powered and controlled by light. Life is light in solid form.

That's why teachers, from ancient mystics to modern physicists, say everything in creation is vibrational in nature. Through the science of cymatics, we've learned matter is created by vibration (1). Conventional physics teaches us that electromagnetic energy powers all of creation (2). And we now know from the work of Luc Montagnier that DNA – the very code for life – can be transmitted on photons (3). In other words, the code for life can travel through space on a beam of light. So if it can do so through a fiber optical cable, the code for life can travel through open space.

If you've never seen Hans Jenny's work into the visualization of frequencies (aka "cymatics"), check out the documentary on YouTube entitled "Cymatics full documentary. Bringing matter to life with sound" (4-part series).

Key take-aways

When mitochondriacs say humans are beings of light, it's a literal statement. Our relationship with sunlight… more than ancient reverence or sentimental attachment. On the contrary, we're built, powered and organized around light. At the center of it all, our semiconductor proteins are Nazis when it comes light. Light regulates every aspect of our biology via POMC signaling – specifically, its cleavage products ACTH, CLIP, melanocyte-stimulating hormones (which make melanin), and β-endorphin – plus amino acid precursors, hormones and neurotransmitters.

UV, in particular, influences phenotype – including our biochemistry, brain and behavior, epigenetics, and so much more. Also beyond our conscious awareness, infrared works in myriad ways to power the body: electron-zone water, heat, circulation, ATP and autophagy, as examples. Now you know why Swedish researchers found in a meta-analysis (2016) that sun deficiency was as bad for you as cigarette smoking. It's just that no one ever taught us precisely how real sunlight sculpts our biology OR man-made frequencies poison us… that is, until Uncle Jack led us to networks of interconnected rabbit holes the size of Amazon warehouses.

In review, the primary ways light influences biology are the following:

- kinetic
  - locomotion
  - vibration
- electronic
  - absorption and emission
- ionization
- chemical bond breaking
- photon/phonon transformation
- photoacoustic effects
- photoelectric effects.

Source: Dr. Alexander Wunsch. A full course of his lectures about light can be found on Vimeo: vimeopro.com/alexanderwunsch/alexander-wunsch-in-english/page/1.

Last point: Humans are 80% water when we're born. But 60-year-olds are closer to 55%. That means when you get to that age and beyond, you're down nearly ⅓ on solar yield from where you started. At best, you only have ⅔ the capacity to capture light energy in water that you had when you were a baby. So you need more light just to stay where you're at, before you even think about reversing disease.

Conclusion: Nature had a plan when it used carbon-based semiconductors to control our biology. Over millions of years, it developed a blueprint for the way life is supposed to be run. On the other hand, modern humans invented silicon-based computer chips, light emitting diodes, wireless technologies and their alien light spectra. The inorganic world runs on these silicon semiconductors. But, now that you've learned how biologic semiconductors operate, do you see how silicon semiconductors conflict with the way our bodies are supposed to operate? Of course, technology offers great benefits. But, as foreign frequencies proliferate, health or sickness boils down to one question: *Whose program are you on: that of Nature or that of centralized controllers?*

18

THE *REAL* CAUSES OF COMMON CONDITIONS

Let's put healthcare back on the right track

To begin steering healthcare back in the direction of truth again, this chapter explains the root causes for common conditions that seem to stump modern medicine today. It's a peek into where the field of intentional wellness might be heading in the next few decades should we learn to live the way we were meant to, instead of continually trying to pound Nature into submission with man-made technology.

The views presented here are the best explanations and theories from educators and practitioners on the bleeding edge. Much of it has objective evidence to back it up, while some of it has yet to be demonstrated clinically. But be forewarned: The Establishment, and its clique of accomplices, will likely disagree with these conclusions because, in almost every case, disease is caused by corrupted forms of light, water, magnetism and mitochondria – not by unlucky genes or biochemical mishap.

Indeed, almost every concept shared in this chapter completely contradicts what centralized medicine says, and with what it offers. So don't expect the FDA, NIH, AMA or CDC to have anything nice to say about the descriptions that follow. They don't have good answers themselves. But that doesn't stop them from attacking truth-seekers offering credible explanations and vastly superior options. So which party is in the right, and which party routinely misguides people's health to serve their own interests? We shall see.

First points to contemplate: How many chronic diseases have become commonplace since 1900? There are hundreds. Now, how many conditions can allopathic, evidence-based medicine accurately explain causation for in 70 years of intensive research? How much have they spent in that time? And how many diseases have they actually cured? Not one. I rest my case. You'd think with countless billions spent, and hundreds of thousands of hours into research, they'd have achieved a real breakthrough with at least one disease. But no. Not one cure of any significance in 70 years. Just symptom suppression and empty explanations. In fact, diseases have gotten much worse in prevalence, severity, and damage done. You've got to admit: that's pathetic. You be the judge whose message is closer to the truth: theirs or ours.

A better understanding leads to better results

As this knowledge empowers you into action, start with the premise that the defect rarely begins in your genes and biochemistry. Instead, the defect is almost always external. Exposures in your environment are to blame for your dysfunction and disease. Repeat with me: You are not broken. Your body is working exactly the way it's designed to work. You just need to change your environmental exposures and, like magic, your problems unravel in the reverse order of how they were created and woven together. On the other hand, you can't get fully well when you stay in the environment that made you sick in the first place.

In addition to the principles presented earlier, this chapter gives you the best assessments the leading edge has to offer about common health problems. Using these insights, you'll be better equipped to make decisions on your own, based on the best intel available. You now have a choice. Better information = better decisions = better outcomes. These are the basic understandings you can use to handle health challenges moving forward – some explained at a root level for the first time in print. It's not necessarily a complete picture just yet, but definitely headed in the right direction.

The #1 reason products and practices lose effectiveness over time

The concept to retain for future reference is the mechanism by which acupuncture works. Many healing modalities operate on the principle of creating some sort of stressful disturbance in the body. Think of it as a crisis the body calls in emergency resources to fix. These can make the modality work temporarily for some people... if the body has the resources to draw from an area of lesser need. But then, after some days, weeks or months go by, the benefit starts to wane – until it stops working altogether and you're back to where you started, or even worse.

Meanwhile, you continue to employ the product or practice because it gave you some relief in the beginning. It seemed to help, and so you became attached to the perception it gave you. But the benefit is gone, and may even be hindering your efforts to get better over time. Just as bad, you continue spending your time, money and commitment on something that's not serving you in the present.

I'd venture to guess this is one of the most common reasons people fail to get well when they have a "sick-until-made-well-by-a-'buyable'-product" mentality. In other words, don't assume a product or practice will continue working for you weeks or months after you start using it. Many tap into the body's limited pool of resources and steal from other areas of need in order to deal with a bigger, more urgent threat. Unfortunately, they do this without replenishing the source. So you take two steps forward, and two vacillating steps back, over and over again.

To say it simply, these types of treatments are basically creating a bigger emergency for the body to deal with. As long as you know that, and apply the method judiciously, you can continue making gains without regressing.

Similarly, keeping its biochemical supplies tightly regulated is a bodily priority because quality of life declines rapidly with increasing deficiency or surplus. When you take exogenous supplements, **positive-negative feedback loops** uncouple demand from supply. The body then loses control over its production and recovery of hormones and neurotransmitters because an external supply is being injected into the system, thus it can't trust its own sensors. This creates poor performance in a majority of cases, organ failure when severe.

Sports injuries

For years, I've been low-key obsessing over why professional athletes are sustaining serious injuries far more easily and often than they used to. The most dreaded being: anterior cruciate ligament (ACL) and meniscus tears in the knee, Achilles tendon, muscle tears, joint dislocation, and broken bones. It's been a passion project of mine to investigate and help solve because the explanations offered by analysts are lazy at best, while implications to fans and franchises have been devastating to soul-crushing. Major injuries – particularly those that happen for no apparent reason – have become so common that observers consider them unavoidable. They say they're just "part of the game." I personally think the following factors may be worlds closer to the truth:

Greater demand. First, athletes are demanding more from their bodies than ever. They train harder and compete closer to their limits and beyond. Take baseball: The speed of a good fastball has risen 10 mph in just a few decades. However, I don't think the demand athletes place on their anatomy is that much greater than before, percentage-wise, to warrant special blame. It's just an influence everyone can agree on.

The glyphosate molecule looks similar to glycine (a major building block of collagen). So the body wrongly inserts glyphosate into tendons and cartilage, thereby reducing their terminal elasticity/breaking strength. Mitochondriacs believe this is partly responsible for the striking rise in sports injuries lately.

Glyphosate in place of glycine. My train of reasoning:

1. Glycine is a major building block for proteins.
2. The body mistakes glyphosate for glycine in the construction of many proteins, notably collagen. Its uniformity and strength suffers.
3. Ligaments, tendons and cartilage that hold the body together are made mostly of collagen.
4. Like a rubber band filled with tiny holes, microscopic imperfections in the collagen matrix might give the tendon/ligament extra elasticity, while offering a germination point for a tear to start.
5. These defects disturb the pathway of piezoelectricity through the collagen, resulting in non-uniform tensile strength of strands/fibers that then fail at different rates when stretched to their limit.

6. Muscles heal so much faster than collagen-dense tissue, they become proportionally stronger than tendons and ligaments. If anything, microscopic defects in muscles tear more easily so they can, in days, be rebuilt stronger. Whereas, tendons and ligaments tend to weaken over years, based on the factors below. That may explain why these types of injuries increase in prevalence with age: Muscles are just as strong in a person's thirties as their twenties, along with added weight. However, ligaments and tendons are proportionally weaker. Hence, safety margin goes down, while risk of injury goes up.

Proteins need water and electrical charge to function properly. Recall that proteins, which look like soap loofahs made of ribbons, take on the shape that they do because some of their surfaces are hydrophobic, while others are hydrophilic (i.e., electrical attraction/repulsion to water).

nnEMFs dehydrate proteins. The proteins in collagen hold water to maintain their shape and function. But wireless frequencies and blue light dehydrate us chronically. Just think of how aggressively microwave ovens dehydrate food in seconds. Then adjust that scale to minute exposure over long time periods. That's how wireless frequencies discharge water from collagen at the same time the body's trying to put it back.

Lack of red/infrared light reduces electrical charge in that water. Redox potential (in water) controls tertiary and quaternary protein folding. Red/infrared light exposure builds charge-separated e-zone for proper confirmational shape of proteins in collagen. On the other hand, lack of red causes e-zone deficiency and protein mischief.

Dirty electricity and nnEMFs eject copper atoms from the materials that form bone. When bone loses the copper pegs that hold apatite (the solid, mineral portion of bone) to collagen (the substrate that gives bones some flexibility and shock resistance), the bone weakens. We call clinical cases osteoporosis but, in athletes, fractures are considered normal.

See Dr. Becker's book, The Body Electric, *Chapter Six, for more information about bone degeneration.*

Circadian disruption retards healing. I think this is a significant factor in slower healing of injury hotspots such as joints and tendons.

Conclusion: When you combine these factors into one dynamic, it's hard to tell which influences cause increased failure rates. But, broken down into individual elements, we now have clearer insight into why sports injuries are happening at crazy rates that now seem normal. Basically, demands on connective tissues go up over time, while environmental threats lower the breaking point of collagen over several years. Exceed that threshold by contact or chance and you suffer a major injury.

Fluoride breaks down collagen and makes bones brittle

Fluoride is particularly harmful to human health because it's a **potent dielectric blocker.** That means fluoride inhibits formation of electron-zone water in cells and mitochondria so water can't absorb as much light, and it can't make as strong a battery. Wherever you get fluoride from – whether it's in toothpaste, water or pharmaceuticals – it lowers water's charge-carrying capacity. **Fluoride steals electrons.** When proteins lack electrons, they lose semi-conductive properties. Proteins then lose their shape, piezoelectricity, and ability to carry light. This is one cause of ACL and Achilles tendon tears.

Another reason to avoid fluoride at all costs is it degenerates the triple helix of collagen in both form and function. Athletes listen closely: tendons, ligaments and cartilage are made of collagen that contains water to: lubricate and cushion joints, make them stretchy, and house fluid in the knee capsule. **So by breaking down collagen, fluoride increases the prevalence of cushioning and connective tissue injuries such as ACL tears, Achilles tendon injuries, meniscus and rotator cuff tears, and strained hamstrings.** For example, the antibiotic ciprofloxacin is known to cause Achilles tendon injuries. That's because it and all fluoroquinolones contain fluorine which is extremely electro-negative. With its unpaired valence electron, fluorine poaches collagen's electrons. This blocks electron flow through the semiconductive circuits of collagen, which limits its ability to maintain and repair itself.

Fluorine is an extremely electro-negative element, while fluoride refers to any molecule containing a fluorine ion.

For countless consumers, that means they may be getting benefits from taking a collagen supplement, but then fluoride could wreck it all by causing an ACL, Achilles, or meniscus tear. Similarly, glyphosate weakens the collagen in tendons by replacing its glycine. Damaged collagen also **loses its piezoelectric current,** which slows bone regeneration. You see, when collagen is under tension or compression, it releases tiny electrical currents. You'll never hear this from your orthopedist, but microcurrents of bioelectricity are the fundamental force that activates bone healing. Fluoride also **forms tiny crystals** in soft tissues, which contribute to arthritis and joint problems as cartilage grates on bones like sandpaper. What's more, the Manhattan Project proved 80 years ago that fluoride dumbs people down and makes them complacent.

A fifth (big) reason to unfriend fluoride is it **releases more calcium** into the system. Calcium efflux exaggerates the stress response from nnEMFs that are already harmful to begin with. Remember: calcium activates nerves and muscles. So fluoride takes the never-ending stress our systems are in from sympathetic activators and it keeps the volume turned up. Whether those stressors are from nnEMFs, stimulants, sleep deprivation, toxins or our emotional state, the stress on your system just never lets up… even when you're asleep. So mobilizing more calcium into cells is exactly what you don't need when trying to tilt your biology toward a parasympathetic state of rest and digest.

Lastly, **fluoride displaces iodine.** That's a big deal because one of iodine's superpowers is it gladly shares its electrons with DHA to prevent DHA from becoming oxidized. Thus, iodine protects DHA, while fluoride is powerless to prevent DHA damage. There's nothing good about fluoride – including for teeth. It should be banned. A chemical cousin of fluoride – bromine – is bad as well, because they both belong to a group of elements called "halogens" that displace iodine in the thyroid, impairing its function. Commercial bread products contain bromine – supposedly as a "dough conditioner." It blocks water's ability to form e-zone like fluoride does. That's a more enlightened reason grains are bad for you.

Peak performance for athletes

Dopamine controls muscles and coordination, so a subtle element in how well athletes perform may be their dopamine level. I suspect availability of dopamine, adrenaline and other neurotransmitters such as acetylcholine could make the difference between playing good and playing one's best. You know that feeling when your timing is on, you're anticipating, and it feels like you can't miss? Getting yourself into a flow state has a lot to do with biochemicals and the resulting fluency of communication between components of the nervous system.

Influences that depress your biochemistry include poor sleep/circadian dysregulation, blue light exposure, nnEMFs, social media, stress, dehydration and electrolyte imbalance. Of course, everyone knows sleep deprivation and electrolyte deficiency reduce your sharpness. But imagine for a moment how artificial light, EM pollution and social media detune an athlete's sports performance. I believe modern influences like these can blunt the fluency of nervous system communications to a degree that's barely perceptible because we live at 80–95% of optimal the vast majority of our lives. We're seldom playing at 100%, and no one can tell you why.

I know it's just a theory at this point. But if poor sleep or electrolyte deficiency can impair athletic performance, as we know they do, we must consider the strong possibility that biophysical forces influence athletic performance as much as they do our health, sickness and aging. Thus, athletes can seek those last few percentage points of performance in: electromagnetic shielding for hydration and parasympathetic tone; circadian synchronization such as watching sunrise, blue-blocking glasses and apps; deuterium-depleted water for happier mitochondria; and near infrared sauna for detoxing and healing through photobiomodulation. When you optimize light, water and magnetism in your life, you're effectively doing legal blood doping. Sunlight improves your red cell mass – hemoglobin, in particular – and cardiovascular performance.

Dental fluorosis

For decades now, the American Dental Association and member dentists have told their patients there's nothing you can do about white spots or pitting on teeth caused by fluoride in toothpaste, water and food. They say it's an unfortunate side effect of fluoride you have to live with in order to get the benefits of harder enamel. They are mistaken. You can do something about it.

First, remove the source of ongoing exposure. Stop drinking water with fluoride in it. And avoid processed foods that are made with it. Unfortunately, the water in some regions is naturally high in fluoride – for example, the desert Southwest of the US, specifically, states supplied by the Colorado River. Most food companies don't bother to filter it out either, wherever they're located. That means fluoride is hidden in many packaged foods because their ingredients are grown with fluoridated water, and/or it's added in manufacturing.

Next, stop using toothpaste with fluoride in it. Better yet, instead of toothpaste, switch to a clay, charcoal, and essential oil-based *toothbrush powder* that naturally whitens, detoxifies and cleans better at the same time. (Note: I plan on coming out with my own toothbrush powder in 2025 because I couldn't find a brand I liked in 2024.) Next, try to find prescription drugs that don't have any fluoride in them. This might be difficult because pharmaceutical companies purposely add fluoride to every drug they can because it dulls the intellect (i.e., reduces reasoning ability).

Finally, you can reverse dental fluorosis when you improve your redox potential, because teeth, like other parts of the body, can heal themselves. More precisely, fluorosis can only happen when dental structures are being damaged by fluoride faster than they're being re-mineralized. However, more calcium isn't the answer, as the dairy industry would have us believe. It's only a building material. And most people get plenty of it in milk and milk products such as cheese, yogurt and ice cream. In fact, most of us get too much calcium – especially relative to magnesium.

Cancer

Cancer is made possible when the immune system loses control over growth mechanisms – that is, autophagy and apoptosis are broken (i.e., cell recycling and programmed cell suicide, respectively). In fact, it's impossible for cancer to manifest when apoptosis is working properly because, under normal circumstances, the immune system can tell when cells are so damaged it's not worth fixing them. When that happens, un-repairable cells are instructed to kill themselves in order to make way for healthy, new cells. On the other hand, in a cancerous state, cells aren't getting repaired properly. And they're not dying off like they're supposed to. Quite the opposite: they're growing and dividing as fast as they can because they don't realize they're cancerous and need to be either fixed or destroyed. So what causes the breakdown of apoptosis that leads to cancer?

Initially, before cancer shows up, the electron transport chain is slow. The slower the ETC, the less chance you can lose control over apoptosis. But later, two things happen to make cancer come knocking on your door: The speed of electrons moving across the ETC must be high to support abnormally rapid cell growth, and make huge amounts of free radicals that damage mitochondrial and nuclear DNA. More simply put, free radical DNA damage, and runaway cell growth, are partners in causing cancer. And an overdriven ETC fuels both fires.

So what causes the ETC to run fast? What slows it down? The ETC runs fast when you're solar deficient. Normally, the vitamin D receptor in Cytochrome III slows down the electron transport chain when you get lots of sun. And UV-B makes vitamin D. You see, when you get lots of sun, your ETC doesn't need to run as fast to make energy. You're getting energy directly from the sun. So the ETC purposely slows down oxidative phosphorylation to balance out the body's energy needs.

Solar exposure also dilates blood vessels to enhance UV light absorption and oxygen delivery. Since UV penetrates tissue under 1 mm, nitric oxide dilates vessels to bring blood closer to the surface (dermal pooling). Through this process, Cytochrome III sees higher levels of nitric oxide (a free radical), and uses that free radical signal to know when to turn down the speed of the ETC. In short, UV light puts the brakes on the ETC when you get more energy from the sun, instead of food. Those are some ways that disturbances in energy flow set the stage for cancer.

Impaired threat detection is the other side of the cancer story. Too much man-made (blue) light trashes your melatonin level and mitochondrial energy production, while a lack of full-spectrum sun – UV, in particular – deprives you of vitamin D. Together, these control cell-mediated innate immunity. Killer T cells (aka cytotoxic T cells) need vitamin D to identify rejects, including cancerous cells or viral particles. Vitamin D helps run killer T cell targeting. That means when killer T cells malfunction, the immune system can't take out cancer cells or the SARS virus. Risk for autoimmunity also goes up.

For these reasons, the more sun you get, the less chance you're going to experience runaway energy production that harms healthy cells with free radicals and feeds cancer. **Hence, the sun is truly Nature's vaccine against cancer.** And, as we'd expect, low vitamin D is the #1 risk factor for breast cancer. Deficient autophagy (low UV), low redox, weak detoxing, and inflammation/acidity also help pave the way for cancerous cells to overstay their welcome. They usually go together.

But, counter-intuitively, when you have poor mitochondrial function, low oxygen levels may actually protect you from cancer. That's one of the reasons some people with bad mitochondria develop **sleep apnea**. Their bodies are purposely lowering oxygen levels to protect them from inefficient energy production, which results in lots of free radicals. In some cases, hypoxia is a protective mechanism.

So even though most cancers love hypoxic conditions, supplementing with oxygen – like in a hyperbaric oxygen chamber – may not be a good idea. With some types of cancer, the addition of oxygen can make a person sicker. It can kill them faster because it speeds up the ETC in the absence of apoptosis, making lots more free radicals. Simply put, hyperbaric oxygen can be deadly when your ETC is prone to making free radicals. Conversely, oxygen can inhibit other types of cancer when autophagy and apoptosis are still working and your heteroplasmy is improving.

Reversing cancer the biophysical way revolves around restoring mitochondrial efficiency to reduce the following: (1) free radicals and inflammation; (2) degenerative programming; and (3) the resulting (epi)genetic mutations. You need to expand your redox potential "gas tank" and fill it up with net negative charge, so your immune system wakes up. Cells can then communicate with each other effectively.

Some cancer mitohacks to consider:

- Correct bipolar magnetic fields in your home – especially around your sleep and work spaces.
- Increase energy production and detoxification with strong earth-type magnetism.
- Decontaminate your home of unnatural frequencies – including blue light, dirty electricity, and microwave sources such as Wi-Fi and 5G.
- Get copious amounts of full-spectrum sunlight in the eyes and skin, as close to the equator as possible (or at altitude).
- Cut fluoride out of your water and food.
- Deplete your deuterium level with a seasonal diet or specific protocol.
- Cut down on foods that acidify, while increasing foods that alkalize.
- Do a heavy-duty detox, and continue detoxing daily.

Mold toxicity

Mold toxicity is exacerbated by nnEMFs disrupting mitochondrial metabolism, which makes you more sensitive to mold. The intensified response causes a state of poor mitochondrial function and redox potential that lowers your immune system's ability to handle mold toxins. Therefore, when you get your light, water, magnetism, and seasonal eating exposures right, mild-to-moderate mold reactions go away because your detox systems can handle much greater exposures to mold without issue.

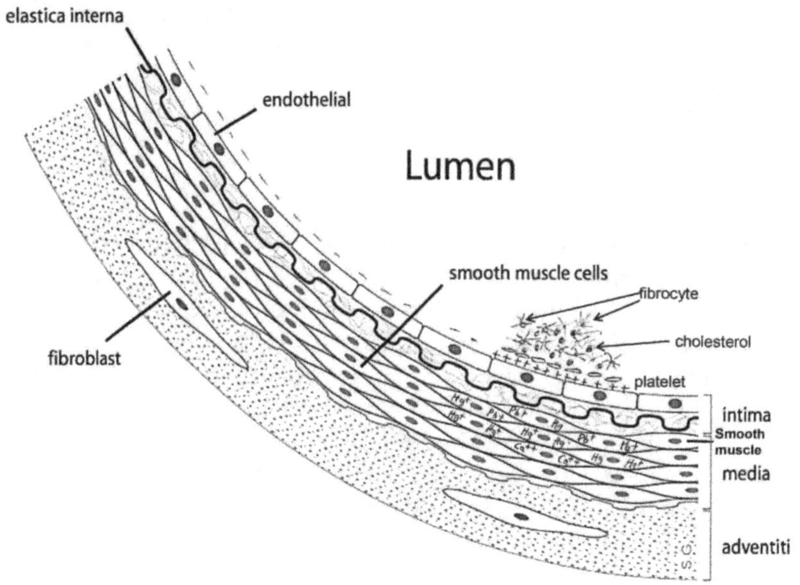

Images of artery clogging used courtesy of Dr. Dean Bonlie, DDS.

Clogged arteries (e.g., atherosclerosis, peripheral artery disease, and coronary heart disease)

Cardiologists and heart associations and tell us cholesterol and inflammation are to blame for blocking arteries and causing cardiovascular disease, because that's what they find there. But, as you'll soon learn, cholesterol and inflammation are both downstream consequences of reversed polarity on the blood vessel wall, which creates a false signal of injury.

Clogged arteries from fatty deposits (including cholesterol) begin as an accumulation of positively-charged heavy metals such as mercury and lead in smooth muscle cells beneath the endothelium (inner vessel wall). You see, blood vessel walls are supposed to be negatively charged when all's well. Negative charge causes negative blood cells to cruise by without a care in the world. More accurately, they're propelled by repulsive, vortical, and expansionary forces. But, when heavy metals (which are positively charged) build up in cells beneath the interior vessel wall, the vessel wall changes polarity. It changes from negative to positive charge locally, which is the polarity of injury, and the signal that calls healing mechanisms to the area. To illustrate, if you pinch yourself, the release of histamine immediately turns the area positive.

Positive charge on the vessel wall then causes (negative) platelets to stick to it. If the injury were a cut, platelets plug the vessel to stop the bleeding. Platelets then tell fibrocytes to attach themselves to the positive endothelium to make a mesh. We recognize the mesh as a scab. Cholesterol then impregnates the mesh to form a soft plaque. If enough bacteria find food and lodging in the plaque, the body tries to wall it off with calcium to minimize the damage done to other tissues. This causes the plaque to solidify into a hard plaque. But that's a story for another day.

Over time, soft plaque restricts blood flow from the artery, which cuts off oxygen and nutrients to cells using that supply. As the plaque gets bigger, not only can lack of blood flow starve organs of oxygen, nutrients and waste removal, but smooth muscle cells of the blood vessel itself can't get enough oxygen and die. That releases (very positive) histamine,

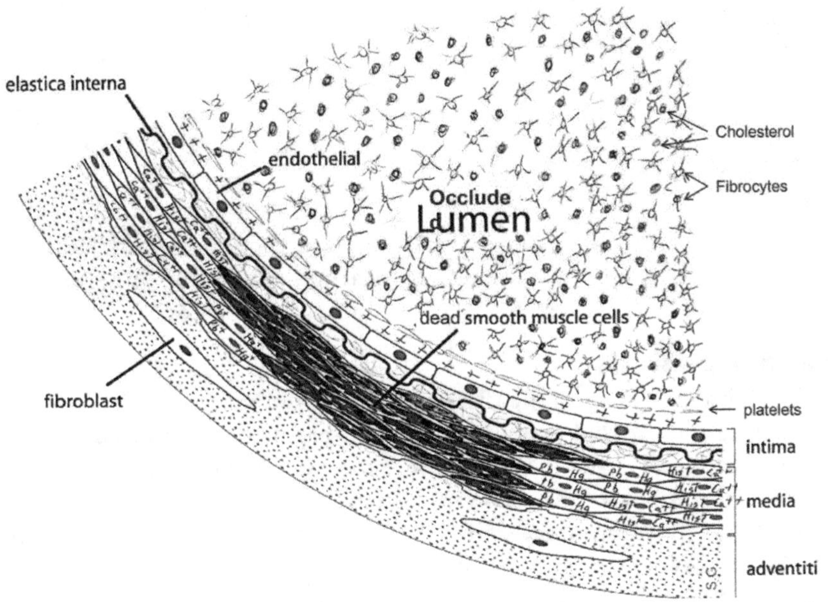

which causes inflammation around the plaque. It also makes the vessel inflexible so it can't dilate when the body needs it to. That's how cholesterol-filled plaques are created in arteries, and why inflammation is found around major blockages. However, cholesterol is not a hairball in your cardiac plumbing. Instead, cholesterol is at the scene trying to help.

So, as you can see, cardiovascular disease is not caused by cholesterol (biochemistry) or resulting inflammation (symptom). Instead, it all begins with biophysics – positive polarity from heavy metals where negative charge should be present – plain and simple. In this case, poor mitochondrial function and ATP deficiency inhibit heavy metal removal – exacerbated by stress, smoking, drinking, poor diet, obesity and diabetes. This sets off a chain of chronic events that has confused medical scientists for more than half a century. They're quick to blame cardiovascular disease on cholesterol, obesity and heredity, because they support Industry's false narrative. But the evidence is piling up that cholesterol and saturated fat are clearly not responsible for causing heart disease.

So, what evidence do we have that heavy metals initiate cardiovascular disease? A pathologist from the University of Wisconsin examined the vessel walls of 120 patients that had died of heart attack, looking for toxic substances they all had in common. To his amazement, he discovered 120 out of 120 had plaque buildup over every spot containing high concentrations of lead and/or mercury in the smooth muscle cells of the vessel. What's more, in locations that didn't have heavy metals, there was no plaque in every case. That means we have 100% positive confirmation in cases AND locations in the body, as well as zero contradictions in location, for this heavy metal-atherosclerosis theory (Dr. Dean Bonlie's). That's what researchers and statisticians call a very high statistical probability and confidence level.

So what can we do about it? Can the process be reversed? A: If you catch the plaque buildup early, while it's still pliable, the Magnetico Sleep Pad and a chelating agent like DMSA can help remove heavy metals in the vessel's smooth muscle cells. With heavy metals gone, negative charge is restored on the vessel wall, the plaque clears up, and everything returns to normal. Though, completely calcified blockages are harder to dissolve.

Autoimmune diseases

The root cause of autoimmune disease is a circadian mismatch between light entering the eye and that absorbed by T-regulator cells in the gut associated lymphoid tissue (GALT) where food is broken down. The GALT is the frontline of the immune system… the Great Wall of the digestive tract. Autoimmunity starts when the circadian clock in the brain (SCN) gets unyoked from circadian clocks in the GALT. In real life, that means your eyes and skin are receiving artificial blue light from smartphones, TVs and LED lights, in the absence of UV and IR from full-spectrum sun… while at the same time your gut lining gets a summertime UV signal from chips, crackers and seed oils you're eating.

For those eating imported foods: when the eye and gut receive conflicting signals about your current light environment, your infradian biology gets confused about which season you're in, and thus which program it should be running. The mismatch increases heteroplasmy rate of mitochondria in T-regulator cells T3 and T4, which impairs their ability to distinguish your own proteins from those you eat frequently, such as gluten. Since T3 and T4 systems control both arms of immunity – innate and acquired – autoimmunity is caused by glitchy "foreign protein detectors" in T3 and T4 cells due to decrepit mitochondria. T-regulator cells basically get blurry vision so they can't tell one protein from another. Many look very similar, so they attack everything.

Now, the science that explains the situation is rooted in evolutionary biology going back 600 million years. Prokaryotic cells (bacteria and archaea) have a different relationship to light than our own cells (especially UV). Prokaryotes don't collect any DHA like our own cells. They can't turn sunlight into DC electricity. So they can't utilize sunlight nearly as well as eukaryotes. Instead, prokaryotes are designed to release light much more liberally than the cells of higher life forms… 5000 times more light than eukaryotic cells. They basically live fast and die young, as far as collecting and "spending" sunlight are concerned.

Prokaryotes: Single-celled organisms lacking a nucleus, including bacteria and archaea.

So, after you eat, bacteria in the gut begin the digestion process. But, not being evolutionarily adapted to retain light, bacteria extract light energy from food electrons as biophotons and other frequencies. They release that light immediately into the gut lumen like a projector. That light is absorbed by cells of the gut lining like a movie screen in this scenario, because our cells *do* have DHA in their cell membranes, so they *are* able to turn light into a DC electric current, as their mitochondria collect seasonal light information from electrons.

The problem is, light absorbed by gut cells turns on an infradian signal for UV in the GALT, but not the eye, because the eye is getting a different signal from your surroundings. That's big, because more light and DC electricity in cells of the GALT makes its clock genes run faster than clock genes in the brain. That raises heteroplasmy rate in the GALT's mitochondria, and confuses communication from brain to immune cells, which then impairs the ability of T-regulator cells to distinguish food proteins from the proteins of your own tissues. That, over time, becomes autoimmunity.

That also means you can't fix an autoimmune condition through diet or repairing a leaky gut alone. Sure, there's plenty of evidence to show that you can alleviate some foreign protein allergies by avoiding foods that give you symptoms, or repairing tight junctions of the gut. However, those are proving to be partial or temporary fixes, because the root cause of autoimmunity is fundamentally circadian clock genes in the GALT running faster than clock genes in the suprachiasmatic nucleus (SCN)... not just avoiding foods that give you symptoms.

Electro-hypersensitivity (EHS)

Electro-hypersensitivity is caused primarily by 50 and 60 Hz electric fields and their dirty electricity, in combination with wireless microwaves from tech devices. These foreign frequencies disturb energy production in the mitochondria. Here's what follows from there:

- The ETC can't burn fats and proteins as well, so mitochondria and cells fall back to glucose and carbs for fuel. That means less energy, lower redox potential and healing, disturbed signaling, and weaker electric and magnetic fields made by mitochondria.
- Inflammation and oxidative stress.
- Increased stress response.
- Elevated histamine.
- Lower melatonin.
- Anti-myelin antibodies.

Which can turn into:

- Brain fog, memory difficulties.
- Disrupted sleep.
- Blood-sugar problems.
- Headaches, dizziness, migraines.
- Asthma/allergies.
- Weakness and fatigue.
- Hearing problems/tinnitus.
- Speech difficulties.
- Depression, anxiety, irritability.

- Skin problems.
- Stress.
- Digestive disorders.
- Heart problems, high blood pressure.
- Flu-like symptoms/breathing problems.
- Sensitivity to light/eye problems.
- Tremors, cramps.
- Joint and muscle pain, numbness.
- Lower sperm motility and testosterone.
- Erectile dysfunction.
- Weight gain.

The solution? You need to remediate your nnEMF environment, or else nothing you do in the way of treatments is going to solve your root problem. Get out of there. Remove yourself from the EMF pollution around you to see how much/how fast your symptoms go away – either by visiting a location undisturbed by nnEMF emissions, or by trying out a well-shielded space. More often than not, you'll be shocked at how well a pristine EMF environment doesn't just soothe your symptoms, but cause them to utterly vanish. They cease to exist, because the true source of the dysregulation and damage is gone. Sometimes symptoms stop immediately – tinnitus, for example – while others can take three to seven days. Susceptibility usually remains, but often there is no permanent damage.

Sleep apnea

A primary cause of sleep apnea (among others) is deuterium overload in the central brain stem that controls breathing. Toxicity compromises mitochondrial energy production and communication of the hypothalamus, thereby confusing our mammalian dive reflex (a remnant of our aquatic ancestors born into water), which is responsible for sensing temperature on the face and telling the autonomic nervous system whether or not the body is underwater. When the hypothalamus sends mixed messages, the autonomic nervous system presumes you're underwater; you need to hold your breath.

To illustrate how complicated sleep problems can be, when the thinking parts of the brain shut down to repair and replenish in sleep, one area of the brain is supposed to stay active so it can keep the airway open. But when deuterium upsets mitochondrial function in that area, the complex coordination required to run sleep breaks down. Among more than half-a-dozen factors supporting or inhibiting easy and effective sleep, the brain center responsible for maintaining the airway shuts down when it's not supposed to. So not only is the urge to breathe depressed in this situation, but the airway is also obstructed until high CO_2 levels in the blood shout at the autonomic nervous system that you're on the verge of death.

This can happen in addition to some combination of poor melatonin levels, excessive cortisol and adrenaline keeping your stress level high, or a bunch of other factors – all biophysical or circadian related, and mostly a result of blue light toxicity, nnEMF exposure, excessive deuterium in the diet, stimulant consumption, and circadian mismatches. This sort of situation also explains why fatty, un-toned tissue around the throat can make it seem as if that's the problem when, in fact, it's only a contributor that would not cause sleep apnea on its own. Conclusion: The physics of the body, and circadian disruptions, answer the complex questions raised by sleep apnea, where conventional medicine offers only non-specific explanations and limited treatment of symptoms.

Thin hair

Have you noticed how many people in their teens and twenties have unusually thin hair these days? It's getting to be epidemic in seemingly normal, healthy adults. You'll really notice the contrast when you go back and examine movies and pictures from the 1970s, '80s, and before. Many women had remarkably thick hair, while you hardly saw any with unexplained thinning problems (and it wasn't just the hairspray). So, what's causing abnormally thin hair of unknown origin today?

It's instigated primarily by too much blue light exposure early in life. This causes high amounts of ACTH and CLIP to be made from POMC. Problem is, excess ACTH and CLIP elevates glucose and insulin, which pressures the hypothalamus to raise sex steroids, including testosterone in males and estrogen in females. These trigger precocious puberty in both sexes. But, in males, high testosterone from infancy to adolescence causes a distorted form of andropause (male menopause) to occur, presenting as hair loss and premature graying, not to mention sexual dysfunction such as low libido, then infertility, then ED and "low T." This pathway gives DHT (dihydrotestosterone) many more years to shrink hair follicles.

What counteracts this mechanism? By nature, UV light is supposed to limit excessive/premature testosterone production. UV light is the virtual 'off' switch for testosterone through this pathway. But modern humans block UV light with clothing, sunglasses, sunscreen, makeup, and never going outside, which disables the off switch. Circadian dysregulation is also to blame. Chronic, systemic stress from all directions – artificial blue light, no natural light, nnEMFs, poor sleep quality, stimulants, psychological stress, toxins such as fluoride, and trauma – these dysregulate thyroid and hormone function, not to mention increasing inflammation and decreasing detoxification. It's basically chronic, subclinical hypothyroidism that makes the thyroid cut back on energy expenditure wherever it can. And hair is an expendable luxury to the thyroid.

On a related note, grey hair happens when skin cells around the hair lose the ability to make V-UV light. Melanocytes, with no V-UV light to stimulate growth, then leave hair follicles to go find another supply. Also, stress produces peroxide in the hair follicle. Conclusion: We've been taught to blame genes, nutrient deficiency, and stress for hair and nail problems. But now you have more credible and actionable reasons for hair loss than medical science ever gave you.

Cataracts

Cataracts are a defense mechanism against too much red light. Red and infrared light are generally beneficial to get. But overexposure to red light, especially in artificial (imbalanced) lighting, causes the body to try and protect itself by turning the lenses of the eye cloudy. This reduces the focus and intensity of non-full spectrum light hitting the retina. Disturbing but all-too true, it also means when eye doctors remove cataracts and replace them with artificial lens implants (which block 100% of UV and 50% of the blue) you actually make the person's health worse, as the ophthalmology literature reports quite clearly in diseases across the board. Quite simply, your health suffers when you don't get UV light. And most lens implants do exactly that. The solution? Of course, it's getting full-spectrum sun during the day, and avoiding artificial light at night.

Pregnenolone steal, libido and gender confusion

Cholesterol is a precursor for pregnenolone. The pregnenolone pathway then goes in one of two directions: progesterone and cortisol, or testosterone and estrogen. Summarized for simplicity, pregnenolone converts into either the stress hormone cortisol or the sex steroid hormones testosterone and estrogen. The problem is, breakdowns in the pregnenolone pathway ruin the lives of millions of centralized medicine addicts, or at least make you dependent on their drugs and side effects.

Most commonly, cholesterol-lowering drugs (statins) intentionally decrease production of cholesterol that's needed to make hormones we need to feel happy and whole. Cholesterol converts to pregnenolone, so when there's not enough pregnenolone to feed both pathways, you can't make cortisol, testosterone and/or estrogen. Since modern humans are under constant stress from blue light, nnEMFs and social media, an inordinate percentage of pregnenolone goes into cortisol production, instead of sex steroid hormones that control emotions, feelings of wellbeing, and sex traits.

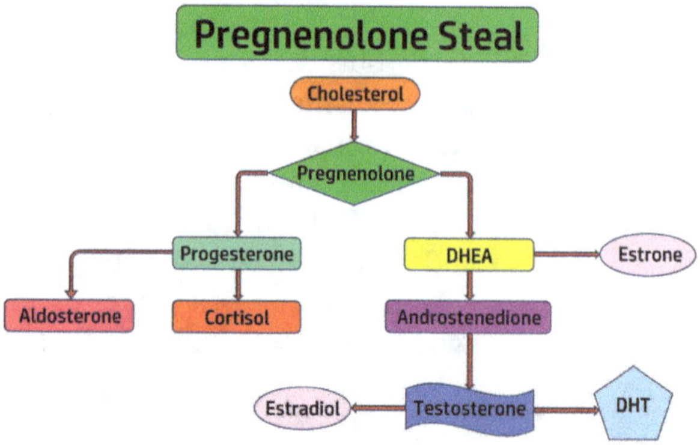

Cortisol goes up to deal with physical and emotional stressors, which deprives you of hormones that control femininity and masculinity. Even worse, when pregnenolone insufficiency gets really bad, you can't produce cortisol either, and your stress tolerance collapses.

If I were to characterize how much I think this effect had on gender confusion/transgenderism in 2025, I'd venture to say (1) chemical crop amendments such as atrazine and glyphosate start a cascade of dysfunction in the womb (as baby's brain is developing), (2) estrogen mimics (BPA) in GMO-soy and plasticizers add fuel to the fire (especially through puberty), and (3) pregnenolone steal keeps the imbalance going after sexual maturity. My guess: Pregnenolone steal syndrome can't cause gender dysphoria on its own. Rather, it reinforces processes already underway – giving people a push that were already headed in that direction, or "merely" causing hormonal deficits in those with established neuro-chemistry and biology.

Psychoactive drugs such as antidepressants, anti-psychotics, anti-anxiety and stimulants may pacify symptoms for some of our young. But are they really helping, or are they hurting? Another factor Uncle Jack mentions: (4) Vitamin A controls sexual dimorphism in hypothalamic neurons. So when vitamin A gets liberated due to blue light stress, a person's brain gets wired – I assume at an early age – so they're not attracted to persons of the opposite sex. That is, retinal alters sexual attraction, away from its natural design.

Sexual dimorphism: Differences between sexes.

Bad skin

Almost everyone thinks dry, wrinkly, unattractive skin is caused by bad genes, too much sun exposure and, in some cases, too much partying. Not true. Not entirely, anyway. I'd say that belief is like a book that's missing chapters, so it doesn't tell the whole story. One of those missing chapters is that unhealthy skin is caused in-part by worn-out mitochondria and intracellular (metabolic) dehydration. Certainly, external creams and treatments can improve the appearance of skin from the outside in. But the thinking woman knows that a better way to improve her skin tone is from the inside out – with hydration, better diet, detoxification, exercise, and healing the microbiome.

Add perkier mitochondria to the mix, and your skin is as beautiful as can be. All those things give you benefits not purely as standalone solutions, but *because* they work on biophysics and the mitochondria. Along with all the other things mitochondria do (or don't do) for us, they make water to keep skin plump, smooth, and youthful-looking from the inside of cells. That's the biggest benefit, because water

1. stores redox energy from the sun and food in its e-zone
2. enables the skin to make vitamin D
3. maintains protein surfaces so mitochondria and cells work properly

4. potentiates magnesium
5. enhances detoxification.

Mitochondria also produce the power (ATP) for skin cells to turn over and resist the aging process – for example, repairing DNA breaks and fixing holes in proteins from sun ionization. Actually, when you think about it, changing your diet, detoxifying, exercising, and healing your microbiome all travel different paths to arrive at common ground, which is enhancing energy production and minimizing free-radical damage by making mitochondria more efficient. The happy by-product is enhanced intracellular hydration. Lively mitochondria also produce more ATP, which helps move oils, toxins, and bacteria efficiently through the pores of the skin, instead of stagnating and causing pimples and boils.

Takeaways from this short course on skin biology: Diet, detoxing, exercise, and ingesting water are one level removed from the real driver of attractive skin, which is mitochondria that rock. Moisturizers and esthetician treatments such as clinical exfoliation and laser treatments are two levels removed from the source of great skin. That means you can take an outside-in approach, or an inside-out approach, to getting great-looking skin. And you might get the results you're looking for. But neither approach will be quite as good, or last as long, as actually *having* younger skin on a cellular level. To put it plainly, when your mitochondria are in great shape, they make your skin look as fabulous as can be.

Loss of smell and taste from COVID-19

The virus is a mitochondrial toxin. In this condition, the virus destroys melanin in that part of the brain responsible for sense of smell, called the olfactory bulb. Getting direct sun exposure on the face, particularly UV and IR, can help renew the melanin there and restore smell and taste. Unfortunately, artificial blue light blocks melanin renovation due to mitochondrial inhibition. Sensory organs for smell and taste can also malfunction because spike proteins clog their receptor sites chronically. Because these sites have far-greater affinity for nicotine, it displaces the spike protein in its presence. Therefore, nicotine (preferably from clean sources in patch or gum form) ejects the spike protein and restores sense of smell and taste. See this book's successor, *The Blood Flow Monograghs*, when it comes out to learn more about the science and solutions to disturbed blood flow and vaccine injuries.

Longevity

Sunlight increases blood flow through nitric oxide and vasodilation. Vasodilation delivers more oxygen per heartbeat, which then makes more ATP. At the same time, infrared charge-separates water to increase zeta potential. And infrared makes more ATP by itself. Combined, these factors help the blood circulate more easily, reducing workload on the heart, so it doesn't have to beat as fast. And what do we know from longevity studies in animals? Animals whose hearts beat faster tend to live shorter lives, while those whose hearts beat slower live longer. Animals basically get a set number of heartbeats in a lifetime, give or take. And when those are used up, you're done. Therefore, sun exposure is the ultimate longevity prescription.

Biochemicals like testosterone enable you to experience the joy for life

The mechanisms that spoil your dopamine and hormone levels also cause you to have low testosterone. These include blue light and microwave EMFs activating your adrenals non-stop, circadian mismatches, signaling breakdowns, and nutrient deficiencies. Environmental factors like these derail testosterone production, as well as its release and recycling. Low testosterone not only inhibits muscle-building, libido, fat distribution and bone mass, but also affects how good or bad you feel physically, which influences occupational compatibility and exercise choice of the sexes. Most significant for this topic, testosterone modulates how men feel physically and mentally.

For example, men are better suited to hard physical jobs because testosterone acts as a natural opioid to the body and mind. It makes men feel good after a strenuous workout (i.e., less exhausted) – producing what we might call a "heavy-workout high" – in addition to actually making muscles and bones stronger. Whereas women tend to feel more rundown, more quickly, doing the same work day after day. That's because testosterone counteracts the feeling of depletion after hard, physical labor.

Testosterone does have body- and mind-enhancing qualities for women, but they have much less available to bounce back after a hard workout. Other biochemicals such as adrenaline and cortisol make up some of the difference. But, fundamentally, lower testosterone is one reason women prefer aerobic exercises, toning practices such as yoga, and training specific body parts such as the butt and abs – instead of working all the major muscles groups, as men do more often. And it's why women do better in jobs that don't require heavy, repetitive movement.

This could be life-changing for you: I bring up testosterone because it demonstrates how biochemicals make you feel as your baseline… your daily norm. When you have an adequate supply of these mood-modulators, you feel vital, energetic, and happy with life. Conversely, when you run low on these chemicals, you go through life feeling dark and dissatisfied physically and mentally. You routinely feel icky or "less than" as your everyday experience.

When your psychological reward centers don't have enough of these pleasure chemicals to give you – whether you're not making them, or because you ran out prematurely – you're not able to feel happy and whole just being… just existing. This is why so many people feel compelled to take stimulants and anti-depressants on a daily basis to boost their biochemicals into a normal range. To their bottomless discontent, they need constant stimulation or else they go through life in a depleted state. Let me tell you: That's no way to live your life. But that doesn't have to be your future from this point forward.

Here are some tips to help *you* get to your happy place:

- First, be acutely aware of how biochemicals such as serotonin, dopamine, and testosterone control your feeling of well-being.
- Next, improve your light and EMF exposures to fix daily and seasonal timing issues that regulate your biochemistry. Most important, see the sun rise, cut out the blue, and increase the UV.
- Fix faults in your lifestyle (e.g., nutrient deficiencies) that interfere with production of biochemicals.
- Reduce your stress level from all sympathetic activators.
- Ground yourself as much as you can and raise your redox potential.

Mitohacks like these raise your mood-makers out of deficiency and into balance, or even positive territory. **Sunlight on the eyes, by itself, makes the feel-good chemicals that enable you to feel pleasure and emotions – both at baseline and during peak experiences.** So if you're feeling depressed or anxious, why not give your body the resources it needs to make its own mood-raising neurotransmitters and hormones from within, instead of taking drugs that force your body into releasing its limited supplies of biochemicals at inappropriate times, causing even greater deficiency and imbalance? That resource is natural, unadulterated sunlight.

Once your body is in firm control of its biochemical levels, you go through life with a natural feeling of well-being, instead of relying on coffee, energy drinks, social media, drama, unhealthy habits, or serotonin-enhancing drugs (SSRIs) that borrow from tomorrow to survive today. Small-to-moderate changes like these can radically improve your perception of living… of life itself. They can change your life experience from *within*, instead of continually looking for external sources to supply you with happiness and satisfaction. Now that's winning.

PARTING WORDS

Some people have the good fortune to be gifted excellent health throughout their life and don't have to work very hard for it. Unfortunately, with our disconnection from Nature so advanced, and our airwaves polluted the way they are, those people are quickly becoming an endangered species. That means most of us will now have to fight throughout our lives to be healthy, and to live the life we desire. But although the health challenges threatening us all may seem daunting, be assured there are solutions to fix the chronic conditions we all know by name. Despite what skeptics say, biophysics, mitochondrial biology and quantum biology have the answers (along with most of the solutions) that mainstream medicine struggles to adequately address, or even acknowledge. I hope that's been made abundantly clear from all the myths dispelled throughout, and mysteries of the body explained as they truly are.

Contrary to what doctors and health agencies say about chronic conditions being incurable, mitophysics *is* a two-way street. That means when you stop contaminating yourself with electromagnetic frequencies that steal your energy and healing capacity. And you start doing the things that deposit wellness into your health bank account. Then your body does what it's designed to do: repair and replenish itself. One more time: You are not broken. It's your environment that's defective. And unless you're down to your last electron of redox potential, you can get better.

The journey to wellness may not be as quick, cheap, or easy as you'd like it to be. But it's almost always quicker, cheaper, and easier to fix health problems than it was creating them. So keep your expectations realistic short term and your outlook optimistic, long term. It's okay to be skeptical, as long as you keep looking for answers and you keep moving forward. Simply refuse to accept underwhelming treatments and lame excuses from a broken system that likes to keep you doped up, disempowered and dependent.

Just remember: The way of the mitochondriac is not a single practice or a set protocol. There's no such thing as a one-size-fits-all, perfect way to achieve immaculate health and resiliency. That's the wrong way to think about the new health paradigm of $n=1$. Instead, think of it as a journey – a new way of life that you build one principle at a time, one practice at a time, and one product at a time. Conversely, relying on a bunch of thoroughly indoctrinated white lab coats to give you a pill for every ill… that's a lazy man's way to accumulate a lifetime of annoying side effects, hidden costs, and declining health. Your crusade to a better place may take a little time, and a little experimentation. However, the results are more all-encompassing, more reliably attained, and far more satisfying, because now you're working with Nature, not against it.

Just test out one or two measures at a time that make sense to you. Start with the quick and easy ones, and see how they fit into your (upgraded) lifestyle and budget. Keep the tips you like. Set aside the ones that don't work well for you. Test and retest (aka "challenge-rechallenge" test) to make sure the measures you adopt continue working for you long-term. Just don't fall victim to complacency. Don't give up for the wrong reasons. Get greedy when it comes to your health. Keep looking to add more mitochondrial capacity, and you'll be rewarded in ways you don't always notice straight away, but will ultimately appreciate when you realize how much they're benefitting you.

Challenge-rechallenge test: Medical testing protocol that purposely stops and restarts a therapeutic effort to confirm its benefits and side effects.

In other words, don't accept anything less than robust energy, a fully-functioning brain, and excellent health for as long as possible, because now you have a choice. It is with these thoughts that I offer this collection of cutting-edge principles and practices to you, your family, and your loved ones.

The End (of the edification for the moment)
…The start of something life-changing for you

GLOSSARY

Note: Some definitions are listed here that are not covered anywhere in the main text.

Archaea microbes that are similar to bacteria, but different enough to be considered another life form.

Aromatic amino acid a building block of neurotransmitters.

ATP adenosine triphosphate is an energy storage molecule made in mitochondria that drives dozens of cellular processes crucial to running the body.

ATP synthase fifth cytochrome 'workstation' of the electron transport chain that completes ATP production.

Autophagy controlled breakdown and replacement of damaged cellular components to keep the cell running well.

Beta-oxidation process by which long-chain fats are broken down to move ATP production forward.

Biophysics the physics that controls biology.

Brown fat a specialized type of fat whose dense mitochondria populations burn it to make heat when you get cold.

Challenge-dechallenge-rechallenge test medical testing protocol that purposely stops and restarts a therapeutic effort to confirm its benefits and side effects.

Chi vital life force, or essential energy, running through all living beings that makes us alive.

Chloroplast primitive symbiotic life form inside plants that uses chlorophyll to convert sunlight into energy. Thought to have evolved from early bacteria, chloroplasts perform photosynthesis.

Chronobiology time-based biological cycles (e.g., circadian, infradian, and ultradian rhythms).

Circadian rhythm biorhythm lasting about 24 hours, such as daily cycles of sleep and waking.

Cold Thermogenesis Protocol Dr. Jack Kruse's way of using cold exposure to increase mitochondrial efficiency and magnetism throughout the body.

Colloid solution with free-floating particles mixed in (e.g., milk).

Coronal mass ejection (CME) the release of large amounts of plasma, magnetic flux, and EMFs from the sun. Large CMEs can shut down power grids, cause fires, and electrocute people touching conductors.

Coulomb force electrostatic attraction or repulsion.

C-reactive protein test a common test measuring inflammation in the body. A highly sensitive C-reactive protein test (hsCRP) measures low levels of C-reactive protein in the blood.

Crop amendment material applied to a crop and/or soil to improve its physical or chemical properties.

Cymatics science of visualizing audio frequencies.

Cytochrome the electron transport chain (that makes ATP) is comprised of five cytochrome complexes, also called respiratory proteins, or just cytochromes for short.

Centralized Medicine (aka mainstream medicine, allopathic medicine, medical Establishment) medical system in which institutions such as the FDA, CDC, NIH, Ivy League medical schools, and healthcare systems control funding for research, definitions for disease, approved treatments, rules and regulations, and insurance reimbursement. It's where a small number of people dictate what's best for everyone.

Central retinal pathway (CRP) helps tell the brain what time of day or night it is. It runs from the retinas at the back of the eyes to the SCN and leptin receptor in the hypothalamus.

DHA docosahexaenoic acid is a very special fat that can convert sunlight into DC electricity, and back again.

Deuterium a hydrogen atom with an extra neutron in its nucleus. Nature uses deuterium's different structure and properties to control biological programs such as energy production, food seasonality, and aging.

Dirty electricity unwanted spikes, surges, and frequencies riding the power lines (below the frequency of wireless communications).

Dimercapto succinic acid (DMSA) is a potent clinical-strength chelator of heavy metals that is over-the-counter and gentle enough to use at home. It grabs hold of metals such as lead and mercury and escorts them out of the body through detox pathways.

Dissipative structure order emerging from chaos. Aggregation and organization as opposed to disorder. Life itself is a good example. Entropy is the opposite.

Ejection fraction percentage of blood pumped out of the heart with each contraction.

Electrosmog (unwanted) electromagnetic fields.

Emergency healing response the body's short-term fix for a problem that borrows resources from another area, instead of making more at the source.

Endogenous made in/by the body.

Epigenetics environmental factors control the way our genes turn into physical traits and behaviors.

Eukaryote multi-celled organism, in contrast to bacteria, which are single-celled.

Exclusion zone water, EZ, e-zone or electron-zone the fourth phase of water, between a liquid and a solid.

Exogenous from outside the body.

Extracellular matrix supportive structures and biochemicals outside of cells (e.g., collagen, enzymes, glycoproteins, and minerals.

Fascia thin, filmy casing surrounding muscles, nerves, organs, blood vessels, and bones that holds each tissue in place, separates it, protects it, and relays information.

GAPS coined by gut-brain health pioneer Dr. Natasha Campbell-McBride, Gut and Psychology Syndrome conditions are gut imbalances, such as leaky gut and gut dysbiosis, that cause impaired brain function, such as ADD, autism, anxiety, and depression.

Ground current stray electricity in the ground underneath you.

Heteroplasmy rate degree to which your respiratory proteins in mitochondria are stretched out vs. condensed. A high heteroplasmy rate is unproductive and unhealthy; low is condensed and productive.

Infradian rhythm biorhythm lasting longer than 24 hours. In this book, infradian rhythm refers to a seasonal cycle. For example, humans tend to gain weight in late fall through winter, and sleep less in summer, while animals breed, hibernate, molt, and grow fur or lose fur seasonally.

Interstitial space the fluid and structural environment between blood vessels and cells.

Inverse Spin Hall Effect discovered around 2008, the ISHE explains that current spin (in this case from magnetic flux) creates an electrical current at a 90° angle.

Isotope elements with a different number of neutrons than their basic variety – often making that isotope radioactive. For example, the radioactive tritium is a hydrogen atom with two extra neutrons.

Kruse, Dr. Jack world's leading mitochondriac educator and neurosurgeon.

Jonesing a strong need, desire, or craving for something – especially by an addict.

Jump conduction transference between conductors of static electricity from nnEMFs.

Kelvin (light) a measurement of a light's color, ranging between 1000°K and 12,000°K or higher. Traditional incandescent light is considered "warm" at about 2700–3000°K (yellowish). Color temperatures around 5500–6500°K approximate daylight (white). Modern, energy-efficient LED lights and screens are 6500–9000°K (blue).

Leptin Prescription Reset Dr. Jack Kruse's protocol to improve the body's response to leptin in order to achieve a healthy weight and fix hormone/endocrine dysfunctions such as hypothyroidism.

Mammalian battery informal, general term describing stores of electric charge and photonic energy that cells can use to do work. (1) E-zone is the biggest cache. (2) ATP holds electrical energy in its chemical bonds. (3) Cell membranes hold electrical charge. (4) Muscle movement releases electrons – as piezoelectricity mostly from bones, ligaments, cartilage and tendons – into the acupuncture meridians. (5) DNA is its own battery, powered by a spiraling coalescence of cosmic energy, often called "scalar energy."

Manhattan Project Top Secret US government project employing 100,000 people to build the first atomic bomb, which ended World War II.

Methylation the transfer of one carbon atom and three hydrogens (CH_3) – called a "methyl group" – to another molecule. Methyl groups control detoxification through glutathione, immunity, inflammation, gene expression, repair of free radical damage, neurotransmitter production for brain function, energy production, the stress response and more. Methylation defects are thought to contribute to autism and many other disorders.

Mitchell, Dr. Peter scientist famous for discovering how mitochondria make energy (ATP): electrons jump along the ETC, pumping protons as they go, finally the fifth cytochrome adds a phosphate group to adenosine diphosphate (ADP) to form adenosine triphosphate (ATP).

Mitochondria microscopic powerplants of the cell. They convert sugar, fat and protein from food into energy that the body can use (ATP).

Mitochondriac a fan/follower of mitochondria, biophysics, and seasonal cycles of the body.

mTOR (mammalian/mechanistic Target of Rapamycin) protein kinase that controls whether cells go into an anabolic state of survival, growth and proliferation, or a catabolic state of breakdown with autophagy and apoptosis.

MuMetal magnetic shielding material made of nickel, iron, copper, chromium, and molybdenum.

n=1 'n' is shorthand for the number of people in a study. So instead of a test group representing the general population, *n=1*, in this context, means you are your own test subject. Mitochondriacs use the term "*n=1*" informally to mean your response to a given diet, supplement, drug, practice, or treatment is unique to you.

Orbital path in which an electron might be found orbiting around its nucleus – not necessarily a linear path, but a path of probability.

Ott, John researcher and author who developed time-lapse photography techniques to study how light affects plant growth, in addition to how their chloroplasts utilize light in photosynthesis, and how children became hyperactive when exposed to artificial light in classrooms.

Oxidative phosphorylation aka the electron transport chain (ETC). The main process by which ATP (energy) is made in mitochondria.

Paramagnetism substance that is weakly attracted to magnetic fields because of its unpaired electron(s).

Parasympathetic state of rest, digest, and calmness.

Paraventricular nucleus (PVN) of the hypothalamus, in collaboration with the suprachiasmatic nucleus (SCN), controls homeostasis (normal operational state) by regulating a broad range of autonomic functions – including cardiovascular, thermoregulatory, metabolic, circadian and stress responses.

Phonon mathematical simplification of sound or heat moving through a liquid, gas or solid. Sound or heat is represented as a single wave or unit, instead of innumerable vibrations and reverberations upon trillions of atoms in a piece of metal, for example. For a great explanation, watch the YouTube video entitled "This is a SOUND PARTICLE - Phonon and Quasiparticle Physics Explained by Parth G" https://www.youtube.com/watch?v=_axrpVnGHpk.

Price, Dr. Weston A. dentist and researcher who travelled the world in the 1930s to study how diet affected the health of indigenous populations.

Prokaryote single-celled organism lacking a nucleus, including bacteria and archaea.

Pulsed electro-magnetic field therapy devices (PEMF) stimulate nerves, muscles, and blood flow using 'on-off' electricity to heal an area.

Pyruvate produced primarily by glycolysis, pyruvate is a chemical compound that fuels the TCA cycle (aka the Krebs cycle, citric acid cycle). See diagram of cellular respiration on page 42, and the citric acid cycle on page 44.

Reactive oxygen species (ROS, aka free radicals, oxygen radicals or redox molecules) metabolism makes dozens of different ROS molecules characterized by oxygen atoms with one or more unpaired electrons.

Redox short for "REDuction-OXidation." Redox reactions involve oxidation and reduction.

Redox potential pools of electrons and their net-negative charge (and, in some cases, pools of positively-charged protons) that the body uses to move materials and perform chemical reactions (think battery power).

Red pill a rude awakening when a person's false beliefs are dispelled after they were programmed to think in another way.

Respiratory protein workstations (1–5) in mitochondria that make ATP, using electrons and protons.

Quantum biology how light photons and electrons influence biology.

Quantum Xeno Effect quantum physics/quantum mechanics tells us that the mere act of observing or measuring subatomic particles, such as electrons or photons, invariably changes the results.

Schumann resonance earth's natural resonant frequency, which is slow and gentle at 7.83 cycles per sec.

Serum (of blood) the liquidy part of blood (i.e., not the cells or clotting factors).

Standards of care normal, acceptable ways to treat disease in the medical system, according to medical boards and public health agencies. When conventional doctors step outside these guidelines, they risk being sanctioned by their boards.

Suprachiasmatic nucleus (SCN) one of the brain's primary control centers that run daily and seasonal cycles (circadian and infradian, respectively), based on the information it receives about your environment.

Sympathetic alert, active, and focused state.

TCA cycle (aka Krebs cycle, citric acid cycle) a preparatory process that makes precursors for the ETC, as well as producing a small amount of ATP on its own.

Time crystal repeating atomic structure (i.e., a crystal) whose electrons and protons perpetually oscillate in spin state and magnetic moment (think tick-tock of a metronome). They flip-flop at regular intervals between two different states, alternating their response. This makes them crystalline in four dimensions (3 in space, 1 in time) and therefore a distinct phase of matter distinguishable from a liquid, sold, gas or plasma. Influences such as light can make them favor one state over the other. Time crystals use light to regulate our biochemistry and circadian programming.

Topologic insulator the surface can conduct an electric current, without affecting the interior. Mitochondria, for example.

Ubiquitination process by which ubiquitin tags defective proteins so they can be degraded (before being repaired or replaced).

Ultradian rhythm biorhythm lasting less than 24 hours. For instance, sleep cycles last about 90–120 minutes.

Voltage-gated calcium channels (VGCCs) extremely sensitive, electrically-powered valves that let calcium into the cell.

Wallace, Dr. Doug world's leading mitochondria researcher. He taught the field most of what it knows about mitochondria, including how they are generationally inherited, and how local food and climate controls metabolism, epigenetic expression, and aging (through free radicals).

INDEX

F

G

H

I

J

K

ABOUT

Randy (The Mito Man) is a citizen scientist, author, and entrepreneur. His past work includes *Gut-Brain Secrets*, which is all about the many factors that go into corruption of the gut microbiome, which then affects a person's mental and physical state – including attention deficit disorder, autism and OCD. His current work, *The Mitochondriac Manifesto,* aims to overturn our old beliefs about where health or sickness comes from, in light of what we now know about mitochondria, seasonal cycles, and energies in and around the body.

His upcoming book, *The Blood Flow Monographs,* explains how blood delivers life to the body when operating efficiently… or poor circulation chokes the life out of organs and tissues through clustering, clotting and constricted vessels. Without blood flow, nothing else matters. Unfortunately, millions now suffer the circulatory side effects of vaccines, non-native EMFs, and poor food choices. Yet, consumers of centralized medicine still don't understand why turbo cancers are exploding, heart problems are destroying the fittest athletes in the world, people's brains don't work as well as they used to, and people are dropping dead at record rates. Coming in 2025: The real reasons blood is able to circulate around the body when it does, or what's actually happening on a biophysical level when it can't (because it's not what they tell you).

NOW WHAT?

1. Help spread the word

If you found this information valuable to your health and life(style), I encourage you to go to Amazon.com (or other review site) and leave an honest review. Total number of reviews helps Amazon and prospective readers determine a book's worth.

2. Recommended reading

(Preceded by my description of subject matter.)

- The science (and solutions) of blood flow: *The Blood Flow Monographs,* by R. D. Lee. Projected 2025 release.
- How a corrupted microbiome causes psychological disorders (GAPS), good food and bad food, GMOs and glyphosate, vaccines, and detoxification: *Gut-Brain Secrets,* by R. D. Lee.
- Dangers of nnEMFs: *The Invisible Rainbow: A History of Electricity and Life,* by Arthur Firstenberg.
- Why the polarity of water matters: *The Fourth Phase of Water,* by Professor Gerald Pollack.
- Healing and polarity: *The Body Electric,* by Robert O. Becker, MD.
- How mitochondria came to power multi-celled organisms: *Power, Sex, and Suicide: Mitochondria and the Meaning of Life,* by Nick Lane.
- Effects of light on human health: *Health and Light,* by John Ott.
- Viktor Schauberger's work on the movement and energy of water: *Living Energies,* by Callum Coats.

RECOMMENDED RESOURCES

Magnetico Sleep Pad

Full disclosure: I have an affiliate relation with The Magnetico Company, and may earn a commission on purchases.

- www.magneticosleep.com
- 1.800.265.1119 (North America only). Direct: 1.702.952.5243.
- Use promo code **TheMitoMan** to receive a small price break.

DMSA Synergy

Looks like some government agency is trying to cut off the supply of DMSA Synergy. Their website is shut down, but it looks like you can still get it from other sites as of late 2024.

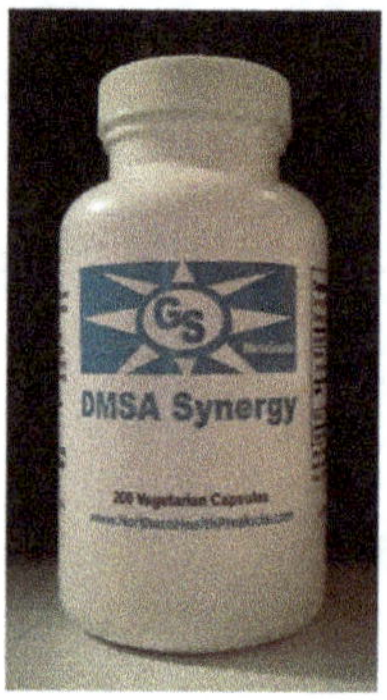

SaunaSpace faraday saunas and personal "Glow" light

- Use the URL "https://sauna.space/MITOMAN" to get 5% off.
- 573.667.2862
- hello@sauna.space

The Biomat

- www.thebiomatstore.com
- 866.952.8111
- info@thebiomatstore.com

www.ingramcontent.com/pod-product-compliance
Lightning Source LLC
LaVergne TN
LVHW061357100726
842905LV00065B/81
* 9 7 9 8 9 9 2 3 4 8 5 2 1 *